INTRODUCTION TO OIL AND GAS TECHNOLOGY

Third Edition

Francis A. Giuliano, *editor*

Scientific Software - Intercomp

Denver, Colorado

PRENTICE HALL

Englewood Cliffs, New Jersey 07632

Library of Congress Cataloging-in-Publication Data

Introduction to oil and gas technology / edited by Francis A.
   Giuliano.—3rd ed.
      p.   cm.
   Includes bibliographies and index.
   ISBN 0-13-474354-7
   1. Petroleum engineering.   2. Gas, Natural.   I. Giuliano, Francis, A.
TN870.I64 1989
622'.338—dc20   89-32315
                                                             CIP

Editorial/production supervision
   and interior design: Gertrude Szyferblatt
Cover design: Diane Conner
Manufacturing buyer: Mary Ann Gloriande

© 1989 by Scientific Software - Intercomp, Inc.
Denver, Colorado

Previously published in 1985, 1981, and 1979

The publisher offers discounts on this book when ordered
in bulk quantities. For more information, write:
   Special Sales/College Marketing
   Prentice-Hall, Inc.
   College Technical and Reference Division
   Englewood Cliffs, NJ 07632

All rights reserved. No part of this book may be
reproduced, in any form or by any means,
without permission in writing from the publisher.

Printed in the United States of America

10 9 8 7 6 5 4 3 2 1

ISBN 0-13-474354-7

PRENTICE-HALL INTERNATIONAL (UK) LIMITED, *London*
PRENTICE-HALL OF AUSTRALIA PTY. LIMITED, *Sydney*
PRENTICE-HALL CANADA INC., *Toronto*
PRENTICE-HALL HISPANOAMERICANA, S.A., *Mexico*
PRENTICE-HALL OF INDIA PRIVATE LIMITED, *New Delhi*
PRENTICE-HALL OF JAPAN, INC., *Tokyo*
SIMON & SCHUSTER ASIA PTE. LTD., *Singapore*
EDITORA PRENTICE-HALL DO BRASIL, LTDA., *Rio de Janeiro*

CONTENTS

PREFACE v

ACKNOWLEDGEMENTS vii

CHAPTER 1. THE PETROLEUM INDUSTRY 1
    I. Introduction 1
    II. Early History of Petroleum Use 1
    III. Development of the Oil Industry in the U.S. .. 2
    IV. Aspects of the Modern Petroleum Industry ... 7
    Bibliography 17

CHAPTER 2. GEOLOGY 19
    I. Introduction 19
    II. Physical Geology 19
    III. Historical Geology 24
    IV. Petroleum Geology 27
    Appendix 35
    Glossary 40
    Bibliography 44

CHAPTER 3. EXPLORATION 45
    I. Introduction 45
    II. Exploration Methods 45
    III. The Exploration Cycle 56
    Glossary 58
    Bibliography 60

CHAPTER 4. LAND 61
    I. Introduction 61
    II. Land Description 61
    III. Land Ownership 63
    IV. Acquiring Land Rights 67
    V. Sequence of Land Work 73
    VI. Regulations 74
    VII. Business Arrangements 75
    Bibliography 78

CHAPTER 5. DRILLING AND COMPLETION 79
    I. Introduction 79
    II. History of Drilling 79
    III. Contemporary Rotary Drilling 81
    IV. Drilling Operations 84
    V. Types of Drilling Rigs 87
    VI. Special Drilling Operations 89
    VII. Completion Methods 92
    VIII. Well Stimulation 94
    IX. Drilling and Completion Reports 96
    Glossary 103
    Bibliography 113

CHAPTER 6. PRODUCTION 115
    I. Introduction 115
    II. Flowing Wells 115
    III. Artificial Lift 116
    IV. Oil Treating 118
    V. Oil Storage and Sales 121
    VI. Saltwater Disposal 122
    Glossary 123
    Bibliography 127

CHAPTER 7. RECOVERY 129
    I. Introduction 129
    II. Primary Recovery Methods 129
    III. Fluid Injection Processes 134
    Glossary 139
    Bibliography 140

CHAPTER 8. TRANSPORTATION 141
    I. Introduction 141
    II. History 141
    III. Modern Transportation Methods 142
    Glossary 148
    Bibliography 150

CHAPTER 9. ECONOMICS 151
    I. Introduction 151
    II. Basic Economics 151
    III. Profitability Analysis 156
    Bibliography 159

CHAPTER 10. GOVERNMENT 161
    I. Introduction 161
    II. Federal Agencies 161
    III. State Oil and Gas Commissions and Reports . 163
    Bibliography 167

CHAPTER 11. ENERGY OUTLOOK 169
    I. Introduction 169
    II. Present Energy Situation 169
    III. Alternative Sources of Energy 174
    Appendix 179
    Bibliography 181

INDEX 183

PREFACE

Introduction to Oil and Gas Technology was initially intended to serve only as a supplement and a textbook for Scientific Software - Intercomp's seminar on "Basic Petroleum Technology". "Basic Petroleum Technology" is the original and most widely attended seminar of its kind, having provided thousands of individuals in the United States and Canada with a fundamental knowledge of today's oil industry. The seminar is an intensive two-day learning experience integrating lectures with slides, films, exhibits and audience involvement. **Introduction to Oil and Gas Technology** is distributed to seminar participants as a comprehensive and permanent reference on the information presented in the seminar. During the course of its development and use, however, the text has evolved into a work that also can stand alone as an introduction to the terminology, operations, and equipment associated with the oil and gas industry.

This book concentrates on the "upstream" functions of the industry: exploration, land work, drilling, production, recovery, and transportation; but it also includes discussions on topics that are related to, or have a bearing on, these operations: geology, economics, and government. The final chapter examines the present energy situation in the United States, current projections about our future energy resources, and possible alternative sources of energy for the country.

Introduction to Oil and Gas Technology has been designed to provide this broad overview of the technology and functions of the petroleum industry for several groups within the industry—new employees, non-technical personnel, and individuals involved in specialized areas, who have not had the opportunity to gain a perspective on the overall industry. In addition, many people in non-petroleum fields, such as banking, insurance, investment, government, computer services, etc., have business and professional dealings with the industry. For these people, this text can serve as a general introduction to the operations and procedures of oil and gas companies. Finally, it is hoped that the book will be of service to anyone who is simply interested in learning how petroleum is found and produced.

For those readers who are familiar with the original version of **Introduction to Oil and Gas Technology**, the first difference you have probably noticed in this edition is the size of the book. By having the entire book typeset, and by presenting the illustrations in a more efficient and consistent manner, we have been able to prepare a text that is considerably more manageable and easier to use.

It should be noted, however, that this second edition has been reduced in physical size only, not in content. The entire book has been revised and rewritten to include additional and up-to-date information on the individual topics. This revision has resulted in an expansion of the material contained in each chapter, material that is now more complete and current.

A significant addition to the book is the inclusion of an index, which serves as a reference guide for terms found both in the glossaries and in the text.

Another major effort undertaken for this edition has been the improvement of the quality and detail of the illustrations in the text. Many of the figures have been redrafted or altered to reflect more clearly the actual conditions or procedures they portray.

The petroleum industry touches practically every segment and facet of our modern society. Because petroleum has such an influence on our lives, people both inside and outside the industry have become interested in learning about the techniques and processes that are used to discover and produce this important resource. We sincerely hope that the reader, for whatever specific reason he or she may have in using the text, will find **Introduction to Oil and Gas Technology** informative and helpful.

Francis A. Giuliano
Editor

ACKNOWLEDGEMENTS

We first would like to acknowledge those people who coordinated and wrote the original version of **Introduction to Oil and Gas Technology,** and upon whose work this edition has been built: Carol F. Chesnut, Dwayne A. Chesnut, Dave O. Cox, Jane E. Hoback, Vicki P. McConnell, David M. Newell, and Richard D. Reese.

We are deeply indebted to those individuals who gave so much of their time and effort to produce this second edition. The one common trait that each person exhibited during this revision was a concern for the quality and accuracy of the material. The following individuals were to serve initially as technical reviewers and consultants, but they ended up contributing very much more: Charles W. Bloomquist, who reviewed the chapter on Recovery, and rewrote and updated the material on Economics; Richard S. Bryson and Richard D. Holt, who revised and reorganized the chapters on Geology and Exploration; Floyd D. Burnside, who provided extensive revisions for the illustrations and text on Drilling, Production, and Transportation, and who was of invaluable assistance on so many matters; and Allison M. Moore, who completely rewrote the chapter on Land, incorporating much more information and detail on the subject.

Special thanks is due to William P. Versteeg, who redrafted and modified many of the illustrations in the text, and whose work exhibits an excellence that has enhanced the quality of the entire book. We also want to acknowledge Doris J. Jurica, who so selflessly provided printed text for various illustrations, and A. Gene Giuliano, Jr., who proofread the material and contributed valuable suggestions on its construction and content.

We want to express our sincere gratitude to the following companies and organizations that either contributed illustrations, or granted us permission to use their material in this text:

American Petroleum Institute
Gulf Publishing Company
Hughes Tool Company
Kaneb Services, Inc.
McGraw-Hill Book Company
Oil and Gas Journal (PennWell Publishing Company)
Petroleum Extension Service, The University of Texas at Austin (PETEX)
Petro-Lewis Corporation
Society of Petroleum Engineers of AIME
U.S. Department of Energy

CHAPTER 1:
THE PETROLEUM INDUSTRY

I. INTRODUCTION

Petroleum plays a significant, even critical, role in the modern industrial world. Petroleum products and petrochemicals affect almost every aspect of our civilization and of our lives as individuals, including the areas of transportation, food, clothing, shelter, and recreation. The supply, production, and consumption of petroleum influence national economies, and have a bearing on global politics and international relationships.

The petroleum industry, however, has attained this present position of importance and prominence in a relatively short span of time. For example, the history of the industry in the United States only dates back to 1859, when the first well drilled specifically for oil proved successful. Since that time, technological advances and concurrent social changes have contributed to the emergence and growth of the industry.

Even though the petroleum industry has a relatively short history, the use of petroleum itself extends much further into the past. Although the word **petroleum** is derived from the Greek words meaning "rock oil," no one knows when it was discovered or what culture first used it. The knowledge and use of petroleum can be traced back for centuries, and most probably precede recorded history.

II. EARLY HISTORY OF PETROLEUM USE

A. Petroleum in Ancient Civilizations

Petroleum was utilized for a variety of purposes by ancient civilizations in Asia and Europe. The records of the Mesopotamians, Elamites, Chaldeans, Akkadians, Sumerians, Babylonians, Egyptians, Chinese, Japanese, Romans, and Greeks show that they found sources of petroleum in surface springs and tar pits, as well as in natural gas seepages. These ancient peoples used petroleum—in the form of asphalt, pitch, bitumen, and oil—in a variety of ways: as a waterproofing substance to caulk their boats, as cement and mortar in construction, in religious ceremonies, as a military weapon, for medicinal purposes, and in everyday use as a lubricant, illuminant, or fuel.

The Book of Genesis makes several references to asphalt or bitumen. Bitumen was used as mortar in the construction of the Tower of Babel, and Noah used the substance to waterproof the Ark. The valley of Siddim was reported to be filled with "tar pits," or bitumen wells.

Archeologists have found drawings from Egyptian tombs, dating as far back as 4000 B.C., showing that these people used petroleum as cooking oil to prepare their meals. The ancient Egyptians used natural asphalt for their process of mummification. The axles of the pharaohs' chariots were greased with pitch. "Syrian oil" or "green oil" was also used by these people in a variety of medicinal preparations.

Asphalt and pitch were widely used by the ancient peoples of Mesopotamia as glue, cement, mortar, and as a waterproofing substance. The Sumerians in 3000 B.C. were using asphalt as glue under inlaid mosaic walls and floors, as well as for bonding and jewel setting. Early Mesopotamian civilizations dug pits with their hands for asphalt, which they used for building cement and for ornamental purposes. Bitumen, a substance similar to asphalt, was used to fasten the blades of implements and weapons in their handles, and to caulk ships and build roads. The Chaldeans used this substance as tile cement and for waterproofing. In 1500 B.C., they were burning a petroleum liquid in censers, or fire pans, to give off light.

Sometime between 1000 and 900 B.C., the Chinese became the first people to utilize natural gas. They transported natural gas from hand-dug wells, through simple but functional bamboo pipelines, to their homes for heat and fuel. Later, in the sixth century B.C., the philosopher Confucius wrote about wells along the Tibetan border of China that were several hundred feet deep. The Chinese used picks and shovels to dig these shafts in search of the natural gas and petroleum lying beneath the earth.

During the same period, the followers of the religious leader, Zoroaster, worshipped before fires that were continuously fueled by natural gas seepages. Also during this time, King Nebuchadnezzar used asphalt to build the walls and pave the streets of Babylon.

In the fifth century B.C., Herodotus wrote in his **History** how oil and bitumen were procured by the people of his time. He also described in his writings how the Greeks used petroleum as a weapon during the siege of Athens in 480 B.C. The Greeks destroyed a Scythian fleet by pouring oil on the sea and setting it on fire, directly in the path of the Scythian ships. The Persians were also beginning to use petroleum for military purposes, by developing incendiary arrows wrapped in oil-soaked cloth.

In 331 B.C., Alexander the Great, according to the first century writings of Plutarch, encountered a continuous flame from the earth that was fed by a natural gas seepage, in what is now the Kirkuk Field.

The Chinese are credited with being the first civilization to drill for oil. It is generally believed that sometime during the third century B.C., they developed a method of pounding holes into the earth using bamboo tubes and bronze bits, originally in order to procure brine. The Chinese could drill approximately two feet a day by alternately lifting and dropping the metal drilling tool (sometimes weighing 300 pounds) in the hole. Estimates as to the depth of these wells vary from 200 to 3,000 feet, but it does seem probable that they reached depths of several hundred feet. Sometimes it would take as long as three years to complete a single well. Early accounts also reported that the Chinese used water to soften the rock formation in the hole, making it easier both to drill and to remove the cuttings.

In the first century B.C., Cicero, the Roman orator, was reported to have used crude oil lamps. Meanwhile, the Arabs of that time began to distill crude oil as fuel for lighting. The distillation of naphtha, especially in centers such as Damascus, became a thriving business.

B. Petroleum in Medieval Times

As time passed, early civilizations began to exploit the potential military uses of petroleum. In 673 A.D., during the naval siege of Constantinople, the Romans destroyed the wooden Saracen fleet by launching flaming containers of naphtha and sulfur at it. Approximately a century later, the Arabs and Mongols were using "grenades," small pots containing burning oil, as weapons against their enemies. The Arabs are said to have developed the first distilling process for obtaining flammable products for military purposes. In the eighth and ninth centuries, the Byzantines used a highly secret formula, called "Greek fire," against the Muslims, which most likely consisted of ignited paraffin.

In the tenth century, in Baku, Persia (now the Soviet Union), there occurred a spectacular spontaneous combustion of gases from a naphtha deposit, prompting the people of the area to worship before this "eternal flame." During his travels in 1271, Marco Polo noted that the people around the area of Baku and the Apsheron Peninsula in the Caspian Sea region were collecting oil from seepages, and that petroleum was even being exploited on a commercial scale, at least for that time. Later, in the eighteenth century, when Peter the Great of Russia conquered this area, he would allow private businessmen to procure, refine, and ship the oil located there, in return for a percentage of their income.

In Medieval times, oil was being used in a variety of ways by the Europeans, while European explorers also were reporting the use of petroleum by the peoples they encountered on their journeys. Off the coast of Sicily, oil was gathered as fuel for lamps. Springs throughout Europe provided natural gathering places for seepages of oil, which was used for fuel and medicinal purposes. In the fourteenth century, court records indicated that King Edward III of England received "eight pounds of petroleum" from one of his expeditions. In the early 1500s, Columbus sent asphalt samples from Trinidad to Spain; while later in the same century, Sir Walter Raleigh visited the great Trinidad pitch and asphalt lake, noting that its contents were excellent for oiling the ships' rigging. In 1650, the Rumanians were digging oil shafts by hand.

European scientists had been discovering and examining the properties in petroleum. In 1556, the German mineralogist, Agricola, became the first person to use the word "petroleum," in a treatise describing its recovery and refining. The term "gas" was coined in 1609 by the Belgian chemist, Jan Von Helmont.

C. Petroleum in Early America

The existence and use of petroleum in America was noted by the early explorers and settlers of this country. In 1542, Juan Rodriguez, a Spanish sailor, wrote about the presence of oil near Santa Barbara, California. The next year, the survivors of the DeSoto expedition found oil residue from seepages near Nacogdoches, Texas, which they used as waterproofing to repair the bottoms of their boats. Joseph de la Roche, a Franciscan priest-explorer, wrote in 1627 that he found a "fountain of oil" in what is now upper state New York. In the early 1700s, a Swedish traveller in America, by the name of Peter Kalm, included a map in the report of his travels that located "oil springs" near Oil Creek, Pennsylvania. In 1753, George Washington and General Andrew Lewis gained title to a section of land near Charleston, West Virginia, because they thought the presence of a gas seepage could make the land valuable.

Long before these discoveries, however, American Indians had already been using petroleum for a variety of purposes. Remnants of wells found in Pennsylvania and elsewhere indicate that the Indians were drilling for oil as early as 1300. More commonly though, they skimmed crude oil from surface springs and streams to be used as fuel, waterproofing for their canoes, and medicine. The medicine men, or tribal doctors, revered for what were thought to be supernatural powers, used oil for healing purposes. They later taught the pioneers to make medicine from the oil they had discovered. Taking their cue from the tribal medicine men, many "doctors" began to appear who bottled and sold "Seneca oil" or "rock oil" that was guaranteed to "cure what ails you."

III. DEVELOPMENT OF THE OIL INDUSTRY IN THE U.S.

A. Introduction

In the first half of the eighteenth century, a series of distinct events were to take place that laid the foundation for what was to happen at Titusville in 1859. Although coal gas had been introduced for street lighting in a few places, lamps fueled by whale oil provided the main source of light in the early 1800s. Whaling became a major industry; but by the mid-nineteenth century, as the demand for the oil increased, whales were becoming scarce, and the cost of the once plentiful whale oil was rapidly increasing. Americans began to search for a cheaper and more plentiful substitute.

In the early 1850s, a Canadian doctor and geologist, Abraham Gesner, patented a distilling process for refining oil from coal. By this process, he produced an improved lighting oil, which he called "kerosene." Within a short time, the kerosene lamp had been invented and was soon being marketed.

At approximately the same time, Samuel Kier, a Pittsburgh druggist and salt merchant, struck oil while drilling for salt. He set up the first refinery, consisting of a one barrel still with a daily capacity of about five barrels. The majority of the production from this refinery was sold by Kier as "rock oil" for medicinal purposes.

The occurrence of oil was being more carefully noted, and its presence in certain areas could be anticipated. Oil was most likely to be found in surface springs, gas seeps, slicks, and in low places along creek banks. In 1842, Sir William Logan, the director of the Geological Survey in Canada, noted that a series of oil seepages occurred along anticlines (large convex folds in rock beds) paralleling the St. Lawrence River. Other geologists also were noting the occurrence of oil in anticlinal structures. In addition, oil was being accidentally discovered in Pennsylvania, West Virginia, and Kentucky in wells that were being drilled for salt deposits.

All of these events—the diminishing supply of whale oil together with the growing need for a suitable replacement, the distillation of kerosene from coal and the introduction of the kerosene lamp, the establishment of a refinery for processing crude oil, and the recognition of potential oil sources—provided the background for the beginning of the oil industry.

The final impetus was supplied by a chemist at Yale University, Professor Benjamin Silliman. Through tests at the University, he discovered that crude oil could be refined into kerosene, producing a smokeless, odorless illuminant that was superior even to whale oil.

Recognizing the rising demand for oil as an illuminant and in medicinal preparations, and spurred by Professor Silliman's report, George H. Bissell and Jonathan G. Eveleth, on December 30, 1854, incorporated the Pennsylvania Rock Oil Company. This was the first company organized for the express purpose of producing oil. The company was later reorganized under the leadership of James M. Townsend, and on March 23, 1858, became known as the Seneca Oil Company. The result of this enterprise would change the face and character of America.

B. The Drake Well

The specific objective of the Seneca organization was to recover oil in commercial quantities. They decided that drilling for the oil would be more effective than the more common practices of ditching (digging shallow trenches and letting the oil seep into them) or skimming the oil from surface springs or streams. "Colonel" Edwin L. Drake, a retired railroad conductor, was hired by Townsend to supervise a drilling project along the banks of Oil Creek near an old oil spring in Titusville, Pennsylvania.

Drake's drilling rig was typical of those used by salt drillers, consisting of an old steam engine and an iron bit attached by a rope to a wooden windlass. One of the innovative procedures that he was to use was to sink a pipe casing into the hole in order to prevent cave-ins, and to seal off the water as he drilled.

At a depth of thirty-nine feet, Drake encountered solid rock, and the drilling was slowed to no more than three feet a day. Soon the project became known as "Drake's Folly." But on August 27, at a depth of 69½ feet, "Uncle Billy" Smith, an employee of Drake's, discovered that oil had appeared in the hole a few feet below the surface. "Drake's Folly," which yielded 2,000 barrels of oil in that first year, marked the beginning of the petroleum industry in the United States (Figure 1.1).

C. Early Expansion and Development

A rush of activity followed Drake's success. People came into the area to seek their fortunes, and towns mushroomed around drilling sites. By 1869, oil wells had sprung up all along the Allegheny River, as well as in West Virginia, Kentucky, and Ohio. Within that same decade, oil had been found in Colorado, California, and Texas. By 1900, oil was being produced commercially in thirteen states, with Ohio, West Virginia, and Pennsylvania producing over sixty percent of the nation's petroleum.

The rapid growth of oil production led to related developments and changes in other industries. Refining methods were improved to more effectively utilize the increased supply of oil as an illuminant. The first large refinery, which produced mainly kerosene, was completed in 1861, about a mile from the Drake well.

One of the first problems the new industry had to deal with was storage. Coopers, who made the barrels to store and haul oil, could not keep up with the increased production. Soon, there were not enough barrels to meet the demand, and what barrels there were had become too expensive to use. Large holes were scraped out of the ground to provide storage areas, but seepage proved a problem, and vapors from the oil tended to ignite. Storage facilities progressed from square wooden boxes, built both above the ground and embedded in the ground, to the round tank that was to predominate. This tank consisted of a round, wooden structure held together by iron bands and secured to a wood floor. The original version held about twelve barrels of crude oil.

The increased production also led to development in methods of transportation. The teamsters devised a massive system of boats, barges, and horse-drawn carts, but the slippery roads of the oil fields, together with escalating prices, soon proved problematic. Seeing this, the railroads began to build spurs into the producing regions to transport oil to the refineries. At first, barrels of oil were simply carried on regular flatcars. In 1865, a flatcar with two round vertical wooden tanks, each holding forty to forty-five barrels, became the first tank car built expressly for transporting oil. In 1871, a railroad tank car was developed with a horizontal, cylindrical tank mounted on two four-wheeled platforms that closely resembled the tank cars used today.

In 1865, Samuel Van Syckel built the first oil pipeline. The pipeline was constructed of two-inch iron pipe, and ran five miles from Pithole City in Pennsylvania to the Oil Creek Railroad. In 1879, the first major pipeline was completed, covering a distance of 110 miles from the oil fields around Bradford, across the Allegheny Mountains, to the railroad depot at Williamsport, Pennsylvania.

Barges and rafts were used to carry barrels of oil down Oil Creek to Oil City, where the barrels were transferred to steamboats, which then carried the oil to market. In November of 1861, the schooner, "Elizabeth

Figure 1.1. THE DRAKE WELL.

Watts," became the first ship to carry petroleum overseas, transporting three thousand barrels of oil to London. In 1869, the "S.S. Charles" of Antwerp, Belgium was the first vessel fitted for bulk shipment of oil, when all of the area in its holds was converted to carry tanks of oil, and nothing else. The first tanker set sail in 1886.

The economic aspects of the fledgling industry presented almost as great a challenge as finding and producing the oil. Prices haphazardly rose and fell, depending on shortages or surpluses on the market. John D. Rockefeller took advantage of the chaos when he formed the Standard Oil Company in 1870. Rockefeller dominated marketing and refining for almost fifty years. By signing contracts with the railroads, and by controlling 80 percent of the refining capacity of the United States and 90 percent of the pipelines, Rockefeller was marketing 95 percent of the nation's refined oil in 1879. By the 1880s, the Standard Oil Company had become the first important integrated oil company in the United States, with involvement in refining, transportation, marketing, and production. The company was the foremost purchaser of crude oil and the largest shipper of petroleum products. The only area the company excluded was exploration, since Rockefeller felt that this end of the business presented too great a financial risk.

In 1882, Rockefeller formed the Standard Oil Trust, but ten years later a suit was filed under the Sherman Anti-Trust Act against Standard Oil. Litigation continued until 1911, when the Supreme Court ordered a dissolution of the trust, and required the company to divest itself of interests in 33 subsidiary companies.

Although technical advances and operational improvements were made in the areas of drilling, storage, transportation, refining, and marketing, there was little progress initially in the field of exploration. Knowledge of subsurface geology was increasing, but oil people paid little attention to the professionals. Instead, they often would rely on such unscientific methods as divining rods and "doodle bugs," as well as spiritualists and fortune tellers, to assist them in their search for oil.

Geologists, nevertheless, continued their research, and in 1863, Henry D. Rogers, a geologist at the University of Glasgow, stated without qualification that in the United States oil was only found in anticlines. Rogers' theory was later proved incorrect, but the importance of geological knowledge in exploration was gaining acceptance, and by 1920, every major oil company had a department of geology.

D. The Lucas Gusher and Gasoline

The Lucas Gusher in the Spindletop Field near Beaumont, Texas, in 1901, has been called the most important event in the history of the oil industry since Drake's well (Figure 1.2). It marked the first major use of the rotary rig, and its proved effectiveness was to launch the oil industry in a big way. This well was also the first exploration of a geological formation that had never been drilled for oil: a salt dome. But perhaps most significantly, the Lucas Gusher produced so much petroleum that it far exceeded the demand for oil in medicines and as an illuminant.

Other uses for this now abundant resource were explored, and as a result, a once useless byproduct of oil now became an important source of fuel—gasoline. When scientists discovered that three barrels of oil

Figure 1.2. THE LUCAS GUSHER.

produced as much energy as one ton of coal, fuel oil soon replaced coal as fuel for boilers, and gasoline replaced steam and electricity as the source of power for the new horseless carriages. In 1901, sales of gasoline exceeded sales of kerosene for the first time. The internal combustion engine had come of age.

In 1892, Charles and Frank Duryea built the first American gasoline-powered automobile. Gasoline had proved to be a more efficient fuel than either steam or electricity, because when it burned, it delivered direct power in the form of a high amount of heat energy in relation to its weight. In 1900, Ransom E. Olds started mass-producing cars for the first time, by standardizing models and using interchangeable parts. When Henry Ford introduced assembly line production with his Model T in 1908, the automobile became practical for everyone. In 1910, there were fewer than a half million automobiles in the United States; by 1920, over nine million motor vehicles were in existence.

Just as the increased supply of oil, marked by the Lucas Gusher, spurred the development of the internal combustion engine, the practical development of the automobile provided the oil industry with its greatest impetus. There was not only a need to produce more crude oil to meet the rising demand for gasoline, but improvement in the refining process was needed to yield more gasoline from the crude. In 1913, Dr. William Burton, a chemist, developed the thermal cracking process. By applying intense heat and pressure to the heavy oil, the larger molecules were broken down into smaller ones, which augmented the gasoline production. This process increased the yield of gasoline from eleven barrels per hundred barrels of crude, to twenty-five barrels per hundred. The process also produced a better burning fuel, which in turn, allowed automotive engineers to develop higher compression ratio engines with greater efficiency. This relationship between the oil and automotive industries was to continue: improvements in the one allowed, and even demanded, further advancements in the other.

E. Judicial and Governmental Activity

The frequent discoveries of petroleum in many parts of the United States were heady stuff for prospectors and speculators. Hundreds of people learned of the discoveries and swarmed into oil-rich areas, rivalling the gold rushes in the west. Families lived in make-shift shacks right in the oil fields, and packed so tightly into a single area that they seemed to live on top of each other. Oil rigs were built in such close proximity to each other that there was little room for anything in between. With no sense of modern "community development planning," buildings and roads were constructed haphazardly, and towns seemed to spring up, or disappear, overnight. Pithole, Pennsylvania reached a population of 15,000 in five months; a few months later it was a ghost town. In Beaumont, Texas, where the great Spindletop oil field was discovered, the population grew from 9,000 to 50,000 within two years time. The town of Borger, Texas was founded in February, 1926, and by December of that same year, boasted a population of 20,000.

This explosive growth was greatly encouraged by what was known as the "Rule of Capture." Oil pools under the ground can easily extend beyond the surface property boundaries of an individual owner. When a well is produced, the pressure in the underground formation is lowered at that spot, causing the oil from surrounding areas of higher pressure to move toward the well. These areas of higher pressure can extend beyond the property lines of the person owning the producing well. The pressing question in the early years of the industry was: who legally owned the recovered oil?

In 1889, the Supreme Court, basing its decision on common law governing the capture of wild game, ruled that, regardless of where it originated, gas and oil which was recovered belonged to the person on whose land it was "captured." As a result, owners or leaseholders of neighboring properties would drill oil wells as rapidly as possible, not only to recover as much oil as they could, but also to prevent the oil beneath their land from being drained by someone else.

The courts also ruled that people who leased land for oil production had implied obligations to protect the landowners' interests. If one company leased two neighboring parcels of land, it had to drill on both leases, to assure that both owners received their fair share of the royalties. The result was a tremendous waste of oil and other resources, and a decrease in the market price of oil because of overproduction. Boom towns soon became ghost towns when fields were "drilled out," and their inhabitants packed up and moved on to more productive areas.

Because of the growing concern over wasted resources and unstable prices, the states of Oklahoma and Texas began to pass prorated production laws to encourage more orderly, and less "cutthroat," production of oil. The laws were based on the doctrine of "correlative rights," which required landowners and oil producers whose oil came from the same source to respect each others' rights. The laws, however, were not effective until the 1930s, after oil from the two states had flooded the market, causing a drastic drop in prices. The governors from the two states ordered an emergency call-up of the National Guard to enforce shutdown orders. The courts ruled that waste could be legitimately prevented by limiting production to just meet market demand. The federal government subsequently enacted laws that would regulate the production of oil, eliminate excess, and protect the rights of property owners.

F. Petroleum Use in War and Peace

Petroleum played an important role in World War I. Oil fueled the ships that brought American troops and munitions to Europe; troops were transported to the battlefields in gasoline-powered vehicles; and the first armored tanks were used by the military in this conflict. This was also the first war fought in the air, and it saw the emergence of the airplane as a significant military weapon.

After World War I, the petroleum industry in the United States continued to expand. New highways, surfaced with petroleum asphalt, increased in number across the country, allowing simultaneous expansion in the areas of commerce and transportation. In the 1920s and 1930s, new oil fields were discovered, including the great East Texas field. In 1937, the catalytic cracker was introduced, which further increased the quality and yield of gasoline.

During the second World War, the oil industry responded to President Roosevelt's request to develop fuel for the 50,000 planes that were being built each year. The annual production of crude oil in the United States increased from 1.4 billion barrels in 1941 to 1.7 billion barrels in 1945, serving to fuel the trucks, tanks, jeeps, ships, and railroad cars that transported men and supplies both here and overseas. The production of 100 octane aviation gasoline increased from 75,000 barrels a day in 1942 to almost 600,000 barrels a day in 1945. The government also subsidized the construction of more pipelines to transport the increased petroleum supply. The "Big Inch" (24-inch) and the "Little Big Inch" (20-inch) pipelines carried crude oil and refined products from Texas to the Atlantic seaboard.

If the war caused a shortage of many items at home, it also made possible the rapid development of the new petrochemical industry. This industry produced toluene for TNT, butadiene for synthetic rubber, and created or

Figure 1.3. MOBILE OFFSHORE RIGS. Left: semisubmersible rig. Right: jack-up rig.
(Photos courtesy of Kaneb Services, Inc.)

developed new lubricants, medicinal oils, paints, and varnishes.

After the war was over, peace brought even greater demands for petroleum products. The number of motor vehicles registered in the United States increased from just over thirty-one million in 1945, to fifty million in 1950. The number of diesel locomotives and farm tractors also increased dramatically, while more homes became equipped with oil-burning, rather than coal, furnaces.

During this post-war period, natural gas developed into the major fuel for industry and home heating. The "Big Inch" and "Little Big Inch" pipelines of war fame were converted to natural gas transmission, while other transmission lines were constructed to link remote areas with the population centers of the country.

The consumption of petroleum rose from 1.8 billion barrels in 1946 to 2.4 billion barrels in 1950. In 1910, coal had provided approximately 85 percent of the energy consumed in the United States, while oil and natural gas accounted for 12 percent. In 1946, petroleum, for the first time, supplied more of the nation's energy needs than coal; and by 1950, oil and natural gas provided almost 58 percent of the energy consumed in this country.

G. Offshore Development

One particular phase of the oil industry that has grown rapidly since the end of the second World War is the exploration and development of petroleum reserves offshore. Offshore drilling actually began in the United States, off the coast of California, as early as the end of the last century. However, the increased demand for petroleum products during the second half of this century has provided the real impetus for accelerated offshore activity.

Drilling offshore presents many more difficulties and is considerably more expensive than onshore operations. The less stable environment and the sometimes violent weather conditions pose unique problems for offshore activities, which has led to the development of specialized equipment and techniques. For example, mobile rigs are used for drilling exploratory wells in offshore areas. The jack-up rig, which is towed to the drilling location, can be raised to a safe height above the water after its elevator legs have been lowered to the sea floor. The semisubmersible, which is partially submerged and moored by anchors, can be used for drilling in deep water (Figure 1.3). The drill ship, a modified sea vessel outfitted with specialized equipment, is also used for deep sea drilling operations.

Once producible amounts of oil or gas have been discovered, permanent platforms are installed. These are huge, fixed structures, large enough to contain living quarters for the crew, storage space for the supplies, and all the necessary drilling and production equipment for the development wells.

Despite the expense and potential hazards, the chances for finding large new fields are much better offshore, where much of the area is untested, than in the well-drilled basins located onshore. Current estimates suggest that nearly one-half of the undiscovered recoverable oil resources, and more than one-fourth of the undiscovered recoverable gas resources in the United States, may occur in offshore regions. Although the challenges and risks are significant, the country's need for energy has increasingly directed the search for oil and gas to offshore areas.

Table 1.1. PARTIAL LIST OF OIL INDUSTRY CAREERS

Accountant	Engineering Technician	Paralegal
Administrator	File Clerk	Physicist
Bookkeeper	Geological Technician	Plant Operator
Chemical Engineer	Geologist	Programmer
Chemist	Geophysicist	Pumper
Civil Engineer	Land Clerk	Roughneck
Computer Scientist	Land Man/Woman	Roustabout
Drafter	Lawyer	Salesman/Woman
Driller	Machinist	Secretary
Economist	Manager	Service Representative
Electrician	Mechanical Engineer	Toolpusher
Electrical Engineer	Paleontologist	Truck Driver

H. The People of the Petroleum Industry

Although we will speak of companies, industries, firms, etc., throughout this text, and we will discuss in detail equipment and processes, people still remain at the heart of the petroleum industry. The petroleum industry provides a wide variety of career opportunities for its individuals, as shown by the partial list of positions in Table 1.1. Salaries paid by the industry are among the highest in the country, and benefits provided by oil and gas companies for their employees top every other industry.

In addition, changes are happening within the industry that reflect the overall social restructuring and revitalization of the entire country. For example, the role of women in the industry has changed considerably over the last ten years. Until that time, women were usually confined to the "traditional" secretarial and clerical positions. Now, they can be found working in a number of different capacities and levels. As opportunities open and horizons expand, women managers, geophysicists, geologists, heavy equipment operators, investment officers, paleontologists, oil field laborers, and semi-skilled workers of all types are contributing to the advancement of the industry.

This diversification of opportunity applies to all groups and sectors within the industry. The success of oil and gas companies depends on the efforts and capabilities of the millions of individuals who are the flesh and blood, both literally and figuratively, of the petroleum industry.

I. Summary

Today, petroleum has become a vital part of our everyday life. Americans have, and use daily, any number of the over 3,000 products that are derived from petroleum. Even with conservation efforts, our consumption of petroleum averages 15 to 17 million barrels of oil per day. Meanwhile, the oil industry has grown to service this need and demand for petroleum in our society. In 1859, the total production of crude oil in the United States for the year was 2,000 barrels. Annual production of crude oil and condensate in this country now surpasses 3 billion barrels. The petroleum industry has come a long way from that single, steam-powered, percussion-drilled well in Titusville, Pennsylvania.

IV. ASPECTS OF THE MODERN PETROLEUM INDUSTRY

A. Introduction

The focus of the present text is on the "upstream" functions of the oil and gas industry—exploration, land, drilling and development, production, and transportation. An overview of these areas is presented in parts B through G of this section, while more detailed discussions on each of these topics can be found in the remaining chapters of the text. The "downstream" functions, including refining, the petrochemical industry, and marketing, are not within the scope of this text, but a brief look at each of these is contained in parts H and I of this section to complete the overall picture of the petroleum industry's operations.

The petroleum industry ranks as one of the largest and most prominent industries in the United States. Recently, four of the five largest corporations in the U.S., and thirteen of the top twenty, were oil companies. Thousands of companies, and millions of individuals, are engaged in the business of finding, producing, refining, and marketing oil and gas to meet this country's demands for energy. The industry's size and complexity are such that few people are ever involved in more than one phase.

Integrated oil companies are those firms involved in the entire spectrum of the industry's operations, from exploration to marketing. There are over 50 integrated firms in the United States.

The greater number of oil companies usually engage in only one, or maybe two or three, of the different functions in the industry. These smaller firms that specialize in a particular area are often called **independents.**

B. Raw Materials

The basic raw materials used by the modern petroleum industry are **crude oil** and **natural gas,** otherwise known as **hydrocarbons** or **petroleum.** While we are familiar with many of the products derived from crude oil and natural gas, such as gasoline, motor oil, and fuel oil, we often know little about the raw materials themselves. Considerable processing and refining are necessary before the raw materials can be converted from their original form into finished products.

Petroleum consists of hydrogen and carbon atoms linked together to form chains of molecules. The basic hydrocarbon unit is the methane molecule, which consists of one carbon atom and four hydrogen atoms:

$$\begin{array}{c} H \\ | \\ H - C - H \\ | \\ H \end{array}$$

This molecule may be combined with itself to form different chains of molecules, such as propane:

$$\begin{array}{ccc} H & H & H \\ | & | & | \\ H - C - C - C - H \\ | & | & | \\ H & H & H \end{array}$$

In general, short hydrocarbon chains are in the form of gases, whereas liquid petroleum is composed of longer, heavier chains as indicated below:

$$\begin{array}{cccccccccc} H & H & H & H & H & H & H & H & H & H \\ | & | & | & | & | & | & | & | & | & | \\ H - C - C - C - C - C - C - C - C - C - C - H \\ | & | & | & | & | & | & | & | & | & | \\ H & H & H & H & H & H & H & H & H & H \end{array}$$

Crude oil is the natural form of oil as it is found in the earth. It is a chemically complex substance, as witnessed by the fact that some crude oils may have over one thousand different organic compounds. Two of the fundamental classifications of crude oil extracted from the earth are **paraffin base crudes** (crudes having high wax content) and **asphalt base crudes** (crudes having a high asphalt content).

Crudes are also differentiated according to their sulfur content. Sulfur may be present in crude oil as dissolved gaseous hydrogen sulfide or as sulfur-containing organic compounds. Because of the odor of hydrogen sulfide, crudes are referred to as "sweet" (oil which contains very little sulfur) and "sour" (oil that has high sulfur content). The sweet crudes, which usually contain less than one percent sulfur, are preferable because sulfur has a significant impact on air quality and causes many pollution problems.

Crude oil varies in color from black to green or yellow, and also varies in texture and density. Some crude, called **heavy oil,** may be so thick that it is solid at room temperature, and requires extensive refining before any useful product can be made from it. Other crude, called **light oil,** is lighter in weight and is easier to refine. There are some crude oils that need no processing at all. For example, in the early days, penzoil, which was a lightweight crude oil found in the Appalachians and in the northern United States, could be used in its native form as a machine lubricant.

Most crude oils, however, do require processing and refining. Crude usually comes from the earth containing foreign materials, such as water or fine, particulate, solid matter, which must be removed before the oil can be sold. The crude is usually processed immediately on the production site, so that only clean oil is sent through the pipelines to the refineries. There, the oil is converted into gasoline, fuel oil, lubricants, synthetics, and the many other petroleum products so familiar to us today.

Natural gas—any of the gaseous hydrocarbons generated beneath the earth's surface—can exist as either "dry" gas or "wet" gas. **Dry gas** is the term used to designate natural gas containing little or no liquid hydrocarbons, while **wet gas** contains liquid hydrocarbons, or **condensate.** Gasolines and other liquid products may be made from this natural gas condensate. To distinguish it from gas produced by gas wells, gas produced with oil from oil wells is called **casinghead gas,** since it is drawn from the casinghead at the top of the oil well.

Generally, natural gas does not require as much processing as crude oil. However, it frequently must be dehydrated, which means that water must be removed, and the liquid hydrocarbons (or condensate) must be separated from the gas. Natural gas also can be "sour," and may contain hydrogen sulfide, a highly poisonous compound that must be removed before the gas can be used for commercial or domestic purposes.

Once the gas is dehydrated, and the condensate liquids and hydrogen sulfide have been removed, natural gas is suitable for use in homes and factories as a fuel or for heating. Mercaptans, compounds containing sulfur in place of oxygen, are usually added to give the gas its familiar odor, since natural gas is generally odorless in its native form. Propane and butane, two forms of easily liquefied natural gas, are used for fuel in remote areas, where other gaseous fuels are not available.

Oil and gas are found beneath the earth's surface in zones called **petroleum reservoirs.** These reservoirs

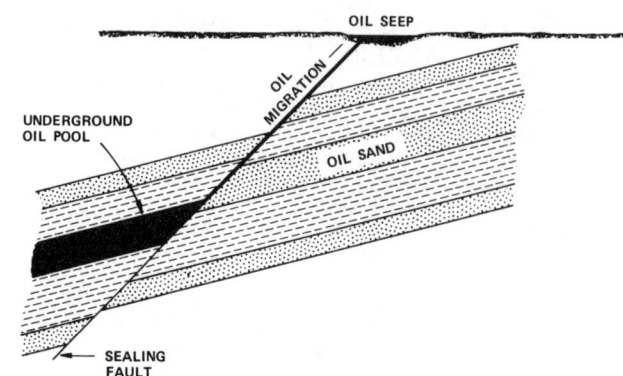

Figure 1.4. A PETROLEUM RESERVOIR OR "POOL." The fault has formed a trap for the petroleum.

consist of rock formations which trap the petroleum deposits in the rock, and prevent the oil and gas from dissipating or migrating into surrounding areas (Figure 1.4). Underground reservoirs are also called **pools,** although this term is often misunderstood. Oil and gas pools do not consist of large, tank-like accumulations of petroleum underground. Rather, oil and gas actually occur within the rock itself. Not all rock is impervious; several types are porous, and contain many small holes, or **pore spaces,** within their structure that may serve to hold quantities of petroleum (Figure 1.5). Therefore, oil and gas pools, or hydrocarbon reservoirs, refer to porous rock formations in which oil and gas have been trapped.

The term, **petroleum reserves,** refers to the estimated amount of oil and gas which is contained in the earth, and is capable of being recovered or produced. On the other hand, the terms, **oil in place** and **gas in place,** refer to the gross amount of hydrocarbons in the earth, whether they can be extracted or not. The amount of oil actually recovered is often only 25 to 30 percent of the oil in place, since the natural energy available is usually insufficient to produce any greater amounts. Much of the oil remains trapped in the small channels and pore spaces of the rock, and resists being extracted despite various recovery techniques. Certain enhanced recovery

Figure 1.5. POROUS ROCK CONTAINING OIL.

Table 1.2. PRESENT WORLD RESERVES OF CRUDE OIL
(as of January 1, 1984)

Country or Area	Oil Reserves (Thousands of Barrels)	
United States	27,735,000	
Canada	6,730,000	
Latin America	81,675,900	
TOTAL WESTERN HEMISPHERE		116,140,900
Middle East	370,100,800	
Africa	56,907,020	
Asia-Pacific	18,969,400	
Western Europe	23,019,480	
TOTAL FREE WORLD		585,137,600
Communist Nations	84,600,000	
TOTAL WORLD		669,737,600

(Source: American Petroleum Institute)

methods including **fluid injection processes**, have made it possible to recover some of the remaining oil. (This will be discussed in the chapter on Recovery.)

The amount of recoverable raw materials, or petroleum reserves, becomes important when our nation's energy needs are considered. Today, the United States consumes approximately 15 to 17 million barrels of oil every day. Yet, our country is capable of producing only a portion of this amount, and as a result, must import large quantities of oil. In recent years, we have been importing five to seven million barrels per day of both crude oil and refined products, or about 30 to 35 percent of our total consumption.

Table 1.2 provides some useful information about the present world energy situation in terms of available petroleum reserves. As the Table indicates, total known reserves of oil are estimated to be approximately 670 billion barrels. Of this amount, the United States has reserves of about 28 billion barrels, which represents only a little more than 4 percent of the world's total reserves. Should the United States have to rely solely on its own supply of oil, and if it could be produced fast enough to meet daily requirements, the nation would only have enough oil to last five to six years, at the current rate of consumption. Even if the current level of oil imports were maintained, these reserves would be sufficient to last only about eleven years.

Of course, these figures do not take into consideration the new discoveries which might occur, and the revisions that are constantly made regarding reserve estimates. Nevertheless, the amount of new oil and gas being discovered has not kept pace with the amount being produced. For example, during 1982, new discoveries of oil and other reserve revisions added 1.38 billion barrels to U.S. oil reserves, while the nation's oil production during the same year was 2.95 billion barrels. This resulted in a net reserve loss of 1.57 billion barrels of oil by the end of the year. The five year period from 1978 to 1983 showed discoveries and revisions adding to 12.03 billion barrels, while production during that time was 14.86 billion barrels, for a net reserve loss of 2.83 billion barrels.

The supply of natural gas in the United States today is in the same predicament. Total U.S. gas reserves at the end of 1983 were approximately 200 trillion cubic feet, or about 6 percent of the world's total known supply. While this seems to be a large amount of natural gas, the United States consumed approximately 17 trillion cubic feet of gas during that same year. At this rate of consumption, U.S. reserves of gas are sufficient to last only about 12 years. No significant amount of gas is imported to the United States because of the difficulty in transporting it from overseas. Since only five percent of total U.S. consumption is imported, the country must rely primarily on its own resources. The gas industry is giving special emphasis to the discovery of new gas fields at home to increase our present gas reserves.

As our discussion indicates, the long range supply of raw materials in the United States is steadily decreasing. Although there is constant speculation and argument back and forth about the extent of the "energy crisis," the fact remains that we are dealing with a finite resource, whose yearly consumption eats away at the total supply. As a result, the petroleum industry is putting more effort into developing new reserves, recovering additional amounts of petroleum from old fields, as well as finding alternate sources of energy to meet our needs. With the decreasing supply of raw materials accompanied by an increasing cost of petroleum products for the consumer, energy conservation has also become very important.

C. Exploration

The raw materials used by the oil industry are not easily found. They lie hidden underground, at depths often several miles beneath the surface. Because there are so few clues to indicate the precise location of petroleum reservoirs, detailed exploratory work is necessary before their discovery is possible. The responsibility for locating these new reservoirs belongs to the geologist and geophysicist, whose efforts are vital to today's petroleum industry. These individuals work in the exploration departments of most major oil companies, as well as in many of the smaller firms.

Geologists, who study the history and transformation of the materials that make up the earth, search for clues in the earth's crust which may indicate the presence of petroleum. The science of petroleum geology began in the mid-1800s in the United States, as oil exploration became important. Today, a great deal of scientific data, together with many scientific and technical procedures, is available to geologists to aid them in their search for petroleum. (These activities will be discussed in detail in the following chapters.)

Geophysicists study the physics of the earth, including earthquakes, magnetism, gravity, and radioactivity. The use of geophysics in petroleum exploration developed in the early 1900s. Modern geophysicists are primarily involved in seismology, or the study of induced vibrations in the earth to determine its subsurface structure and nature. In a sense, seismology enables geophysicists to "see" into the earth's crust. With this information, these scientists can determine in what areas petroleum might be trapped beneath the surface of the earth.

Geophysicists also use data from gravitational and magnetic studies in their search for oil and gas. These methods are often combined with those used by geologists to insure that exploration activities are conducted with maximum efficiency and productivity.

As a result of the exploration efforts in the oil industry, approximately 170 major oil and gas fields have been discovered in the United States since 1947.

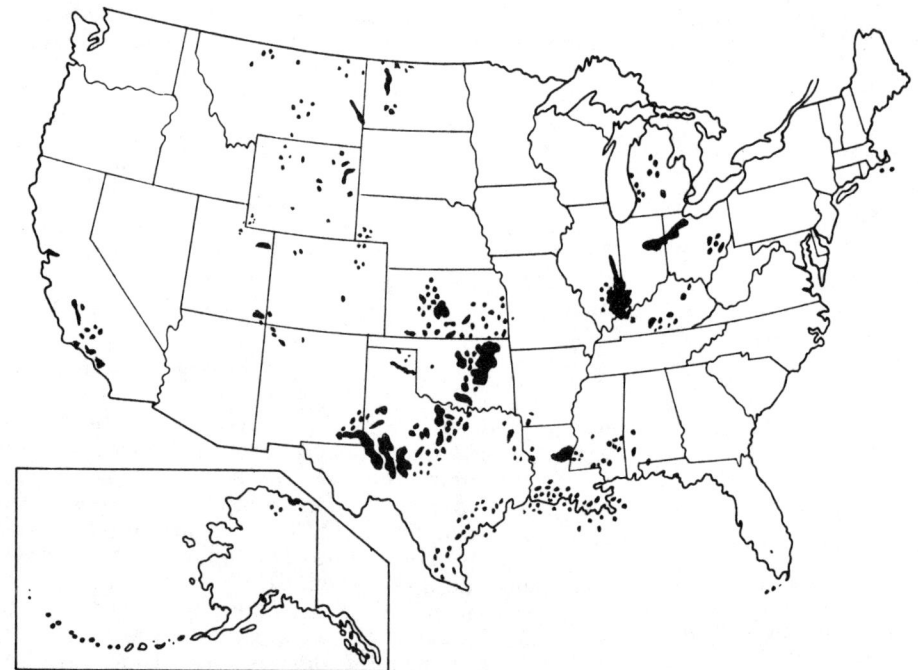

Figure 1.6. OIL PRODUCING AREAS IN THE UNITED STATES.

over the last twenty years in order to meet the challenge. As the search for oil and gas moves to more demanding environments, such as offshore or the Arctic regions, the science of petroleum exploration will become even more important to the overall success of the industry.

D. Land

After the explorationist has identified areas which may contain accumulations of oil and gas, proper legal rights to the land must be secured, so that exploration and development can proceed.

Men and women, called "landmen," are responsible for obtaining the appropriate rights from landowners. Most commonly, the necessary rights are acquired through an **oil and gas lease**, a legal document authorizing exploration, drilling, and production of oil and gas on a particular tract of land. In return for signing the lease, the owner receives various forms of compensation. (Lease arrangements will be discussed in the chapter on Land.) Once a lease has been acquired, landmen are further responsible for insuring that all of the obligations in the lease are properly met.

In addition to acquiring leases or other property rights, landmen are often involved in negotiating and preparing **farmins, joint ventures,** and other agreements between individuals and companies in the industry.

(Major oil fields include those which have reserves of over 50 million barrels, while major gas fields are those with over 300 billion cubic feet of reserves.) In addition to these major finds, over 3,000 fields of significant size have been discovered in the same period. (Significant fields contain more than one million barrels of oil or more than six billion cubic feet of gas.)

Petroleum is located in several areas throughout the United States. Oil and gas fields are found in the Appalachian Mountains, the Midwest, the Gulf Coast of Texas and Louisiana, the Rocky Mountains, California, and Alaska (Figures 1.6 and 1.7). In recent years, many offshore oil and gas fields have been discovered on the shallow continental shelf surrounding the United States. The Gulf Coast is the most prolific source of such fields at the present time. New fields are also being discovered off the western coast, primarily in California. In addition to these regions, exploration work is being conducted on the eastern coast, and along the offshore continental shelf of Alaska.

Major discoveries, however, are becoming fewer all the time. The last major oil discovery in the United States was made along the northern slope of Alaska. Because petroleum is becoming increasingly more difficult to find, modern exploration techniques have necessarily become more sophisticated

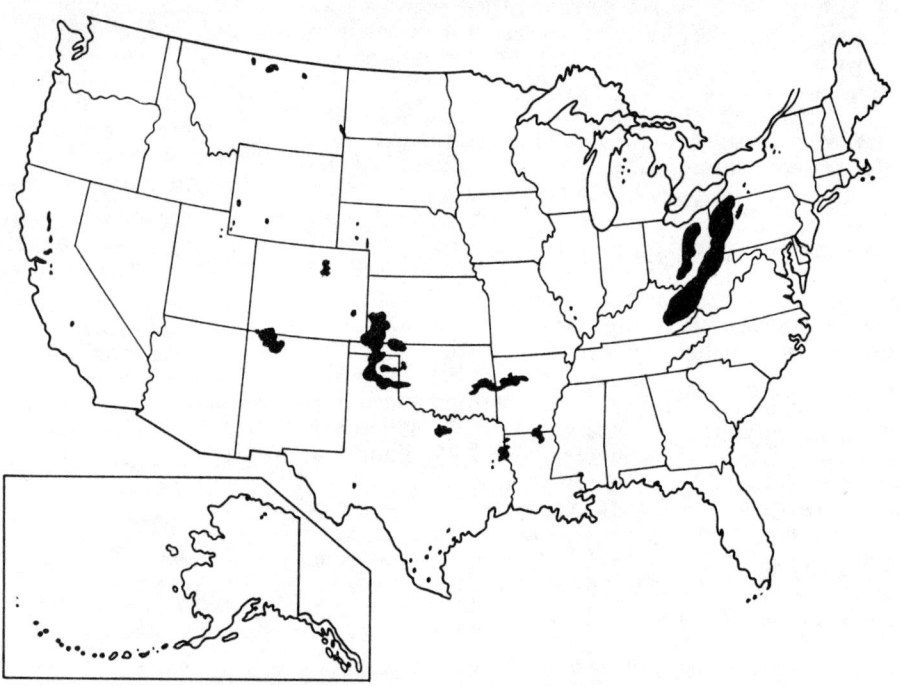

Figure 1.7. GAS PRODUCING AREAS IN THE UNITED STATES.

People in land work also coordinate their efforts with the other groups within an oil company. For example, they often work closely with the exploration department in obtaining rights-of-way for surface development work. It is also their responsibility to secure the necessary land rights for roads, power lines, and other facilities, whenever new fields are developed.

E. Drilling and Completion

After the exploration department has identified an area as having sufficient oil or gas potential to warrant further exploration, and once the land department has obtained the proper leases and agreements to accomplish this work, drilling operations may begin. To some degree, all of the exploration and land work is speculative; only by drilling a hole in the ground can the existence of petroleum actually be verified. Oil companies are constantly involved in testing new prospects and developing new fields. Consequently, a great deal of drilling activity is continually going on in the United States. Even though economic factors may affect activity, both onshore and offshore rigs drill tens of thousands of wells each year in this country. In 1982 alone, approximately 85,000 wells were completed.

Although the process of drilling can be complex, its goal is rather simple: to make a small hole in the earth, often measuring no more than six to eight inches in diameter. In the early years of the oil industry, drilling

Figure 1.9. ROTARY RIG. (Photo courtesy of Petro-Lewis Corporation)

was originally accomplished by using a **cable tool rig** (Figure 1.8). This method consisted of pounding a hole in the earth, by continually raising and dropping a chisel-like bit attached to the end of a weighted piece of pipe. The process was slow, however, and could not be used to drill deep wells or pressured formations.

Although there are some instances in which it is still used, the cable tool rig has largely been replaced by the modern **rotary rig** (Figure 1.9). Rotary drilling involves rotating the drill bit, which is attached to a long string of drill pipe. This rotating action allows for faster and more efficient cutting of the rock. The use of rotary drilling techniques has made possible the exploration of many new areas.

Another significant improvement with this system is the utilization of a **fluid circulating system.** A fluid, called "drilling mud," is pumped down the inside of the drill pipe and out through the bit at the bottom of the hole. The mud carries the fragments of broken rock, cut by the drill bit, to the surface. It also counteracts any high pressure zones encountered in the well, and cools and lubricates the bit.

Following the drilling of a production well, the well must be completed and stimulated to enhance the flow of oil from the formation into the wellbore. Generally, steel pipe, called **casing**, is run into the hole to keep it from collapsing (Figure 1.10). The casing is cemented in place to make it stationary, and to prevent the migration of fluids between permeable zones.

After a well has been cased, it must be perforated. Holes are shot in the casing, opposite the producing formation, so oil or gas can enter the well. Sometimes, the well must be stimulated to either enlarge already existing pore channels or to create new ones in the formation. This allows the fluids to flow more efficient-

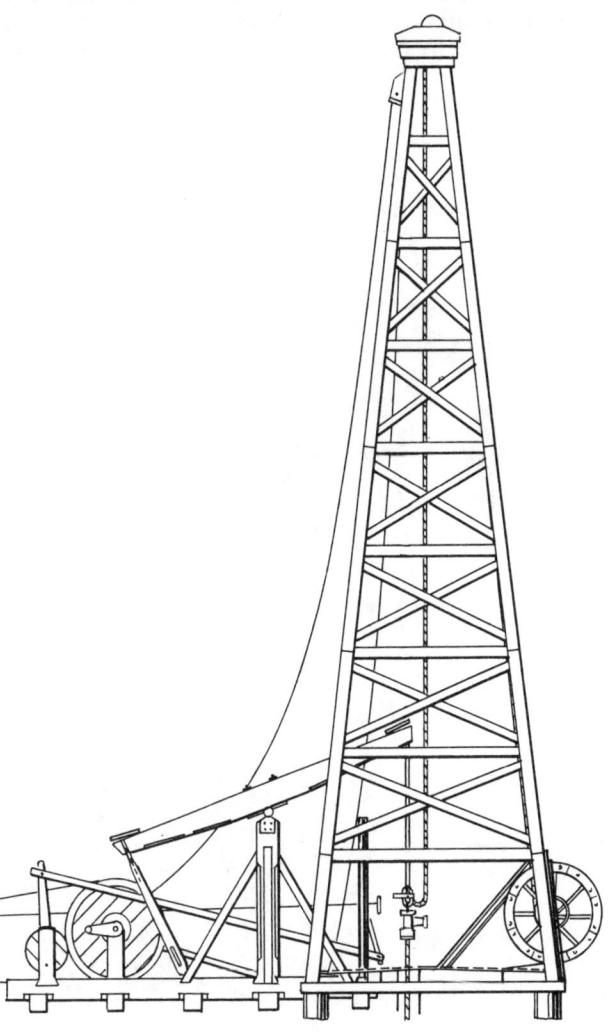

Figure 1.8. CABLE TOOL RIG.

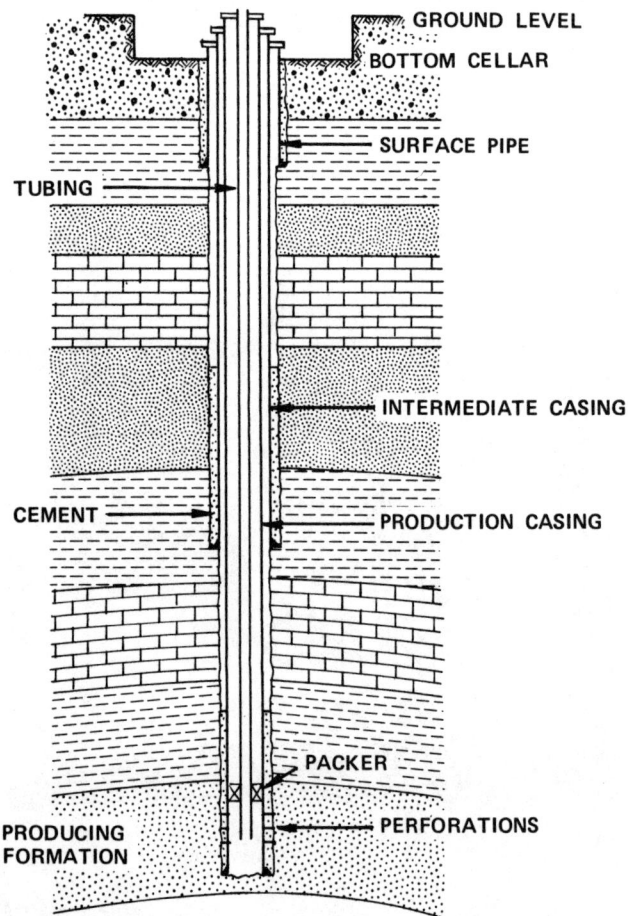

Figure 1.10. TYPICAL WELL COMPLETION.

ly into the well. Stimulation involves treating the formation with acid or other substances to dissolve any restrictive material, to improve permeability, and to clean the area around the well.

F. Production

If exploratory drilling and testing of the well reveal the presence of petroleum in commercial quantities, production operations are set into motion. At this stage, facilities are constructed to recover the oil or gas, and prepare it for shipping to the refinery.

After the well has been completed, **tubing** is inserted in the casing to serve as a flow path for formation fluids. Oil and gas reservoirs often have sufficient pressure to cause the petroleum to flow naturally into the wellbore and up through the tubing to the surface. These are called **flowing wells,** and include almost all gas wells and many oil wells (Figure 1.11).

In many cases, however, reservoir pressure is not sufficient to drive the fluids from the producing zone to the surface. Artificial lift equipment must then be installed. **Pumping units,** which are similar to the old-fashioned, hand-operated water pumps, lift the fluids to the surface (Figure 1.12). Other types of pumping equipment and artificial lift techniques will be discussed in the chapter on Production.

When oil comes from the ground, it often contains water and other impurities, called BS&W (basic sediment and water), which are not acceptable to purchasing companies for delivery in a pipeline. The oil must first be treated at the production site. One common method of cleaning the oil is through the use of **free-water knockouts**—large vessels in which water and sediments settle to the bottom and are drawn off by a disposal line. **Heater-treaters** are other common oil-treating devices, in which heat is applied to the oil-water mixture, causing the water to separate from the oil. As

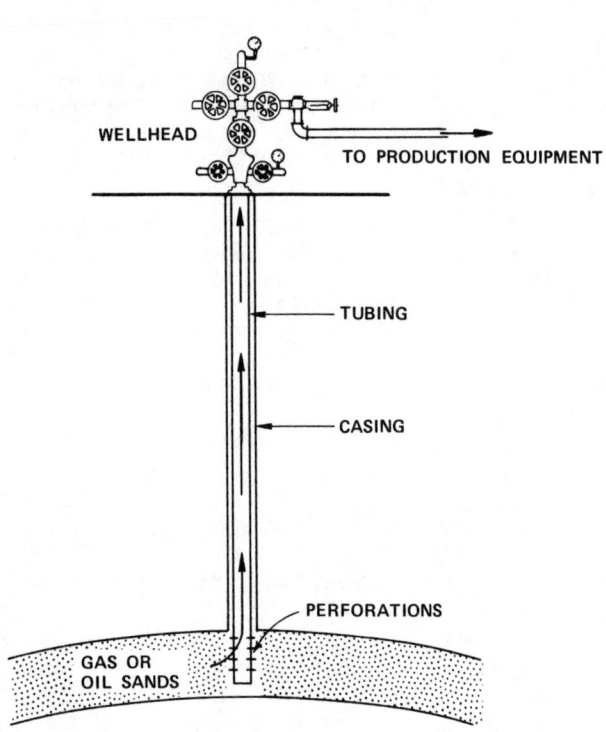

Figure 1.11. FLOWING GAS OR OIL WELL.

Figure 1.12. ARTIFICIALLY LIFTED WELL.
(Photo courtesy of Kaneb Services, Inc.)

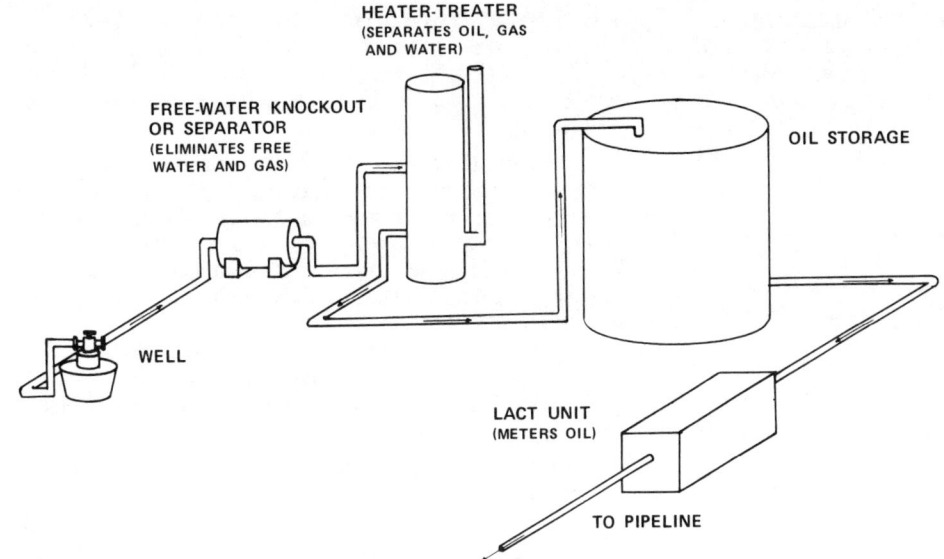

Figure 1.13. SURFACE PRODUCTION FACILITIES.

H. Refining and Petrochemicals

The final step in the process of converting crude oil and natural gas into useable products is refining and petrochemical processing. There are more than 300 refineries in the United States today. Although some are near producing oil fields, most refineries are located in heavily populated industrial centers, which have access to water transportation facilities. The refining and petrochemical industries will not be discussed in detail in this text; however, some of the basic processes which are common to all refineries are described briefly here.

illustrated in Figure 1.13, this equipment is often located right at the wellhead.

Oil **storage tanks** are used to accumulate the oil at the well site. These tanks are usually connected to a pipeline or trucking outlet, through which the oil can be transported to a refinery. A **LACT unit** is often included with this equipment (Figure 1.13). LACT, or Lease Automatic Custody Transfer, is a sophisticated metering device that measures the quantity, and monitors the quality, of the oil shipped from the production site.

Detailed accounting and record-keeping is maintained on all production operations. Records are kept on the amount of oil and gas sold and their market prices, while the costs of the production operations themselves are carefully monitored. In addition, oil and gas companies must file several different government reports on various aspects of their oil and gas production.

G. Transportation

An extensive transportation network has been developed in this country for transporting oil and gas from remote areas or offshore fields to refining facilities located in populous, industrial centers. The earliest forms of transportation were wagons and barges, which hauled the oil from the fields to rail centers and water shipping points. Today, oil is shipped in a variety of ways: **tank trucks, railroad tank cars, tankships, barges,** and **pipelines,** with each method capable of meeting certain unique problems and circumstances.

Most petroleum in this country, including natural gas, crude oil, and refined products, is transported by **pipeline.** Approximately 250,000 miles of crude oil and refined products lines cross and recross the United States. There are over a million miles of major gas pipelines, excluding the small distribution lines running through many cities.

In the United States, over 165,000 **railroad tank cars** and approximately 150,000 **tank trucks** haul oil to many locations that pipelines and tanker vessels cannot reach. **Tankers** have emerged as important vehicles of petroleum transportation because of the high volume of international trade between oil producing areas and importing countries. There are now more than 5,100 ocean-going tanker vessels throughout the world.

Refining is a combination of processes and operations that convert crude oil into its many products. The fundamental process in refining is called **fractional distillation,** in which the crude oil, or **feedstock,** is heated until it becomes a mixture of hot vapors and liquid. It is then run through a **fractionating tower,** which typically may be 15 feet in diameter and more than 100 feet high. As the vapors travel upward through the tower, they cool and condense at different levels on horizontal trays (Figure 1.14).

The rising vapors must bubble up through the liquid formed in each successive tray. Since the various components in petroleum have different boiling points, they condense at different temperatures. As shown in the illustration, the heavier components, such as asphalt and heavy fuel oil, are separated out first, because they condense at the highest temperatures. The very light hydrocarbons, such as gasoline, condense near the top of the tower. At each level, the separated components are drawn off by pipes at the sides of the tower. Gases that do not condense are carried from the top of the column. Through the use of the fractionating tower, a variety of products can be distilled from the crude oil used as feedstock.

With the growing demand for gasoline in the early

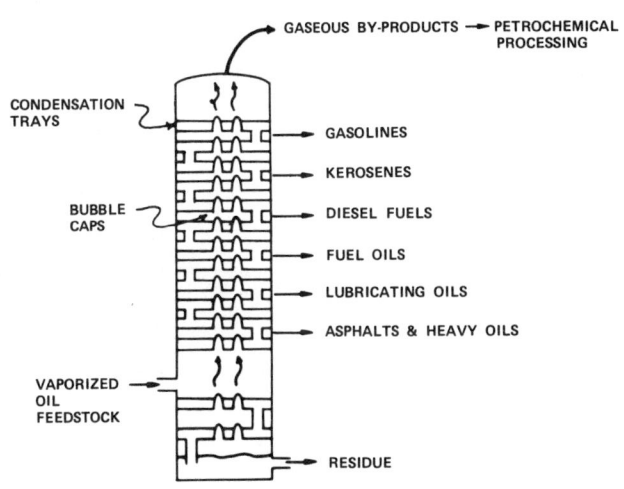

Figure 1.14. FRACTIONATING TOWER.

1900s, a method had to be found for increasing the amount of gasoline that could be made from a barrel of crude oil. Simple fractional distillation only yields a small portion of gasoline from a typical barrel of crude oil. In 1913, William Burton, a young chemist, discovered a new method for producing more gasoline—**thermal cracking.** Burton discovered that by heating the crude oil to high temperatures and increasing the pressure, the large molecules were forced to break down into smaller ones, thereby producing gasoline. Thermal cracking not only increased the quantity of gasoline produced, but also improved the quality, since the resulting gasoline exhibited considerably better antiknock characteristics.

Today, **catalytic cracking** and **hydrocracking** have replaced thermal cracking in modern refineries. In **catalytic cracking,** a **catalyst** (a substance that accelerates chemical reaction without undergoing chemical change) is used to break up the molecules at a lower pressure. This is the most common cracking process used today.

Hydrocracking requires high pressures and temperatures, and utilizes hydrogen to increase the yield of gasoline. For this reason, it is a relatively expensive process. Its advantage for the future, however, is that it produces higher octane products, and does not leave a carbon residue. **Reforming, alkylation,** and **isomerization** are refining processes that also increase the octane level of gasoline by rearranging the molecular structure of the hydrocarbons.

An additional function of refining is to **treat** the oil, specifically to remove the sulfur compounds. Removal of these compounds increases the quality and effectiveness of the different fuels, but more importantly, it reduces the sulfur levels in the atmosphere.

Over 3,000 different petroleum products are now available as a result of the refining process. These include gasoline, jet fuels, kerosene, diesel fuels, fuel oils, petroleum coke, liquefied petroleum gas (LPG), lubricating oils, greases, waxes, and asphalt. Table 1.3 illustrates what percentages of the different products have been made in our refineries over the last few years. Note that the majority of crude oil is used to make gasoline, fuel oils, and jet fuel. Table 1.4 gives a more detailed listing of various products derived from crude petroleum.

The **petrochemical industry** is a relatively recent outgrowth of the refining industry. Although there was some petrochemical manufacturing as early as the 1920s, the greatest impetus for the industry came during World War II, when the war effort demanded a ready supply of certain products. Petrochemical processing was to expand even more during the post-war years, as many of these products were transferred to peacetime uses. There are over 700 petrochemical plants in the United States, mainly centered along the Gulf Coast in Texas and Louisiana.

Petrochemicals are pure chemical substances derived from crude oil or natural gas, which serve as the feedstock for the manufacture of a variety of products. There are three major groups of petrochemicals: **aliphatics, aromatics,** and **inorganics.** The **aliphatics** are organic compounds in which the carbon atoms are arranged in a straight row. These include ethylene, butylene, propylene, and ethyl alcohol. The **aromatic** petrochemicals are organic compounds with their carbon atoms in a circular configuration. The principal aromatics are benzene, toluene, and xylene. The **inorganic** petrochemicals, which contain no carbon atoms, include ammonia, sulfur, and carbon black.

There are approximately 3,000 different chemicals that can be derived from petroleum. They are used in the manufacture of a wide range of products such as plastics, synthetic rubber, detergents, solvents, cleaning compounds, films, paints, adhesives, fibers, herbicides, explosives, saccharin, aspirin, insulation, and pharmaceuticals.

I. Marketing

After the raw materials have been refined and processed, large-scale marketing operations are necessary to distribute all the consumer products made from petroleum. Most people do not see the petroleum

**Table 1.3. PERCENTAGE YIELDS OF REFINED PETROLEUM PRODUCTS FROM CRUDE OIL[1]
IN THE UNITED STATES—ANNUALLY 1964-1979**

	1964	1965	1966	1967	1968	1969	1970	1971	1972	1973	1974	1975	1976	1977	1978(r)	1979(p)
Gasoline	44.1	44.0	44.4	44.0	43.9	44.8	45.3	46.2	46.2	45.6	45.9	46.5	45.5	43.4	44.1	43.0
Jet Fuel	5.6	5.7	6.2	7.5	8.3	8.2	7.5	7.4	7.2	6.8	6.8	7.0	6.8	6.6	6.6	6.9
Ethane (including Ethylene)	—[2]	—[2]	—[2]	—[2]	—[2]	0.2	0.2	0.2	0.2	0.2	0.1	0.1	0.1	0.1	0.1	0.1
Liquefied Gases	3.3	3.2	3.0	3.1	3.1	2.9	3.0	2.9	2.8	2.8	2.6	2.4	2.4	2.3	2.3	2.3
Kerosine	2.9	2.8	2.9	2.7	2.7	2.6	2.3	2.1	1.8	1.7	1.3	1.2	1.1	1.2	1.0	1.3
Distillate Fuel Oil	22.8	22.9	22.5	22.2	22.1	21.7	22.4	22.0	22.2	22.5	21.8	21.3	21.8	22.4	21.4	21.5
Residual Fuel Oil	8.2	8.1	7.6	7.7	7.2	6.8	6.4	6.6	6.8	7.7	8.7	9.9	10.3	12.0	11.3	11.5
Petrochemical Feedstocks	1.8	1.7	2.1	2.4	2.5	2.5	2.5	2.7	2.9	2.9	3.0	2.7	3.3	3.6	4.1	4.7
Special Naphthas	0.8	0.9	0.9	0.8	0.7	0.7	0.8	0.7	0.7	0.7	0.8	0.6	0.7	0.6	0.7	0.6
Lubricants	2.0	1.9	1.8	1.8	1.7	1.7	1.6	1.6	1.5	1.5	1.6	1.2	1.3	1.2	1.3	1.3
Wax	0.2	0.2	0.2	0.2	0.2	0.2	0.2	0.2	0.1	0.2	0.2	0.1	0.1	0.1	0.1	0.1
Coke	2.6	2.5	2.5	2.5	2.5	2.6	2.7	2.6	2.8	2.9	2.8	2.8	2.6	2.5	2.5	2.6
Asphalt	3.5	3.7	3.8	3.5	3.6	3.5	3.6	3.8	3.6	3.6	3.7	3.2	2.8	2.9	3.2	3.1
Road Oil	0.2	0.2	0.2	0.2	0.1	0.2	0.3	0.2	0.2	0.2	0.2	0.1	0.0	0.1	0.1	—
Still Gas	4.0	4.1	3.9	3.9	4.0	4.1	4.1	3.8	3.9	3.9	3.9	3.9	3.7	3.6	3.7	3.8
Miscellaneous	0.4	0.5	0.5	0.4	0.4	0.4	0.3	0.4	0.4	0.4	0.5	0.7	1.0	1.0	0.9	0.8
Shortage[3]	-2.4	-2.4	-2.5	-2.9	-3.0	-3.1	-3.2	-3.4	-3.3	-3.6	-3.9	-3.7	-3.5	-3.6	-3.4	-3.6
TOTAL	100.0	100.0	100.0	100.0	100.0	100.0	100.0	100.0	100.0	100.0	100.0	100.0	100.0	100.0	100.0	100.0

[1]Other unfinished oils added to crude oil in computing yield
[2]Included with liquefied gases
[3]Processing Gain (-) or Loss (+)
(r) Revised
(p) Preliminary

(Source: U.S. Energy Information Administration, Annual Petroleum Statements.
Quoted in **Basic Petroleum Data Book,** American Petroleum Institute.)

Table 1.4. CRUDE PETROLEUM AND SOME OF ITS PRODUCTS

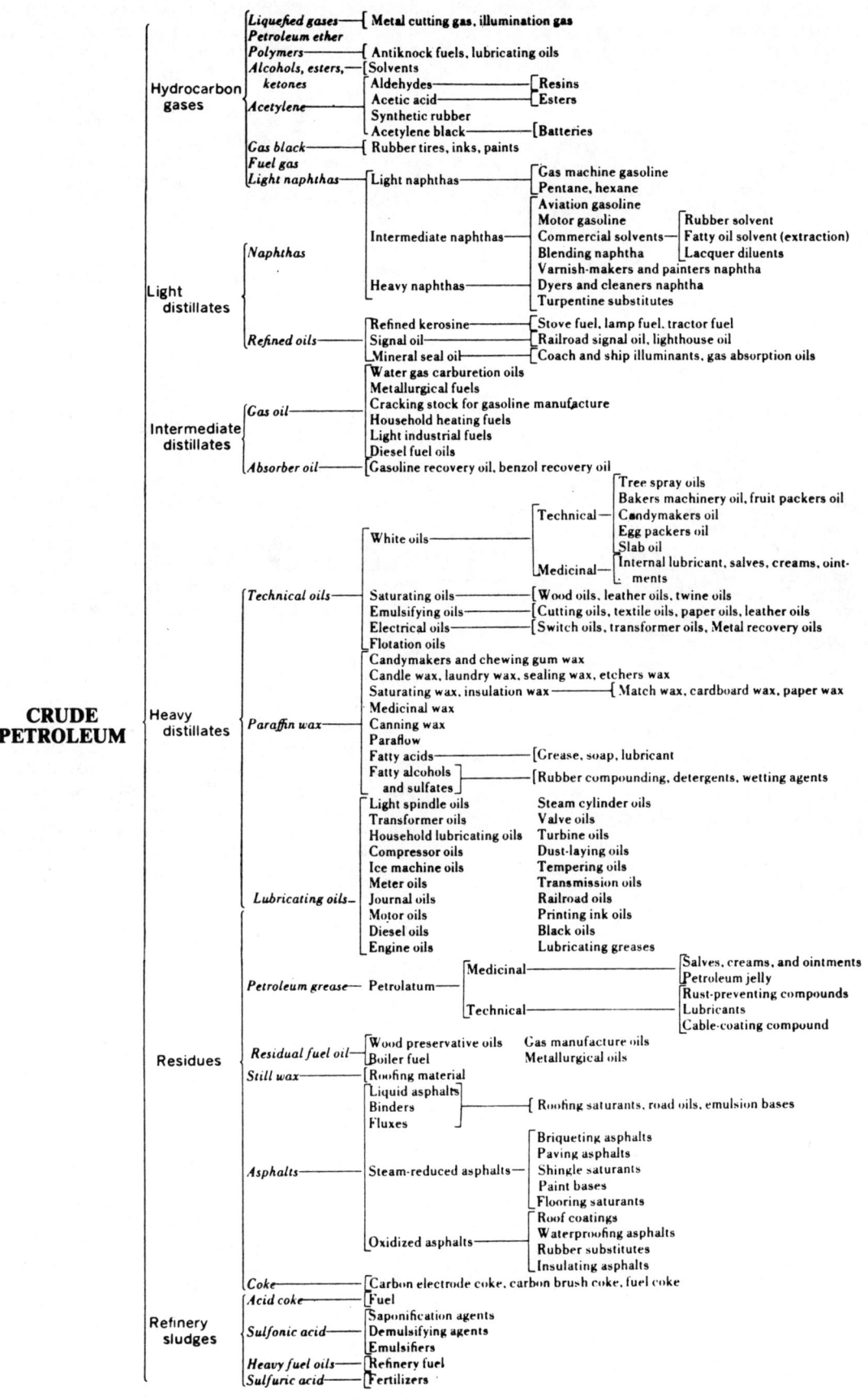

From P. Albert Washer, Texas A. and M. College Extension Division (First Session). Printed in **Encyclopedia of Energy.** McGraw-Hill (New York) 1976.

products as they are transported in pipelines, tank cars, or trucks from field, to refinery, to retail outlet. Consumers are generally aware of petroleum only in its refined, marketed condition, such as gasoline for automobiles or as fuel oil for their homes or businesses.

Marketing of gasoline and fuel oil is most important in this country, since the majority of crude petroleum is used to make these products. Petroleum products are shipped from refineries to **terminals** or **bulk plants** by pipeline, tanker, barge, rail, and truck. Petroleum **terminals** are large centers that redistribute the products to bulk plants, wholesale distributors, dealers, service stations, and large customer accounts. **Bulk plants** are smaller redistribution points servicing large commercial accounts, homeowners, farmers, and service stations.

The transferral of products from these distribution points is handled by **refiner-marketers** or **distributors**. A **refiner-marketer** usually sells the products under its own brand to distributors, dealers, and consumers, but may also sell products without a particular brand to distributors or dealers, who then sell the products under their brand.

Generally, fuel oil is marketed directly from the wholesale distributor to the consumer. Gasoline, however, is usually marketed by a retail marketer through independent service stations. Service station dealers receive their products either from refiner-marketers or distributors. The dealer may sell these products under the brand of the refiner-marketer, the distributor, or may use his own private brand, depending on the particular arrangement.

In years past, the typical service station was a branded operation, supplied by a major brand refiner and run by an independent dealer. The station would not only offer gasoline, but also a variety of auto-related products and services. Over the last two decades, however, significant changes have taken place in gasoline outlets throughout the country, as they have been affected by inflationary costs and periods of tight supply.

First, there has been an overall decline in the number of service stations and gasoline outlets available, due to a reduction in new station construction and the closure of marginal outlets. From a survey of 28 companies conducted by API, the number of new stations declined from 2,500 in 1970 to a little over 300 in 1978. At the same time, deactivations increased from approximately 3,600 in 1970 to over 5,000 during 1978. A U.S. Department of Commerce report revealed a similar reduction in gasoline outlets (defined as retail establishments receiving over half of their gross income from gasoline sales) from a little over 225,000 in 1972 to approximately 175,000 in 1978. (Source: **Recent Changes in Retail Gasoline Marketing**, API.)

Secondly, the character and type of retail gasoline sales outlets have changed considerably. Independent, private branded gasoline marketers have provided strong competition to major brand operations, with an emphasis on high gasoline volume and a significant reduction in the availability of additional services. Using this approach, these independent marketers have reduced the cost to the consumer.

The number of self-service stations has increased dramatically over the past decade, usually because they offer only gasoline at the cheapest possible price. Full-service stations, which provide automobile service and products besides gasoline, still exist and function, but a considerable number of them now provide self-service options also. In addition, other retail outlets, such as convenience stores, are offering self-service gasoline facilities.

Although gasoline and fuel oil are those commodities most familiar to the average customer, the marketing field is not limited to these two products. There are more than 3,000 products made from petroleum, including diesel fuel, jet fuel, lubricants, liquefied petroleum gas, fertilizers, waxes, asphalts, industrial oils, etc., that must make their way from the refinery to the appropriate consumer. Thus, the petroleum marketplace is a complex system providing goods to a number of different sectors in our society.

PETROLEUM INDUSTRY—BIBLIOGRAPHY

GENERAL

American Petroleum Institute: **Facts About Oil,** API (Washington, D.C.) no date.

American Petroleum Institute: **Recent Changes in Retail Gasoline Marketing,** API (Washington, D.C.) 1981.

Berger, Bill D. (ed.): **Facts About Oil,** Technology Extension, Oklahoma State University (Stillwater) 1975.

Berger, Bill D. and Kenneth E. Anderson: **Modern Petroleum: A Basic Primer of the Industry,** The Petroleum Publishing Company (Tulsa) 1978.

Boatright, Mody C. and William A. Owens: **Tales from the Derrick Floor,** Doubleday and Company, Inc. (Garden City, N.Y.) 1970.

Carter, D.V. (ed.): **History of Petroleum Engineering,** American Petroleum Institute (Dallas) 1961.

Clark, J. Stanley: **The Oil Century: From the Drake Well to the Conservation Era,** University of Oklahoma Press (Norman) 1958.

Exxon Corporation: **The Offshore Search for Oil and Gas,** Exxon Background Series (New York) 1980.

Leffler, William L.: **Petroleum Refining for the Non-Technical Person,** PennWell Books (Tulsa) 1979.

Petroleum Extension Service: **Fundamentals of Petroleum,** The University of Texas at Austin (Austin) 1979.

The Petroleum Resources Communication Foundation: **Our Petroleum Challenge: The New Era,** 2nd Ed. (Calgary, Alberta) 1980.

Wheeler, Robert R. and Maurine Whited: **Oil: From Prospect to Pipeline,** 3rd Ed., Gulf Publishing Company (Houston) 1975.

REFERENCE

American Petroleum Institute: **Basic Petroleum Data Book,** API (Washington, D.C.) published three times a year.

Desk and Derrick Clubs of North America: **D & D Standard Oil Abbreviator,** 2nd Ed., The Petroleum Publishing Company (Tulsa) 1973.

Langenkamp, Robert D.: **Handbook of Oil Industry Terms and Phrases,** The Petroleum Publishing Company (Tulsa) 1974.

Langenkamp, Robert D. (ed.): **The Illustrated Petroleum Reference Dictionary,** The Petroleum Publishing Company (Tulsa) 1980.

Lapedes, Daniel (ed.): **Encyclopedia of Energy,** McGraw-Hill Book Company (New York) 1976.

PennWell Publishing Company: **International Petroleum Encyclopedia** (Tulsa) published yearly.

Petroleum Extension Service: **A Dictionary of Petroleum Terms,** 2nd Ed., The University of Texas at Austin (Austin) 1979.

Thrush, Paul W. (ed.): **A Dictionary of Mining, Mineral, and Related Terms,** Bureau of Mines, Department of the Interior, 1968.

Williams, Howard R. and Charles J. Meyers: **Manual of Oil and Gas Terms,** Matthew Bender and Company (New York) 1976.

TRADE JOURNALS

AAPG Bulletin (monthly), American Association of Petroleum Geologists, Inc. (P.O. Box 979, Tulsa, OK 74101).

Drilling Contractor (monthly), International Association of Drilling Contractors (3737 Westcenter, Houston, TX 77042).

Geophysics (monthly), Society of Exploration Geophysicists (P.O. Box 3098, Tulsa, OK 74101).

Journal of Petroleum Technology (monthly), Society of Petroleum Engineers of AIME (6200 N. Central Expressway, Dallas, TX 75206).

Oil and Gas Journal (weekly), PennWell Publishing Company (1421 S. Sheridan Road, Tulsa, OK 74101).

Oilweek (weekly), Maclean Hunter Ltd. (200-918-6th Avenue S.W., Calgary, Alberta T2P 0V5).

Petroleum Engineer International (monthly), Energy Publications Division, Harcourt Brace Jovanovich (P.O. Box 1589, Dallas, TX 75221).

Pipeline and Gas Journal (monthly), Energy Publications (P.O. Box 1589, Dallas, TX 75221).

Pipe Line Industry (monthly), Gulf Publishing Co. (P.O. Box 2608, Houston, TX 77001).

Society of Petroleum Engineers Journal (bimonthly), Society of Petroleum Engineers (6200 N. Central Expressway, Dallas, TX 75206).

Western Oil Reporter (monthly), Hart Publications (660 Bannock St., Suite 300, Denver, CO 80203).

World Oil (monthly), Gulf Publishing Company (3301 Allen Parkway, Houston, TX 77019).

CHAPTER 2:
GEOLOGY

I. INTRODUCTION

The word **geology** is derived from two Greek words: "geo," meaning **earth,** and "logos," meaning **word,** or as in English usage, **science.** Geology, then, is the science of the earth: the accumulated body of knowledge concerning the planet we inhabit. Geology is typically divided into two major sections: **physical geology** and **historical geology. Physical geology** deals with the nature, properties, and distribution of the materials composing the earth, and the processes by which these materials are formed, changed, and transported. **Historical geology** is concerned with the history of the planet: its development, physical changes, and evolution.

An understanding of geology is important for the discovery and production of the various minerals and fuels having commercial value. Because most fossil fuels, such as oil, gas, and coal, lie beneath the earth's surface and are hidden from direct observation, geology is a vital part of today's petroleum industry.

The sections in this chapter cover three major areas: physical geology, which discusses basic rock classifications and the geological processes directly related to the formation of petroleum; historical geology, which examines earth history and geologic time; and petroleum geology, which explains the origin and characteristics of petroleum reservoirs.

II. PHYSICAL GEOLOGY

A. Introduction

Physical geology studies the structure, form, and arrangement of the rocks, minerals, and core materials that make up the earth, and the processes and forces by which these materials are developed, changed, and transported. Knowledge of these processes is necessary in petroleum geology. Only certain types of materials in the earth contain oil, gas, or coal, and it is important to be able to distinguish these materials, as well as to understand how they were formed.

This section is divided into two parts. The first, "Basic Rock Classifications," examines the materials themselves. The second, "Geologic Processes," considers the processes that act upon these materials.

It should be pointed out that there is a difference between a **mineral** and a **rock.** A **mineral** is defined as a naturally occurring element or compound with a definite composition and structure. A **rock,** on the other hand, is a naturally occurring solid, composed of one or more minerals. (Biologically-formed carbon compounds, such as coal and asphalt, are sometimes excluded from these definitions.) Legally, petroleum is considered a mineral, just as water and ice are sometimes considered to be minerals. Here are some examples showing the difference between a rock and a mineral: sand (or sandstone) is a rock, normally composed of the minerals quartz and feldspar; shale is a rock which usually consists of a suite of very fine grains of clay minerals; limestone is a rock containing the mineral calcite.

The materials of the earth will generally be described in this book as rocks, although their mineral constituents will also be noted at times.

B. Basic Rock Classifications

Geologists have divided the earth's rocks into three major families, based on the origin of the rock:

1. Igneous Rocks

The word igneous is derived from the Latin word "ignis," meaning **fire.** These rocks were once a hot molten liquid known as **magma,** which subsequently cooled and solidified (Figure 2.1). For example, lava

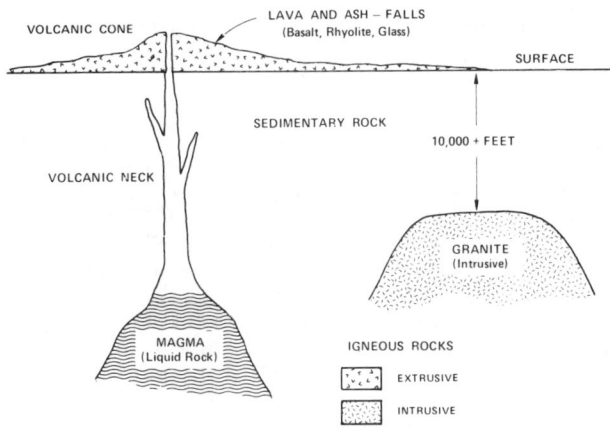

Figure 2.1. FORMATION OF IGNEOUS ROCKS.

from a volcano flows over the earth's surface, then cools and hardens into an igneous rock (**extrusive**). Other igneous rocks actually cool and harden while still beneath the surface (**intrusive**). Common igneous rocks include granite, basalt, obsidian, and pegmatite. Cores of mountain ranges are typically composed of igneous rocks, most usually granite.

2. Sedimentary Rocks

The word sedimentary comes from the Latin, "sedimentum," meaning **settling.** These rocks are commonly composed of particles which have been broken off of pre-existing rocks (Figure 2.2). These particles are transported by wind, water, or ice to a depositional area. After deposition of the particles or sediments, **lithification** (or the process of cementation, compaction, and hardening) occurs and a sedimentary rock is formed. Common sedimentary rocks include sandstone, siltstone, limestone, dolomite, and shale.

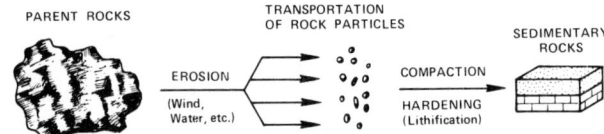

Figure 2.2. FORMATION OF SEDIMENTARY ROCKS.

3. Metamorphic Rocks

The term metamorphic is derived from the Greek words "meta," meaning **change**, and "morphe," meaning **form**. Metamorphic rocks, then, are rocks that have changed their form, or **metamorphosed**, because of pressure, temperature, and accompanying chemical reactions. For example, at high temperatures and pressures, soft coal is changed to anthracite or hard coal and then to graphite (Figure 2.3). Common metamorphic rocks are schist, gneiss, marble, and slate.

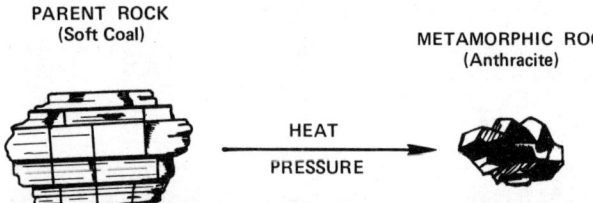

Figure 2.3. FORMATION OF METAMORPHIC ROCKS.

A **rock cycle** may be devised showing the interrelationship of the different types of rock (Figure 2.4). A starting point can be the solidification of igneous rocks from magma. When the igneous rocks, or any other rocks, break down, particles of rock are transported and deposited as sediments. These sediments are lithified and become sedimentary rocks, which in turn, may be transformed into metamorphic rocks when subjected to extreme heat and pressure. Igneous rocks subjected to heat and pressure may also be metamorphosed, or changed, directly to metamorphic rock. The metamorphic rock may undergo additional heat and pressure changes causing melting and re-crystallization back to igneous rock. In nature, the cycle is more complicated than the above, and the processes may be stopped or altered at any point.

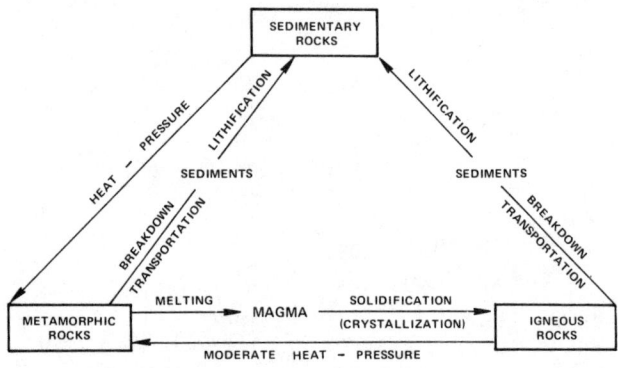

Figure 2.4. ROCK CYCLE.

C. Geologic Processes

There are several geologic processes which act upon rocks (Figure 2.5). Those processes which affect

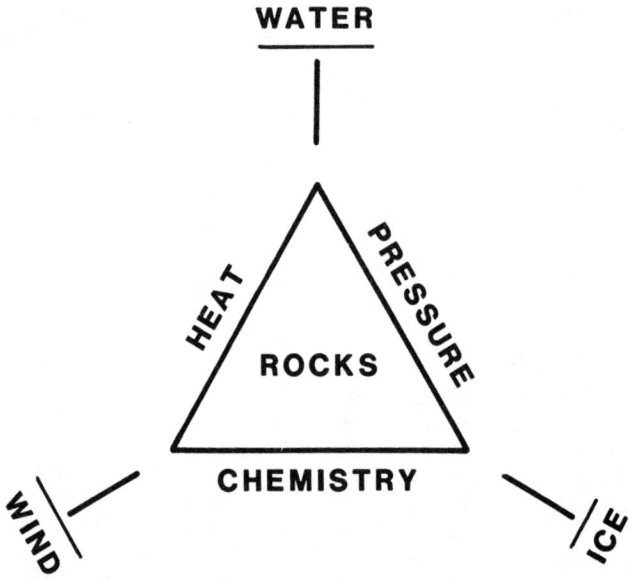

Figure 2.5. GEOLOGIC PROCESSES.

sedimentary rocks are particularly important in our discussion, since most oil and gas bearing rocks are sedimentary in origin.

The fundamental principle governing geologists' study of the earth and its geologic processes is known as the doctrine of **uniformitarianism**. This concept states that the processes which operated in the past to modify the earth's surface are the same processes that act today. This principle was formulated by a Scottish medical man, James Hutton, in 1785, and may be thought of as the birth of modern geology. Geologists have learned much about the past development of the earth by observing present day events.

This section is divided into two parts. The first part, "Genesis," studies the processes which cause the formation of sedimentary rock. The second part, "Post-Depositional Phenomena," examines some of the processes which act on the sedimentary rock after it has been formed.

Some of the major topics which will be considered are **sedimentation**, which denotes the entire process of separation, transportation, deposition, and consolidation of rock particles into sedimentary rock; **stratigraphy**, which deals with recognizing, describing, and defining various sedimentary units; and **structural geology**, which is concerned with the form and arrangement of rocks, including folding and displacement. Because of the vast amount of information on these topics, our treatment is not always technically complete. Instead, we will concentrate on those aspects which pertain to the formation and development of petroleum.

1. Genesis

"Genesis" means the origin or coming into existence of something. In this part, the "coming into existence" of sedimentary rocks will be discussed. Igneous and metamorphic rocks will not be considered, since they generally do not contain petroleum.

As illustrated in Figure 2.2, sedimentary rocks are formed in a multi-step process. First, the **parent rocks**, or original rocks, are broken up into fragments. These pieces are then carried to a new location, where they are deposited and lithified (compacted and hardened) into a new sedimentary rock.

The process of **erosion**, whereby rocks are broken up into many small particles, can be accomplished in several different ways. These different types of erosion are called **weathering processes**, of which three are most common: **physical weathering, chemical weathering,** and **biological weathering.**

Physical weathering occurs when rocks are physically broken up into small fragments, either through the abrasive action of water, wind, ice, and frost, or from the expansion and contraction of the rock due to changes in temperature.

In **chemical weathering**, some minerals in a rock are dissolved by chemical action, as when water containing dissolved carbon dioxide forms a weak acid and dissolves calcite from a rock such as limestone.

Biological weathering is the process in which plants or animals physically or chemically alter rocks. For example, tree roots can grow through a rock, causing the rock to split.

After weathering, or erosion, of the parent rock has occurred, the particles generally are **transported** to a new area by water or wind. Streams and rivers transport large amounts of rock material. For example, every year the Mississippi River carries over one-half billion tons of rock and soil materials which were eroded from the mountains and plains of the central and western United States. Also, wind often carries tiny rock and soil fragments many miles across land. The sand dunes across the western United States are evidence of this ongoing process.

Following erosion and transportation, the sediments are **deposited** in various environments (Figure 2.6). If transportation is by water, the sediments may be deposited in the stream carrying them (**fluvial deposits**), in a lake (**lacustrine deposits**), or in an ocean or sea (**marine deposits**). If transportation is by wind, the deposits are called **eolian deposits.**

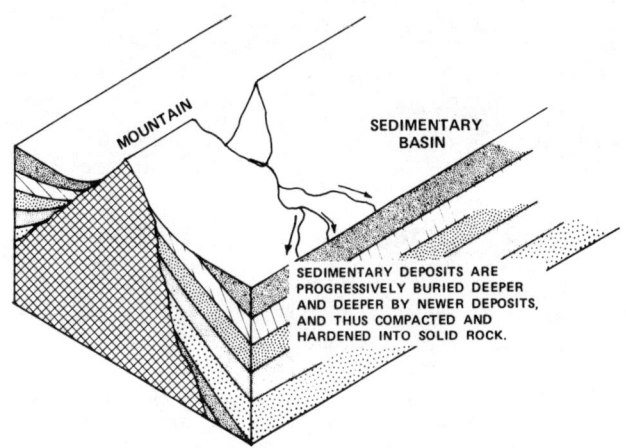

Figure 2.7. LITHIFICATION.

The final step in the formation of a sedimentary rock is the process of **lithification**, which converts unconsolidated rock-forming materials into consolidated rock. The basic methods of lithification are **compaction, desiccation, cementation,** and **crystallization.**

In **compaction,** the volume or size of the sediments is reduced either by the pressure of overlying sediments or by the pressures resulting from earth movement (Figure 2.7).

Desiccation, which means "drying out," occurs when compaction forces water out of the sediments. Desiccation may also be caused by evaporation.

Cementation results when some binding agent forms between particles of the sediment. Silica, calcite, dolomite, and iron oxide are common cementing agents.

Crystallization, or the developing of an orderly mineral structure (a crystal) within the sediments, is a common form of lithification for chemical sedimentary rocks and, also, for some fine-grained sediments.

There are several different types of sedimentary rocks, but only those which commonly contain petroleum deposits will be described here—**clastic** sedimentary rocks and **carbonate** sedimentary rocks. Each plays an important role in petroleum geology.

The term **clastic** refers to those rocks which usually are composed of fragments or particles of rock compacted or cemented together into a new rock. Perhaps the most common example of a clastic rock is sandstone. The clastic nature of the rock is evident when a piece of sandstone is rubbed against a hard surface, causing many small pieces of the rock to crumble off. These grains of sand are small rock particles which have been bound together to form the sandstone.

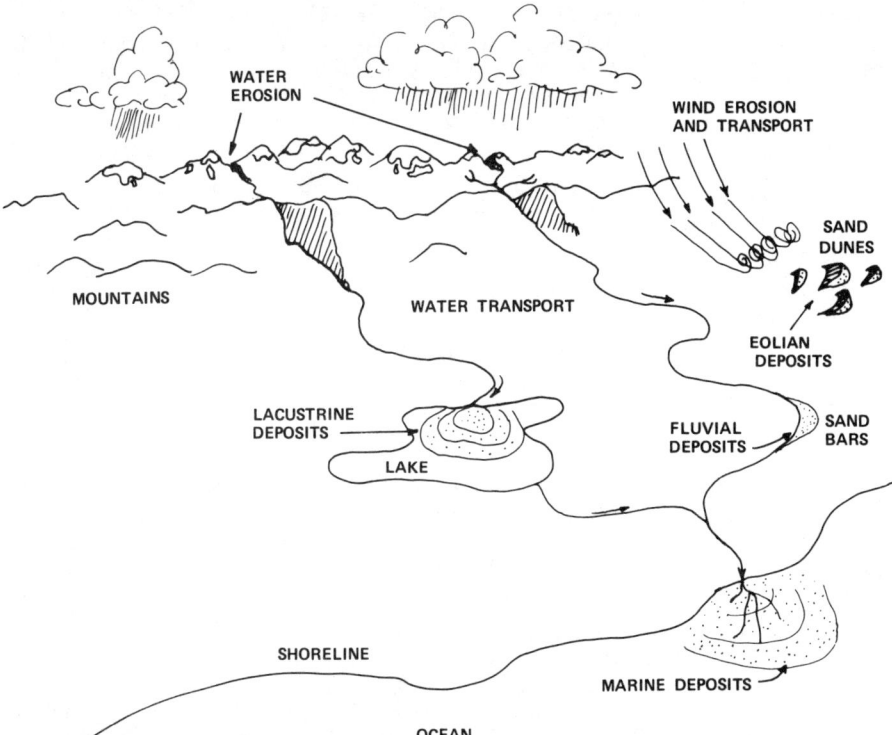

Figure 2.6. EROSION, TRANSPORTATION, AND DEPOSITION OF SEDIMENTARY ROCKS.

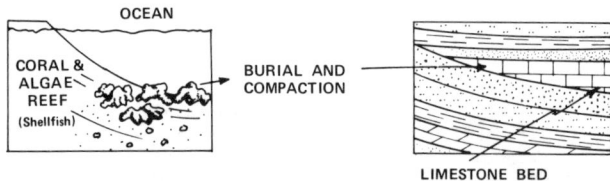

Figure 2.8. FORMATION OF LIMESTONE.

Carbonate rocks are another principal type of sedimentary rock. While clastic rocks are formed because of physical weathering, carbonate rocks result from either lithification of organic debris, such as the hard fragments of coral and algae, or less commonly, by direct precipitation of carbonates from sea water. One of the most common carbonate rocks, limestone, often has a biological origin and is formed from the shells of dead animals or plants (Figure 2.8). Such deposits often are called by the name of the source animal or plant, such as an oyster reef deposit or crinoidal limestones.

Once the sedimentary rock has been formed, it can be classified. Usually, particle size determines the name for clastic sedimentary rocks, as listed in Table 2.1.

Conglomerates and sandstones are coarse enough that individual rock grains can be seen without a microscope, but siltstones and shales require magnification to determine their clastic nature. Generally, recoverable petroleum is found in the more coarsely textured rocks, particularly sandstone. Shales are so finely grained and tightly compressed that normally the petroleum they contain is not recoverable.

Although carbonate rocks and clastic rocks are subject to the same process of lithification, carbonate rocks are classified according to their chemical composition and not by particle size. Two common carbonate rocks associated with petroleum are classified below:

Mineral Present	Rock Name
Calcite, $CaCO_3$	Limestone
Dolomite, $CaMg(CO_3)_2$	Dolomite

Since sedimentary rocks are prime sources of petroleum deposits, they receive special attention from petroleum geologists. By using rock samples taken from wells drilled in sedimentary deposits, geologists can study the **stratigraphy** of the rock sections encountered in the well (Figure 2.9).

Table 2.1. CLASSIFICATION OF SEDIMENTARY ROCKS

Average Particle Size (mm)	Average Particle Size (Approximate) (in.)	Rock Name
Greater than 2 mm	Greater than .08 in.	Conglomerate
Between $\frac{1}{16}$ mm and 2 mm	Between .002 in. and .08 in.	Sandstone
Between $\frac{1}{256}$ mm and $\frac{1}{16}$ mm	Between .0001 in. and .002 in.	Siltstone
Less than $\frac{1}{256}$ mm	Less than .0001 in.	Claystone or Shale

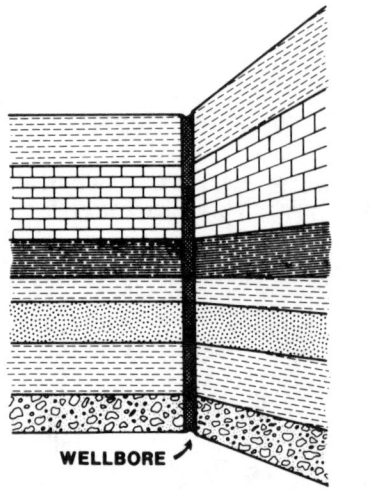
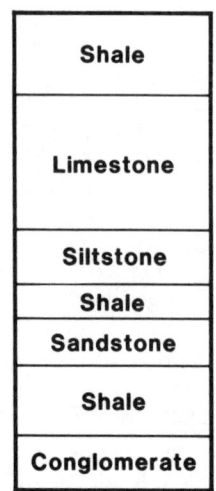

Figure 2.9. STRATIGRAPHY. Left: stratified (layered) sedimentary rocks. Right: stratigraphic column.

The word **strata** refers to the different rock beds or layers which have been laid down during the depositional process. These layers often are clearly divided from each other by texture or composition. In studying these strata, geologists (in this case **stratigraphers**) can construct **stratigraphic cross sections** which show the lateral and vertical variations and the distributions of these beds or formations. These sections provide information on the geologic age of the rocks. A **stratigraphic column** can also be compiled, showing the vertical succession of the strata found in a particular area (Figure 2.9). Sample stratigraphic columns or charts from various parts of the world are shown in the Appendix at the end of the chapter.

2. Post-Depositional Phenomena

Once the sedimentary rock has been deposited and lithified, other processes can occur that will affect the rock. In this section we will look at some of the processes that are particularly relevant to the petroleum industry. The actual results of these processes will be discussed in greater detail in the section on petroleum geology.

The first set of post-depositional phenomena may collectively be labeled **rock movement.** The results of these rock movements are studied by the structural geologist, who is concerned with deformations in the earth's crust and their relation to petroleum.

One major type of rock movement is **subsidence** and **burial.** When an area of deposition subsides, younger sediments may continue to accumulate on the older sediments, as shown earlier in Figure 2.7. In the Ganges River basin of India, sedimentary deposits which once were laid down on the earth's surface, now are estimated to be buried under 45,000 to 60,000 feet (8 to 11 miles) of more recent sediments.

The opposite process is known as **uplift.** Local uplifts often occur as a result of mountain building or formation (Figure 2.10). The Ellenburger formation in west Texas was once buried almost 80,000 feet (15 miles) below the surface of the earth. Today, because of regional uplift and erosion, it lies only 16,000 to 22,000 feet below ground level. When sediments are uplifted to the surface, erosion can take place. The erosion surface formed, when buried by younger sediments, is known as an **unconformity.**

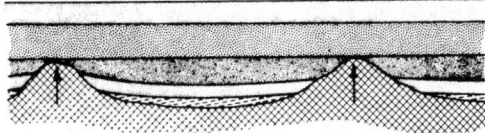
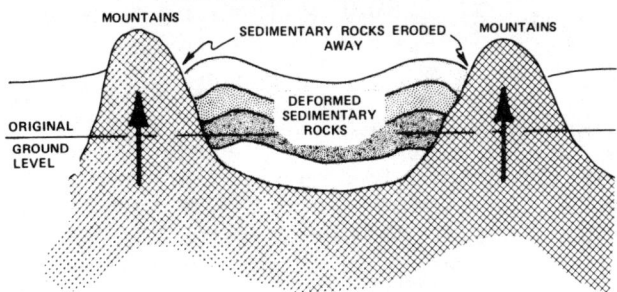

Figure 2.10. UPLIFTING AND FOLDING OF ROCKS.

Under conditions in which a force is applied to the rocks, the rocks may be deformed or fail. **Deformation** of the rocks involves **folding,** whereby the rocks are actually bent like a piece of cardboard (Figure 2.11). Many oil accumulations occur as a result of folding.

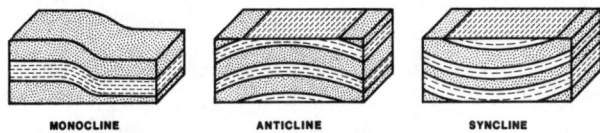

Figure 2.11. FOLDED ROCKS.

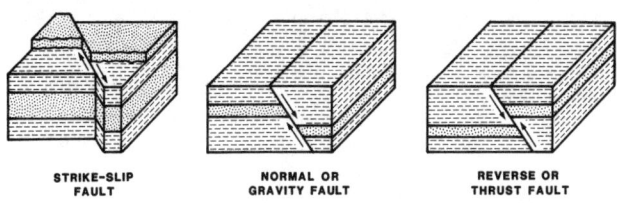

Figure 2.12. FAULTS.

Rock **failure** results when sufficient force is applied in a short period of time to actually break the rock. **Faults** mark the line where the rock has failed (Figure 2.12). Faults also may cause petroleum entrapment. On a smaller scale, **fractures** are likewise the sign of rock failure (Figure 2.13).

A final significant form of rock movement is the process known as **diapirism.** A diapir is caused when a mobile underlying sediment actually "flows" and pierces or ruptures overlying, less mobile layers (Figure 2.14). **Salt domes** are examples of diapiric structures. Folding and faulting often are associated with diapirism. Together, these fractures may form traps for oil and gas.

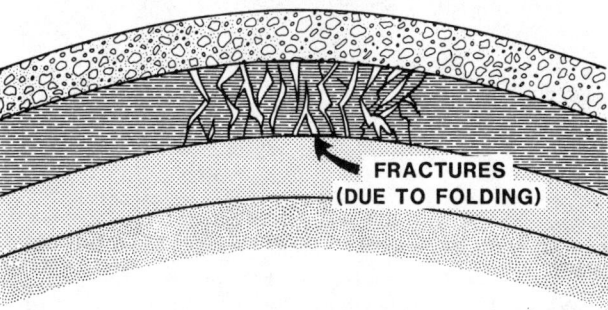

Figure 2.13. FRACTURES.

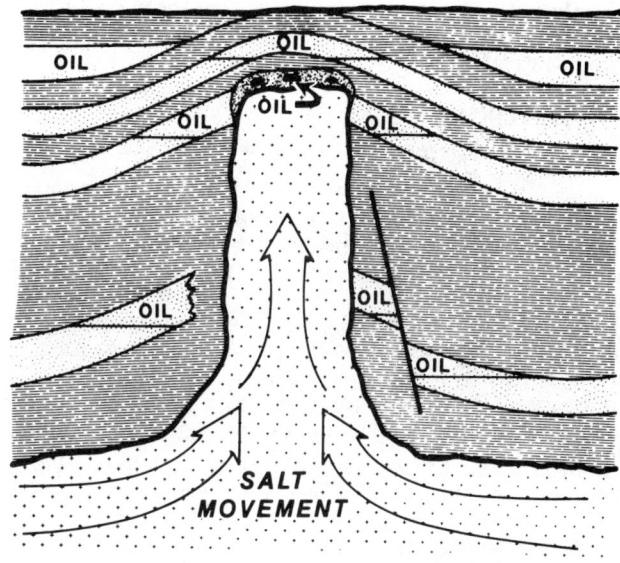

Figure 2.14. DIAPIRISM: SALT DOME.

The second set of post-depositional phenomena is characterized by **fluid movement** within the pore spaces of sedimentary rocks. An obvious case of fluid movement is the migration and accumulation of petroleum in a reservoir. This specific topic will be discussed in detail in the section on petroleum geology.

Water movement constitutes another example of the significance of fluid movement as a post-depositional phenomenon. Once deposition has occurred, water may act as an agent in two rather opposite ways. First, solids may be precipitated out of water into the pore space between the particle grains of the rock, forming a **secondary cement** (Figure 2.15). At times, secondary cement may entirely fill the voids between particles.

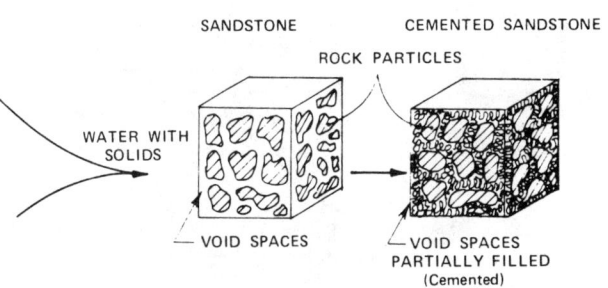

Figure 2.15. CEMENTATION BY FLUID MOVEMENT.

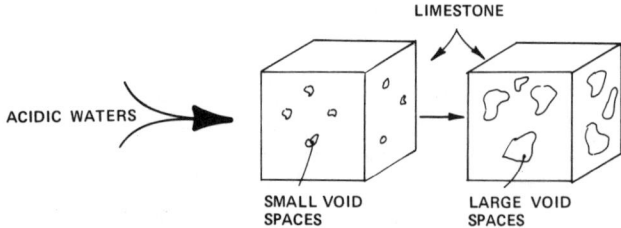

Figure 2.16. WATER AS A DISSOLVING AGENT.

Water, on the other hand, may also act as a dissolving agent, as described earlier in the section on genesis. The dissolution of some of the rock-forming minerals can create more void space in the rock, called **secondary** or **solution porosity** (Figure 2.16). Secondary porosity is very important in some limestone and dolomite reservoir rocks.

III. HISTORICAL GEOLOGY

A. Introduction

To achieve a better understanding of the geologic processes that influence the formation of petroleum, it is necessary to become acquainted with historical geology. Since petroleum was formed over a period of several million years, its history is closely associated with the evolution of the earth. This history covers an enormous span of time. From a lifeless, primordial mass, the earth went through long and complex changes to become the planet that we know today. Such an extensive subject can only be touched upon briefly here.

In this section, some of the major geologic events will be discussed. In addition, the special terminology used to classify the earth's evolutionary events will be related to **geologic time**.

B. Earth History

The history of the earth extends much further into the past than the recorded history of man. Such questions as how the earth was formed, when life first appeared, and what forces caused the physical changes in the earth's surface have long puzzled scientists, and are not easily answered. Detailed study of clues in the earth's rocks and minerals have enabled geologists to piece together this ever changing story of our planet's evolution.

Sometime in the distant past, perhaps 15 to 20 billion years ago, the universe as we know it came into existence. The earth, the moon, and the rest of the solar system, however, were not formed until approximately 4.6 billion years ago, when they condensed out of a large dust cloud, according to the theories of Von Weizacker. By that time, the earth presumably had the same type of form and structure that it has today.

Modern scientists believe that the earth is composed of three distinct zones or layers: the **crust**, the **mantle**, and the **core** (Figure 2.17). The **crust** is the thin outer layer of the earth, ranging in thickness from 3 to 22 miles. Beneath the crust, extending approximately 1,800 miles, lies the section known as the **mantle**. This interval consists of a viscous fluid from which volcanic lava and granitic mountains originate. Finally, the **core,** a dense, molten, inner section, extends to the center of the earth.

Table 2.2 MAJOR GEOLOGIC EVENTS

Number of Years Ago	Event
4,600,000,000	Formation of the earth
3,800,000,000	Oldest rocks found on the earth are formed
3,500,000,000	The atmosphere and the ocean have been formed
3,300,000,000	Oldest fossils of one-celled plants; first appearance of life
2,000,000,000	The ozone layer of the atmosphere, which protects us from the dangerous radiations of the sun, is formed
600,000,000	First abundant fossil record laid down
550,000,000	Mountains appear where the Appalachians now exist, as part of the supercontinent, Pangea
400,000,000	Fish and land plants appear
360,000,000	Insects and land animals appear
300,000,000	Vast swamps exist in what will become the eastern United States; the remains of these swamps form coal
250,000,000	The last major cycle of mountain building in the Appalachians occurs; evidence of an ice age (much of the land's surface is covered with thick sheets of ice)
225,000,000	Dinosaurs appear
200,000,000	The supercontinent of Pangea, which is composed of virtually all the land on earth, begins to break up
180,000,000	Mountains appear where the Rocky Mountains now exist
150,000,000	South America begins to separate from Africa, Australia, India, and Antarctica; birds and mammals appear
120,000,000	The Atlantic Ocean begins to form; seed-bearing plants appear
110,000,000	Africa separates from Australia, India, and Antarctica
70,000,000	Dinosaurs become extinct
50,000,000	Ancestors of the horse appear in North America
40,000,000	Australia and Antarctica break apart
15,000,000	Grasses and manlike creatures appear
2,000,000	Man appears; the Great Ice Age begins, during which ice advances to the middle of the North American continent four times
6,000	Civilization arises in several parts of the world

Figure 2.17. EARTH STRUCTURE.

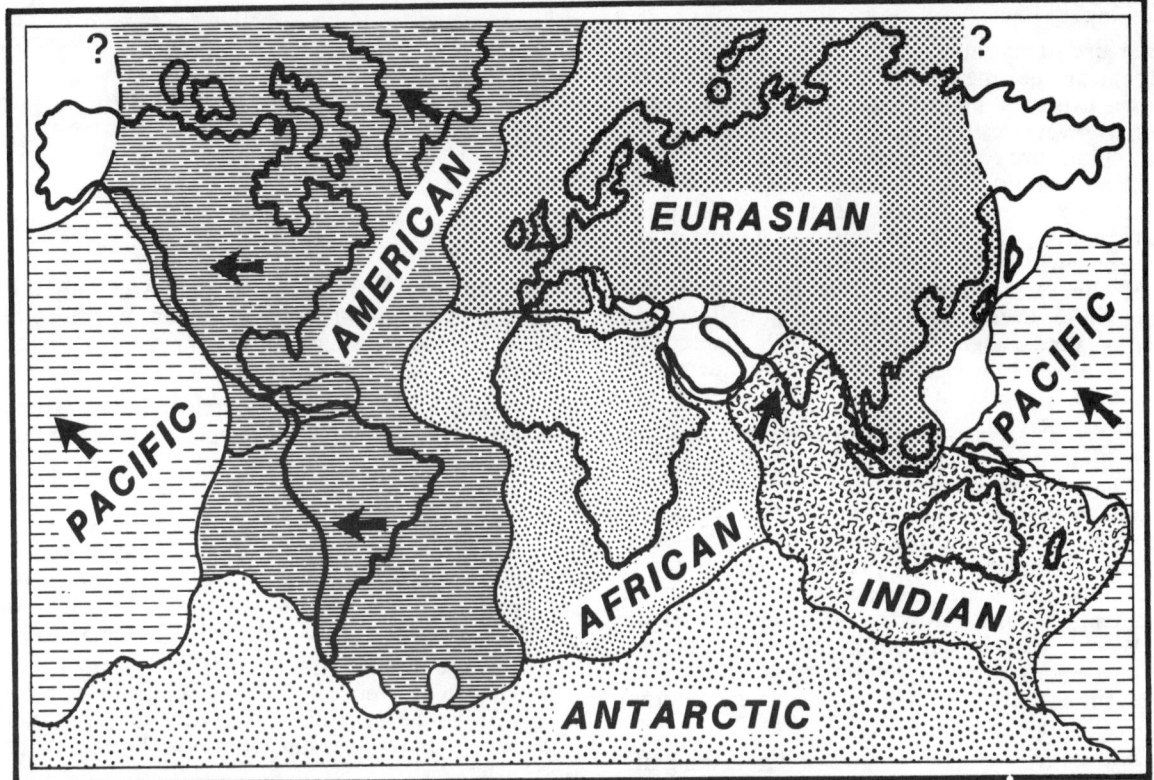

Figure 2.18. PRESENT DAY PLATES. Arrows indicate direction of movement.

This section on historical geology will concentrate on only those events which have occurred on or within the earth's crust. In comparison to the massive interior of the earth, the crust is as thin as the skin of an apple; yet it is within this skin that the majority of the important geologic events have taken place.

Table 2.2 lists the major geologic events that make up the earth's history. The approximate dates associated with each event illustrate the vast span of time over which these events have occurred.

Two of the more intriguing details noted on this "calendar" are the relatively recent appearance of man and the concept that all the continents of the world were once combined into one large "supercontinent."

If all the time that has elapsed since the formation of the earth could be compressed into a single day, then man would have arrived one minute before midnight. Civilization, which is more recent still, would have developed within the final second.

The branch of geology that deals with this concept is now known as **plate tectonics.** Plate tectonics suggests that the continents and major islands of the world are plates floating on the viscous rocks within the mantle (Figure 2.18). **Sea-floor spreading** is the associated idea that the continents are being forced apart or pushed together by magma coming up through mid-ocean fissures.

The theory proposes that at one time a supercontinent, **Pangea,** existed and was made up of all the major land masses of the world (Figure 2.19). About 200 million years ago, Pangea broke up into two bodies of land called **Laurasia** and **Gondwanaland.** Laurasia consisted of North America, Greenland, Europe, and Northern Asia, while Gondwanaland contained Africa, Arabia, Australia, India, Antarctica, and South America. Later, Laurasia and Gondwanaland separated into the continental masses that exist today.

A great deal of evidence, including geologic correlations and studies of magnetism around the earth, supports these theories. Plate tectonics and sea-floor spreading help to explain the mountain chains of the world, deep sea trenches, a number of peculiar geologic occurrences, and even such features as the San Andreas Fault in California. The study of the evidence of these theories is one of the fastest growing and most exciting areas in geology today.

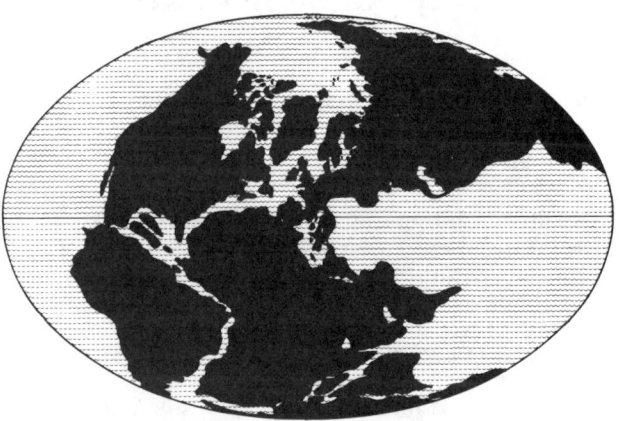

Figure 2.19. PANGEA.

C. Geologic Time

Because of its immense duration, geologic time differs from normal measurements of time. The earth has been in existence approximately 4.6 billion years, according to present estimates. To illustrate the magnitude of a billion, imagine that you earn one dollar every minute. You would need almost 2,000 years to accumulate one billion dollars. An understanding of geologic time is important when considering some of the major events which have occurred throughout the history of the earth. Geologic time is the basis for arranging these events into a usable, well-defined schedule, or **time chart.** Familiarity with the terminology of a geologic time chart is basic to an understanding of geologic information.

Two different scales are used to measure geologic time. The first is **absolute time,** in which a date is specified. The second is **relative time,** in which one event is compared to another. As an example of the difference between the two scales, consider the statement, "Women won the right to vote in 1920." This is dating on an absolute time scale, in which a date is specified. The statement, "Women won the right to vote after the end of World War I, but before the stock market crashed," is an example of relative time, in which a specific event is dated relative to one or more other events.

This distinction between methods of dating is significant in geology, since commonly used methods of absolute dating have only recently been developed. These methods are known as **radioactive dating,** in which the amount of radioactive element in a rock and the amount of its decay products are compared. Radioactive carbon, as well as the uranium isotopes contained in rock compounds, can help determine the ages of rocks. These methods work because the rate at which a radioactive element decays is considered a constant, unaffected by any known physical or chemical process (Figure 2.20).

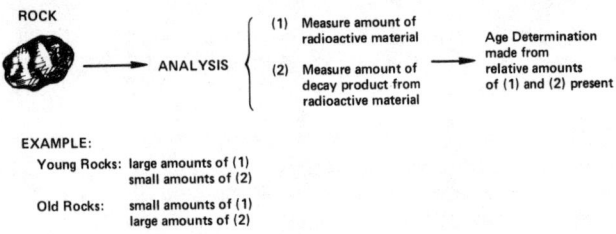

Figure 2.20. RADIOACTIVE DATING.

Radioactive dating is not always possible, however, since not all rocks contain radioactive elements in sufficient amounts to accurately determine their age. It then becomes necessary to use **relative dating.**

Relative dating consists of correlating rock formations between two different areas, using some conspicuous feature common to both areas. **Fossil dating** is one type of relative dating. Fossils are evidence of past life, such as dinosaur bones or footprints. To use the example in Figure 2.21, suppose a rare type of fossil leaf in a certain area is found only in one rock layer; call it formation A. If like fossil leaves are found in another area 50 miles away in a similar formation (call it formation E), fossil dating suggests that formation A

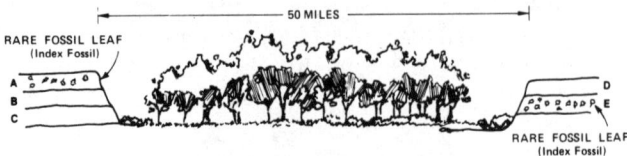

Figure 2.21. FOSSIL DATING. Formation A correlates with Formation E.

and formation E were deposited at essentially the same time. Of course, care must be taken to insure that the fossils in both formations are from the same type of plant.

Fossils from plants or animals that have been found to be restricted to a given time period, and are therefore useful to relative dating, are known as **index fossils.**

A second type of relative dating is by **physical feature correlation.** For example, a volcano distributes a layer of volcanic ash over a large area and does not erupt again for a long time. If no other volcanic eruptions occur in the area, that bed of volcanic ash can be used as a time reference to date the rocks.

In Figure 2.22, the layer of ash from the volcano provides a reference for determining the relative rock ages. It can be determined that rock layers A and D are the same relative age, since they both lie directly beneath the ash layer.

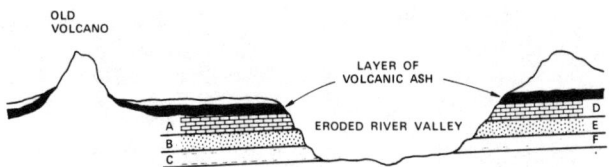

Figure 2.22. PHYSICAL FEATURE CORRELATION.

If several physical features such as this exist, the relative dates of various rock strata can be determined. Occasionally, these relative dates can be tied in with the absolute dates of one or two rock strata in the area to provide a fairly accurate dating of an entire sequence.

Using relative and absolute dating of rocks, geologists have devised a geologic time chart, which is divided into **eras, periods,** and **epochs.** Analagous divisions of "conventional" time might be centuries, years, and days.

There are four major eras:

1. **The Precambrian Era**

   The Precambrian Era includes the time from the beginning of the earth to the time when abundant fossil records exist. Little is known about this era, compared to the more recent ones.

2. **The Paleozoic Era**

   The Paleozoic Era was mainly a time of sea life, and might be called the "Age of Fish." This era lasted from 570 million years ago to 225 million years ago.

3. **The Mesozoic Era**

   The Mesozoic Era might be called the "Age of Reptiles," since numerous dinosaur and other reptilian fossils from this era have been uncovered. In addition, birds and mammals first appear in the Mesozoic Era, which lasted until approximately 65 million years ago.

4. The Cenozoic Era

The Cenozoic Era is the era in which a great evolution of birds, mammals, and seeding plants occurred. It extends from 65 million years ago to the present.

The time chart in Table 2.3 contains the names of the important eras, periods, and epochs, as well as the major events which occurred within them.

IV. PETROLEUM GEOLOGY

A. Introduction

The word petroleum is derived from the Latin words, "petra," meaning **rock,** and "oleum," meaning **oil.** Petroleum, then, is "oil found in rock." Generally, the term petroleum is used to designate a large number of gaseous and liquid **hydrocarbons** found throughout the earth. Liquid petroleum is normally called **crude oil,** to differentiate it from refined products, while gaseous petroleum is called **natural gas,** to differentiate it from gasoline. A classification of hydrocarbons appears in Table 2.4.

For clarification, hydrocarbons are organic compounds containing carbon and hydrogen, which are linked together in certain specified molecular configura-

Table 2.4. CLASSIFICATION OF HYDROCARBONS

GASEOUS	Dry Gas	Marsh Gas
	Wet Gas	Natural Gas
LIQUID	Petroleum	Crude Oil
	Viscous or Semi-Solid	Mineral Tar
	Bitumen	Brea
SOLID	Asphaltite	Gilsonite (Uintahite)
		Lignite
		Subbituminous
	Coal	Bituminous
		Semibituminous
		Anthracite
	Bituminous Shale	Kerogen
		Petroliferous

(Classification presented by Hager, as adapted from Amyx, Bass, and Whiting, p. 3.)

tions. Other elements, such as sulfur, oxygen, or nitrogen, are often present as well. **Bitumen,** which has the adjective form bituminous, refers to hydrocarbons that are soluble in a chemical called carbon disulfide. The gaseous hydrocarbons frequently are considered bituminous as well. **Marsh gas** is a term used for the gases (predominantly methane) given off by the decom-

Table 2.3. GEOLOGIC TIME CHART[a]

ERA	PERIOD	EPOCH	MILLION YEARS AGO	MAJOR EVENTS	ABBREVIATIONS
CENOZOIC	Quaternary	Holocene		Civilization	Q
		Pleistocene	2	Man appears	
	Tertiary	Pliocene			Tpl
		Miocene		Grasses become abundant	Tm
		Oligocene			To
		Eocene		Horses first appear	Te
		Paleocene	65		Tp
MESOZOIC	Cretaceous	(b)		Extinction of Dinosaurs; seed-bearing plants appear	K
	Jurassic			Birds first appear; mammals first appear	J
	Triassic		225	Dinosaurs first appear	$\mathbb{R}$ or TR
PALEOZOIC	Permian	(b)			$\mathbb{P}$ or P
	Pennsylvanian[c]			Coal-forming swamps; reptiles appear	$\mathbb{P}$ or Cp
	Mississippian[c]				M or Cm
	Devonian			First land animals (amphibians) appear; insects appear	D
	Silurian			First vertebrates (fish) appear; first land plants appear	S
	Ordovician				O
	Cambrian		570	First abundant fossil record	E
PRECAMBRIAN	(d)		4600	Formation of ozone layer; oldest rocks; life appears	P_E or PE

[a]Based in part on Leet and Judson, pp. 184, 185.
[b]Each period in the Paleozoic and Mesozoic Eras is commonly divided into Upper, Middle, and Lower Epochs. The term "Middle Cretaceous," however, is generally not used.
[c]The Mississippian and Pennsylvanian periods are often combined into a single Carboniferous period.
[d]Subdivisions of the Precambrian Era are not in general agreement. The Proterozoic and Archeozoic Periods are sometimes used.

position of organic matter in marshes and bogs. **Asphaltum** and **brea,** the tars found in tar sands and tar pits throughout the world, are used to pave roads. **Gilsonite,** also known as **uintahite,** is a solid, black, brittle, glassy hydrocarbon with a conchoidal fracture found almost exclusively in Utah. Bituminous **shales** include the "oil shales" of Wyoming, Colorado, and Utah, as well as the black, petroliferous shales in many other parts of the world. **Ozocerite** is a naturally occurring solid paraffin or wax used in making candles.

Our discussion of petroleum geology will include the origin of petroleum, its migration and accumulation, reservoir traps, reservoir rock types, reservoir properties, and reservoir mechanics.

A **petroleum reservoir** may be defined as a porous and permeable rock containing oil and/or gas which can be produced. The following criteria are necessary for a petroleum reservoir:

1. The rock must be porous; that is, it must have voids or pores that contain fluids.
2. The rock must be permeable, meaning the rock must have connected pores which can transmit fluids.
3. Petroleum must exist in the pores of the rock.
4. The petroleum must theoretically be recoverable.

Each of these criteria is necessary to the definition of a petroleum reservoir. The rock must be porous to contain petroleum, and it must be permeable to allow production. Petroleum must be present for a petroleum reservoir, and the petroleum present must be recoverable, which excludes, for example, most organic shales. As technology improves, many accumulations of petroleum which were not previously considered reservoirs become recoverable and, therefore, become petroleum reservoirs.

An oil or gas **field,** on the other hand, is determined by the surface area under which the petroleum accumulations occur. A field may contain several subsurface petroleum reservoirs (Figure 2.23).

With this background information in mind, some of the aspects of petroleum geology will now be discussed.

B. The Origin of Fossil Fuels

Theories concerning the origin of fossil fuels typically fall into two major categories: **inorganic** theories and **organic** theories. The **inorganic** theories

Table 2.5. THEORIES ADVANCED IN EXPLANATION OF THE ORIGIN OF PETROLEUM

Name of theory or its originator	Salient features	Evidence
Inorganic theories		
Berthelot's alkaline carbide theory	Deep-seated deposits of alkaline metals in the free state react with CO_2 at high temperatures, forming alkaline carbides. These, on contact with water, liberate acetylene, which, through subsequent processes of polymerization and condensation, forms petroleum.	Evidence lacking. Neither free alkaline metals nor carbides found in nature.
Mendeleev's carbide theory	Iron carbides within the earth on contact with percolating waters form acetylene, which escapes through fissures to overlying porous rocks and there condenses.	See above. Magnetic iron oxides would also be formed as a product of these reactions. Magnetic irregularities have been noted in the vicinity of some oil fields.
Moissan's volcanic theory	Moissan suggests that volcanic explosions may be caused by the action of water on subterranean carbides.	Small quantities of petroleum noted in volcanic lavas near Etna and in Japan. Petroleum also associated with volcanic rocks in Mexico and Java.
Sokolov's cosmic theory	Petroleum considered to be an original product resulting from the combination of carbon and hydrogen in the cosmic mass during the consolidation of the earth.	Small quantities of hydrocarbons occasionally found in meteorites.
Limestone, gypsum, and hot-water theory	Reactions between carbonate and sulfate of lime in the presence of water at temperatures sufficient to dissociate the water theoretically may form hydrocarbons.	Practically, it has been found impossible to demonstrate this reaction in the laboratory.
Organic theories		
Engler's animal-origin theory	Petroleum formed by a process of putrefaction of animal remains. Nitrogen thus eliminated and residual fats converted by earth's heat and pressure into petroleum. Activity of anaerobic bacteria thought to play a part in the reactions.	Oils resembling petroleum can be distilled from sediments containing fish remains. Many petroleum deposits associated with marine sediments contain an abundance of foraminifera.
Hofer's vegetable-origin theory	Petroleum formed by decay of accumulated vegetable refuse under conditions which prevent oxidation and evaporation of the liquid products formed.	Deposits of petroleum found in close association with sedimentary deposits containing diatoms, seaweed, peat, lignite, coal, and oil shale of known vegetable origin. Oils closely resembling petroleum can be distilled from these substances.
Hydrogenation of coal or other carbonaceous materials	Solid organic materials converted into liquid hydrocarbons by combination with free hydrogen at high pressures and temperatures in the presence of a suitable catalyzer, such as nickel.	Hydrogenation of coal in the laboratory and in commercial plants. The ash of some petroleums is chiefly nickel. However, the existence of free hydrogen in nature is yet to be demonstrated.

(From **Petroleum Reservoir Engineering** by J. W. Amyx, D. M. Bass, Jr., and R. L. Whiting. Copyright 1960, McGraw-Hill. Used with permission of McGraw-Hill Book Company.)

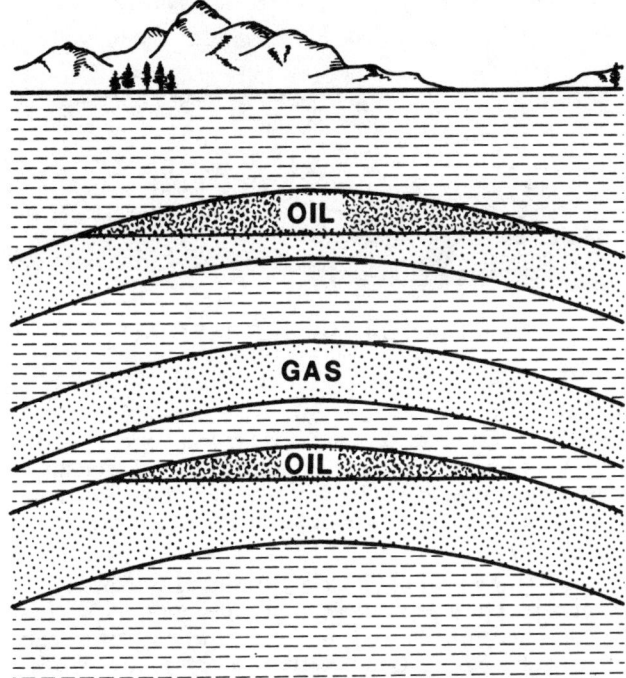

Figure 2.23. FIELD WITH MULTIPLE RESERVOIRS.

assume that some combination of chemical reactions occurs naturally to form crude oil or natural gas. The **organic** theories claim that fossil fuels were formed from the remains of plants and animals that lived long ago. The various theories on the origin of petroleum are summarized in Table 2.5.

The inorganic theories receive little support today, since they are inconsistent with both the amount of petroleum known to exist, and the areas where petroleum occurs. The organic theories, however, predict the existence of oil and gas in sedimentary basins, and are consistent with the form and composition of petroleum as found in nature.

Current theories on the origin of petroleum center around the transformation of plant and animal remains into petroleum by bacterial action, heat, pressure, catalytic reactions (which are chemical reactions aided or accelerated by a substance called a catalyst), and radioactive bombardment (exposure to radioactive elements), as illustrated in Figure 2.24.

Less controversy surrounds the formation of coal, since products ranging from peat (which is simply decomposed plant matter and the first step in coal formation) to anthracite (the hardest coal and the final step in the process of coal formation) are common. Coal is the final product of plant remains from swamps that existed millions of years ago (Figure 2.25).

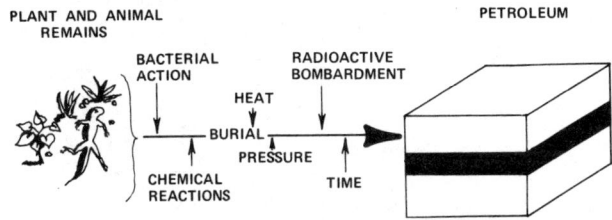

Figure 2.24. ORIGIN OF PETROLEUM: ORGANIC THEORY.

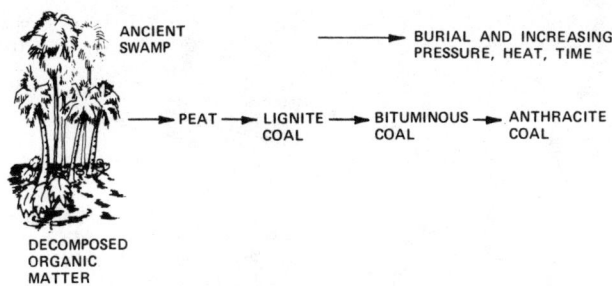

Figure 2.25. ORIGIN OF COAL.

Normally, the dead plant matter from which the coal is formed is approximately 50 percent carbon. As decomposition and decay occur, **peat** is formed. Peat, which is not coal, is about 60 percent carbon. Under mild temperature and pressure conditions and with the passage of time, peat is transformed into the low grade coal known as **lignite**, containing approximately 70 percent carbon. Over a longer period of time, and with increased pressure and temperature, lignite changes into **subbituminous** and, finally, **bituminous** coal, which is about 80 percent carbon. Lignite and bituminous coals are considered organic, sedimentary deposits. The hardest coal, **anthracite**, is a metamorphic rock derived from bituminous coal, and may contain 95 to 98 percent carbon.

Oil shale, another fossil fuel energy source, is neither oil nor shale. Oil shale is a **marlstone**, which is a lithified mixture of clays and calcite. The hydrocarbon contained in the marlstone is known as **kerogen**. Kerogen is a bituminous, mineraloid, waxy solid of organic origin. The "oil shales" of the Green River formation in western Colorado, northeastern Utah, and southwestern Wyoming were deposited in inland lakes over a period of 6 million years or more during the Eocene Epoch. According to estimates, over 1,000 billion barrels of oil-equivalent resources exist in the Green River oil shales. This is more than the entire world's known resources of petroleum. However, much of it is not recoverable where the oil shale is very low grade and is spread over a large area.

C. Migration and Accumulation

Usually petroleum is not found in the same rock where it was first formed. Before it can have any commercial value, petroleum must move from the rock where it originates (the **source rock**), through other rocks (the **carrier rocks**), into the rock where it accumulates (the **reservoir rock**). This movement of the oil is known as **migration**.

The fundamental mechanism involved in migration is **buoyancy**, which is the tendency of the petroleum fluids to rise to the top of water, much as submerged wood floats to the surface of an ocean. Buoyancy occurs because petroleum is lighter or less dense than water (Figure 2.26).

Another mechanism of migration is **hydrodynamics**, or the flow of fluids. Hydrodynamic forces are those by which one fluid flows past another, carrying the second fluid with it (Figure 2.27). An example of this can be seen in the kitchen. When dishes are rinsed, the hydrodynamic force of the water rushing past the dish carries the suds along with the water.

A third type of migration is known as **capillarity**. Capillarity is the tendency of a fluid to rise in a small

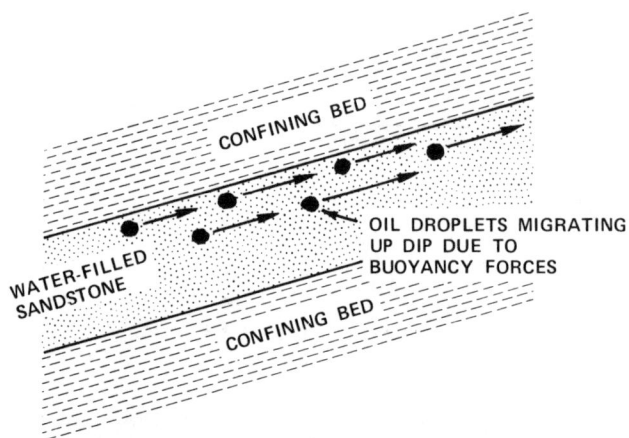

Figure 2.26. MIGRATION BY BUOYANCY.

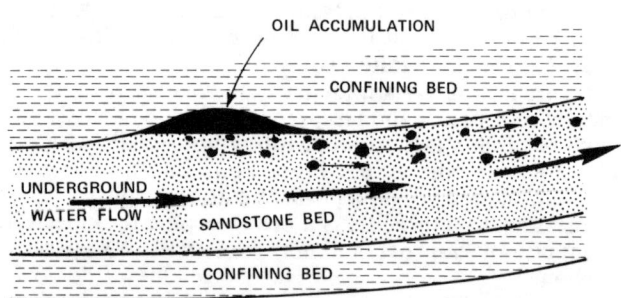

Figure 2.27. MIGRATION BY HYDRODYNAMIC FORCES. Hydrodynamic forces of underground water flow carry oil droplets along with water.

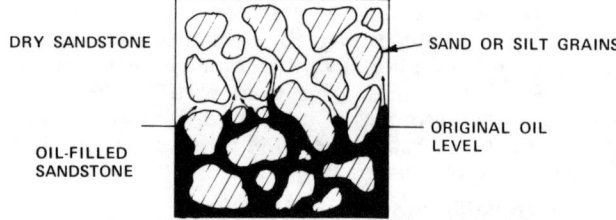

Figure 2.28. MIGRATION BY CAPILLARITY. Oil is absorbed into dry rock by capillary action.

tube, and is a measurable force in rocks (Figure 2.28). A common example of capillarity is the way a sponge soaks up water.

The three forces of buoyancy, hydrodynamics, and capillarity cause the petroleum fluids to move, or migrate, from the source rocks, through the carrier rocks, to the reservoir rock. At this point, a condition must exist to stop the flow of petroleum through the reservoir rock. This condition is known as a **trap,** and many different types of traps exist. In addition, non-porous rock called a **caprock,** or **confining bed,** must exist over the reservoir rock to provide further sealing of the reservoir (Figure 2.29).

D. Reservoir Traps

A reservoir trap consists of a subsurface geologic feature which stops oil migration and, consequently, results in an oil accumulation. Oil and gas have been

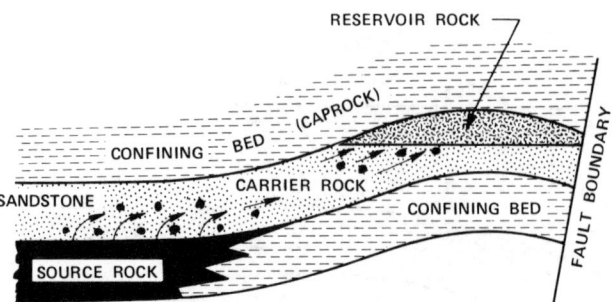

Figure 2.29. OIL MIGRATION.

found in a number of different types of traps or reservoirs. These different reservoir traps are classified and described in the following excerpt from **Elements of Petroleum Reservoirs,** by Norman J. Clark. (For complete reference to this book, see the bibliography at the end of this chapter.)

"Many proposals have been made to classify the different physical shapes of petroleum reservoirs that have been discovered; however, the simplest means of such classification is perhaps a grouping according to the geologic features causing their occurrence. These features are (1) structural folding, (2) structure with faulting, (3) structure with an unconformity, (4) structure caused by some deep-seated movement of earth materials such as salt domes or serpentine plugs, (5) changes in permeability with a formation, and (6) combinations of two or more of the foregoing.

"Reservoirs formed in folded strata usually result in domes or anticlines (Figures 2.30 and 2.31). These traps were filled by oil moving upward through permeable beds to a point where it was stopped by the impermeable beds on top of the reservoir strata. It is common to find traps partially filled with water where the structure is large enough to hold more oil or gas. Examples of reservoirs formed by anticlinal folding of the structure are the Pegasus field, Mid-

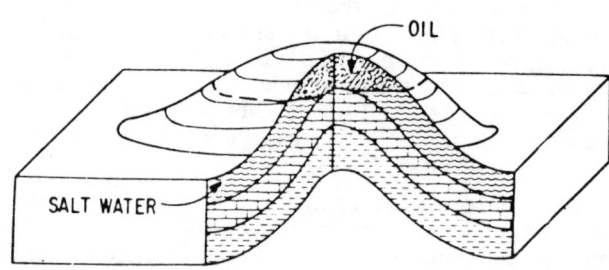

Figure 2.30. OIL ACCUMULATED IN DOMAL STRUCTURE.

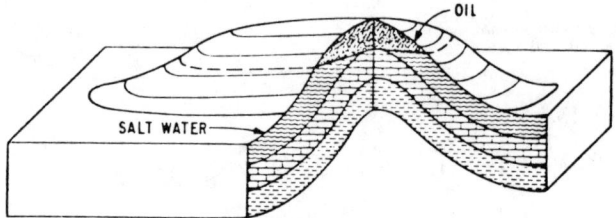

Figure 2.31. OIL ACCUMULATED IN ANTICLINAL STRUCTURE.

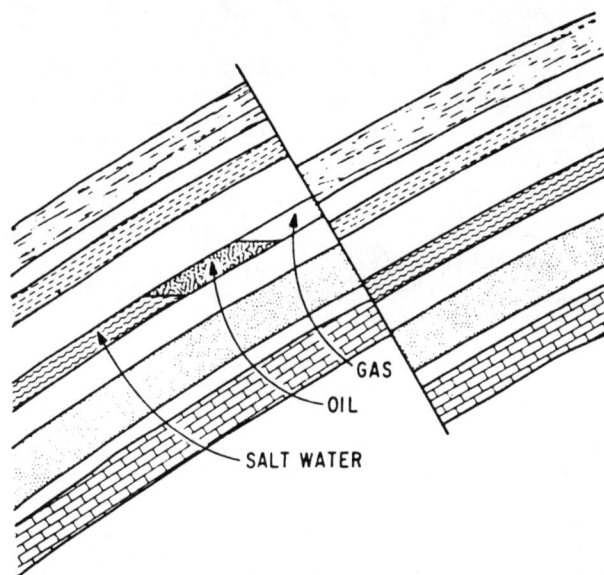

Figure 2.32. STRUCTURAL TRAP RESULTING FROM FAULTING.

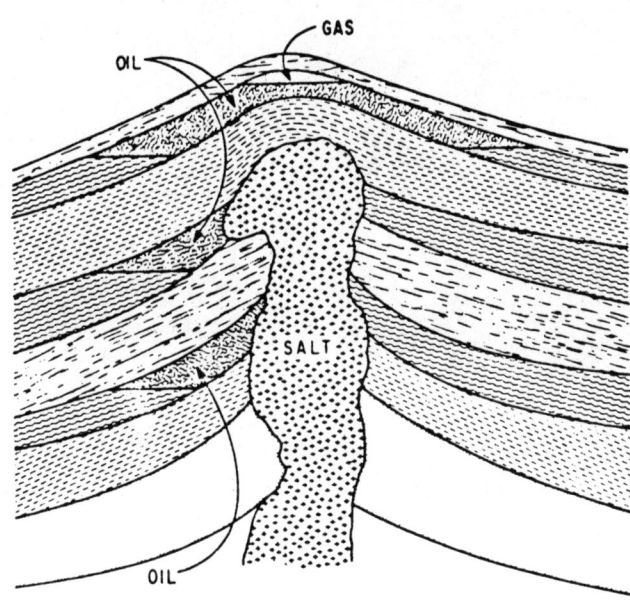

Figure 2.34. OIL ACCUMULATION IN THE VICINITY OF A PIERCEMENT-TYPE SALT DOME.

land and Upton Counties, Tex., and the great Azmari Limestone fields of Iran.

"Reservoirs formed by faulting occur where escape of oil is prevented by impermeable beds moving into position against the oil-bearing rock on the opposite side of the fault plane. The oil is held in traps of this type by the structural dip of the bed and the faulting (Figure 2.32). Reservoirs of this type are typified by the Mt. Poso field, Kern County, Calif., and the Luling, Mexia, and Talco fields in Central Texas.

"Another type of reservoir is one formed as a result of an unconformity where upward escape of oil has been stopped by the impermeable material laid down on the weathered surface of the lower beds (Figure 2.33). The East Texas field is formed in this manner.

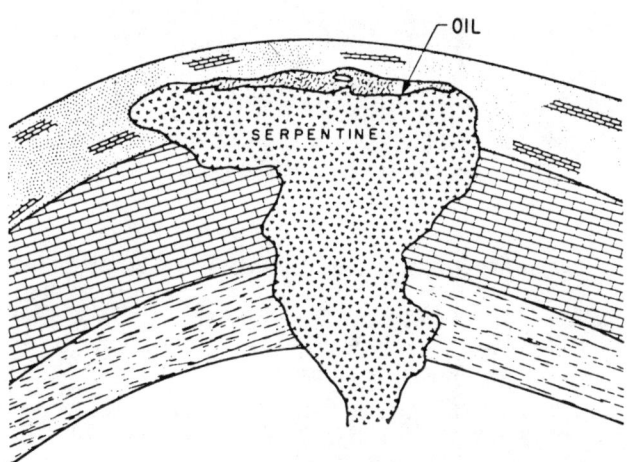

Figure 2.35. OIL ACCUMULATION IN AREAS OF FRACTURE POROSITY ON A SERPENTINE PLUG.

"Accumulations of oil are found in sediments on or surrounding material such as salt or serpentine plugs that have pierced and deformed the overlying strata (Figures 2.34 and 2.35). Examples of salt dome fields are the Spindletop field, Jefferson County, Tex., and the Avery Island field, Iberia Parish, La. Examples of serpentine plug fields are the Lytton Springs field in Caldwell County Tex., and the Hilbig field, Bastrop County Tex.

"Another type of trap is one that is closed by variation in permeability within the strata (Figures 2.36 and 2.37). The size of oil and gas deposits in reservoirs of this type is governed by the manner in which the beds were laid down rather than by structural segments of the bed which are surrounded by impermeable segments of the same bed. Fields having sand reservoirs of this type are the Goose Creek field in Harris County, Tex., and the Spraberry field in West Texas. Limestone reservoirs of this type are

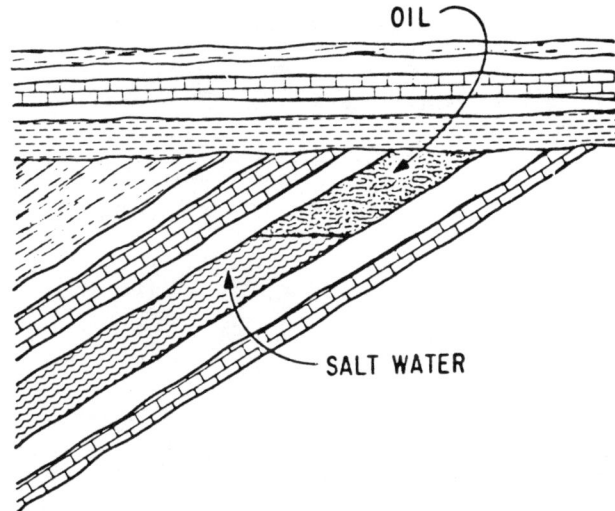

Figure 2.33. OIL ACCUMULATION UNDER AN UNCONFORMITY.

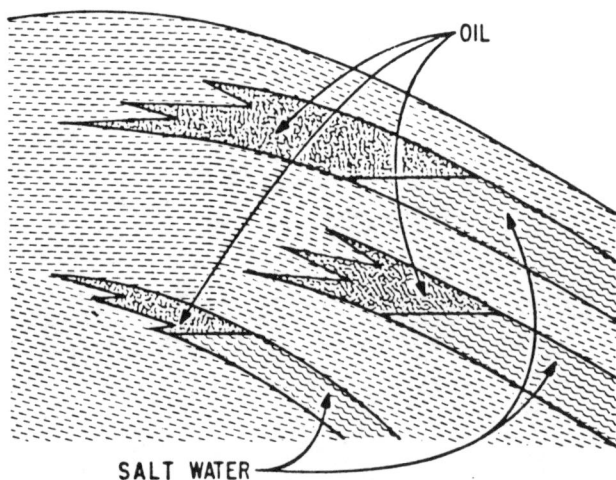

Figure 2.36. OIL ACCUMULATION IN SAND LENSES OF THE SAND BAR TYPE.

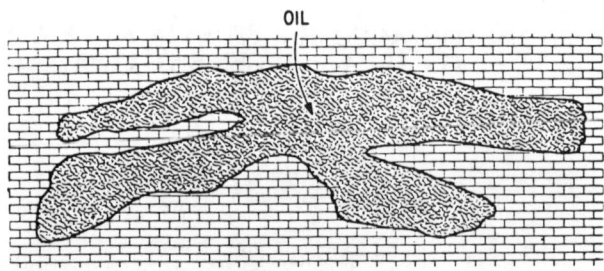

Figure 2.37. OIL ACCUMULATION IN POROUS ZONES IN LIMESTONE.

exemplified by the East Dundas field, Richland and Jasper Counties, Ill. (which produce from McClosky limestone) and the Scurry Area Canyon Reef (SACROC) field in Scurry County, Tex. (which produces from a reef section).

"Perhaps the most common type of reservoir is one that is formed not by a single structural feature alone, but instead, by a combination of folding, faulting, changes in permeability or other conditions. Examples of reservoirs of this nature are the many reservoirs found in the Seeligson-Tijerna Canales Blucher-La Gloria trend of fields in Southwest Texas and the Wilmington field, Los Angeles County, Calif."

E. Reservoir Rock Types

Almost all oil and gas bearing rocks are sedimentary. The three most common rock types associated with petroleum reservoirs are **sandstone, limestone,** and **shale.** Typically, oil reservoirs are either sandstones or limestones, while caprocks, which do not contain recoverable oil but act as confining beds, most often are shale.

Some of the more common reservoir rocks, together with the causes of the rock porosity and the relative frequency of occurrence of the reservoir rock, are listed below. Definitions of geologic terms used in the description of these reservoir rocks can be found in the glossary.

RESERVOIR ROCKS

1. Sand, conglomeratic sand, and gravel in varying states of consolidation; porosity due to fragmental textures; common
   a. Clean sands, etc.: pore space between sand grains uncontaminated
   b. Argillaceous sands, etc.: pore space partly filled with clay material
   c. Silty sands, etc.: pore space partly filled with silt
   d. Lignitic sands, etc.: pore space partly filled with lignitic matter
   e. Bentonitic sands: pore space partly filled with clay formed from volcanic ash
2. Porous calcareous sandstone and siliceous sandstone; porosity due to incomplete cementation; frequent
3. Fractured sandstone and fractured conglomerate; porosity due to fracturing in tight sandstones and hard conglomerates caused by faulting or folding; infrequent
4. Detrital limestone (calcitic and dolomitic); porosity due to fragmental texture and frequently increased by solution effects; common
5. Porous crystalline limestone (calcitic and dolomitic); porosity due mainly to solution and re-crystallization of the limestone; common
6. Cavernous crystalline limestone (calcitic and dolomitic); porosity due to strong solution effects; common
7. Fractured limestone (calcitic and dolomitic, and siliceous); porosity due to open fissures along fracture patterns; frequent
8. Sugary dolomite; porosity possibly due to volume shrinkage in the process of formation of dolomite from calcitic sediment; common
9. Oolitic limestone; porosity due to oolitic or granular texture with uncemented or partially cemented interstices; frequent

(Source: Amyx, J.W., Bass, D.M., and Whiting, R.L., pp. 21,22.)

F. Reservoir Properties

The properties of petroleum reservoirs are related to the characteristics of the rock, the fluids in the rock, and the reservoir itself.

Lithology, porosity, and **permeability** are specific rock properties. **Lithology** describes the mineralogy and rock type present (for example, sandstone). **Porosity** is a measurement of the ratio of the void space in a rock to the bulk volume of the rock. **Total porosity** includes all the voids or pore spaces in the rock, while **effective porosity** considers only interconnected pores, or those pores through which fluids can pass (Figure 2.38). A rock which is composed of 70 percent material and 30 percent void space is said to have a total porosity of 30 percent. The effective porosity of the same rock may be somewhat less, depending on the configuration of the pore spaces.

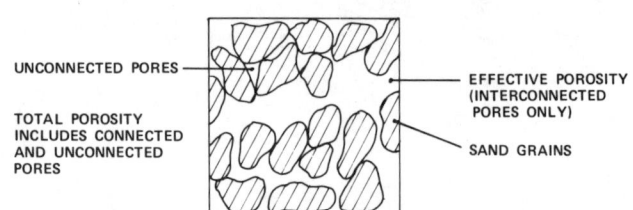

Figure 2.38. POROSITY.

Permeability is a measure of the ability of a rock to transmit fluids. The unit of permeability is known as a **darcy** or a **millidarcy** (1/1000 of a darcy). The higher the permeability of a rock, the easier fluids will flow through it. Unconsolidated sands found in the Gulf Coast region often have permeabilities of several darcies, while much tighter rocks found in the sedimentary basins of the Rocky Mountains may have permeabilities as low as a few millidarcies.

The distribution and type of lithology, porosity, and permeability are important. The term **primary** (or, in carbonates, **matrix**) is used to distinguish the porosity and permeability properties that first existed in the rock from **secondary** or **solution** porosity and permeability which resulted later (Figure 2.16). Fractures, or cracks where the rock has broken, may also have significant porosity and permeability.

Another important rock property is the rock or pore volume **compressibility**. The compressibility of the rock is the measure of its tendency to expand or contract with changes in the forces on the rock, and is a necessary factor in determining the productivity of a formation, or how quickly fluids can be produced. For example, rocks which are highly compressible, such as unconsolidated or loosely packed sandstone, can be very productive. However, if these same rocks are compressed from the weight of overlying rocks and the sand grains are squeezed closer together, the rock's porosity and permeability are reduced and its productivity is decreased.

Fluid properties of the reservoir are related to the type of fluids present as well as their quantities. Oil, water, and natural gas are common fluids found in petroleum reservoirs. The amount of fluid present is indicated by fluid **saturations**, which are determined by measuring the ratio of the volume that a specific fluid occupies in a reservoir to the reservoir's void space. If the void spaces of a porous rock are half filled with water and half filled with oil, the rock is said to have a 50 percent oil saturation and a 50 percent water saturation.

Fluid **densities** determine the location of different fluids in a reservoir. Because it has the lowest density, gas is found in the upper portions of petroleum reservoirs. Water, which is heavier than either oil or gas, is found in the lower segments of a reservoir (Figure 2.39).

Another fluid property, **viscosity**, measures the fluid's resistance to flow. For example, water, which flows easily, has a low viscosity, while a milkshake, which is more resistant to flow, has a much higher viscosity. The term **heavy oil** refers to oil which has a high density and viscosity.

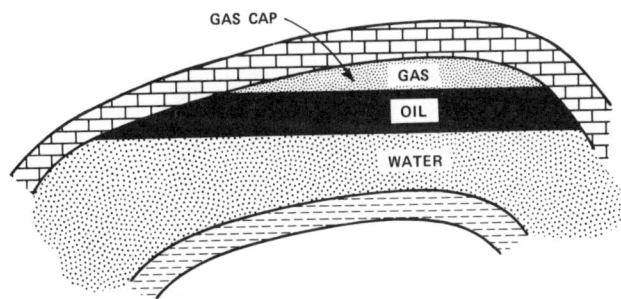

Figure 2.39. EFFECT OF FLUID DENSITIES ON RESERVOIR FLUID DISTRIBUTION.

Two other fluid properties are actually functions of the rock-fluid system; that is, they depend on both the rock present and the type of fluids present. **Capillary pressure**, which was discussed earlier, is one rock-fluid property. The other is **relative permeability**, or the ease with which any given fluid can move through the rock. Relative permeabilities differ with different fluids. This is in contrast to **absolute permeability**, which is a function of the rock itself, and does not change with the fluids present. Absolute permeability is measured by flowing air through rock specimens contained in special laboratory devices. Relative permeability is measured by flowing a particular fluid through the rock in the same manner.

Properties which characterize the oil accumulation include pressure, temperature, thickness, area, structure, trap, and drive mechanisms. The specific combination of these properties distinguishes each reservoir and characterizes the potential for recovering the petroleum present.

G. Reservoir Mechanics

Reservoir mechanics is a study of fluid movement and changes in petroleum reservoirs before, during, and after the production of petroleum. Reservoir mechanics is concerned with the **fluids** and **energy** in a reservoir, as well as its history and its future.

The **reservoir fluids**, which include oil, gas, and water, may be used as a starting point in the classification of reservoirs. **Aquifers**, which are rocks containing water, often contain petroleum accumulations. **Dead oil reservoirs** contain oil but essentially no pressure from solution gas in the oil. **Black oil reservoirs** contain moderate amounts of solution gas in the oil, while **volatile oil reservoirs** contain large amounts of solution gas. **Gas condensate reservoirs** are accumulations of gas with significant amounts of hydrocarbon liquids in solution with the gas; **dry gas reservoirs** contain gas with negligible amounts of liquids in solution with the gas.

The above terminology is not exact since terms tend to overlap. Furthermore, different fluids may exist in conjunction with each other. A volatile oil reservoir may have a gas cap (a gas accumulation directly above it) and an associated aquifer. A dry gas reservoir may have a few oil stringers, or small intervals of oil, within it.

Reservoir energy is the force that causes fluids to flow to the producing wells. Classifying a reservoir by the type of energy present is very important in determining producing rates, locations of infill wells (wells drilled between other pre-existing wells), and the ultimate recovery. The four major types of reservoir energy are **fluid expansion, rock compaction, gravity drainage,** and **artificial energy.**

Fluid expansion occurs when pressure on the fluids is released, and the fluids are allowed to expand and flow. This is significant because pressure in the reservoir may be several thousand pounds per square inch (psi). As an illustration of the magnitude of this pressure, consider that normal air pressure at sea level is only 14.7 psi. As another example, you would have to dive over 2,000 feet deep into the ocean to have 1,000 psi of pressure on your body.

The type of fluid expanding may be used to characterize many reservoirs. An **undersaturated oil reservoir** is one whose major energy comes from oil expansion. A **solution gas drive reservoir** is one whose major energy is derived from gas coming out of solution from oil, much as carbon dioxide fizzes in soda pop. A **gas cap drive reservoir** is one in which a separate accumulation of gas, or a gas cap, exists above the oil and provides the major source of energy for expansion. Finally, the term **water influx** is used when water expansion is the primary energy available to a reservoir. **Shale water influx** (water expanding out of shales), **edge water influx** (water expanding from the edges of the petroleum accumulation), and **bottom water influx** (water expanding from beneath the petroleum accumulation) are special types of water influx.

As pressure is removed from the fluids in the reservoir, the weight of the overlying sediments compresses the rock which forms the reservoir. This compressional energy is known as a **compaction drive.** Compaction drive is an important source of energy for reservoirs in unconsolidated sands (sands that have not lithified). Unconsolidated sand reservoirs are common in California and on the Gulf Coast.

Gravity drainage, which is the tendency of fluids to move downward because of gravity, is also important in some reservoirs. Many of the heavy oil accumulations (oil of less than $20°$ API) have no energy other than gravity drainage.

Artificial energy, such as water injection, may be necessary to produce petroleum at sufficient rates to be of commercial value. These methods will be discussed in the chapter on Recovery.

Several of these forms of energy may act together to allow production in a given reservoir. These interactions will also be discussed in more detail in the chapter on Recovery.

GEOLOGY—APPENDIX

From **Oil—Prospect to Pipeline**, Third Edition by Robert R. Wheeler and Maurine Whited. Copyright 1975 by Gulf Publishing Company, Houston, Texas. Used with permission. All rights reserved.

Regional Stratigraphic Terminology

The following charts are designed to show the general sequence of the more commonly used geologic formation names arranged according to age.

Please understand that these charts were compiled by the authors from many sources and that, although the general age relationships as to system and series designation are generally accepted, the equivalency of many individual formations is the subject of continuing dispute among local authorities.

The first chart shows the subdivision of geologic eras and their subdivisions into systems (representing a *period* of geologic time) and series (representing an *epoch* of time). The series terms used in this chart are the standard nomenclature for the United States.

Subdivision of Geologic Eras

Era	System (Period)	Series (Epoch)
CENOZOIC	Quaternary	Recent / Pleistocene
CENOZOIC	Tertiary	Pliocene / Miocene / Oligocene / Eocene / Paleocene
MESOZOIC	Cretaceous	Upper Cretaceous (Gulf) / Lower Cretaceous (Comanche)
MESOZOIC	Jurassic	Upper Jurassic / Middle Jurassic / Lower Jurassic
MESOZOIC	Triassic	Upper Triassic / Middle Triassic / Lower Triassic
PALEOZOIC	Permian	Ochoa / Guadalupe / Leonard / Wolfcamp
PALEOZOIC	Pennsylvanian	Virgil–Cisco / Missouri–Canyon–Hoxbar / Des Moines–Strawn–Deese / Atoka–Bend ("Lampasas") / Morrow
PALEOZOIC	Mississippian	Chester / Meramec / Osage / Kinderhook
PALEOZOIC	Devonian	Upper Devonian / Middle Devonian / Lower Devonian
PALEOZOIC	Silurian	Cayugan / Niagaran / Medina (Albion)
PALEOZOIC	Ordovician	Cincinnatian / Champlainian / Canadian
PALEOZOIC	Cambrian	Croixian / Albertan / Waucoban
PRE-CAMBRIAN	PROTEROZOIC / ARCHEOZOIC	Undifferentiated "Basement"

Oil and Gas Provinces of Lower 48 States

Eastern Interior Region

System	Series	Group/Formation		
Pennsylvanian	Monongahela	Merom	Dixon	Pittsburgh Coal
Pennsylvanian	Conemaugh	Shoal Creek	Wabash	
Pennsylvanian	Conemaugh	Trivoli		Niagara
Pennsylvanian	Allegheny	Hebrin / Harrisburg Coal / Pleasant View / Oak Grove	West Franklin / Brereton / Alum Cave / Velpen	Madisonville / Pennywinkle / Goshen Coal / Sholtztown Coal
Pennsylvanian	Beaver River	Isabelle / Seahorne / Seville		Sebree / / Curlew
Pennsylvanian	Pottsville	Makanda / Babylon / Wayside	Minshall / Mansfield / Hindostan	Mannington / / Bee Spring
Mississippian	Chester		Kinkaid / Degonia / Clore / Palestine / Menard / Waltersburg / Vienna / Tar Springs / Glendean / Hardinsburg / Golconda / Cypress / Point Creek / Bethel / Benoist-Renault / Aux Vases	Mauch Chunk / / / / / / / / / / / / / / / Greenbriar
Mississippian	Meramec	St. Genevieve / / / St. Louis	McClosky / Roseclaire / Fredonia / St. Louis / Salem / Warsaw	Loyalhanna
Mississippian	Osage		Keokuk / Burlington / Fern Glen	Keener / Burgoon / Weir
Mississippian	Kinderhook	Woodford	Chattanooga	New Albany / Berea
Devonian	Chemung / Hamilton		Lower Chattanooga / Beachwood / Misenheimer / Spieds	Hydraulic
Devonian	Oriskany / Helderberg		Jeffersonville / Geneva / Bailey	"Corniferous" / Dutch Creek
Silurian	Cayugan		Salina / Guelph	
Silurian	Niagaran		Racine / Bellewood / Joliet / Rockdale	Louisville
Silurian	Medinan		Brassfield	Alexandrian
Ordovician	Cincinnatian		Maquoketa	Richmond / Maysville / Eden
Ordovician	Champlainian	Mohawkian	Kimmswick / Decorah / Plattin	Trenton
Ordovician	Champlainian	Chazyan	Joachim / Glenwood / St. Peter	St. Peter
Ordovician	Canadian		Powell / Cotter / Jefferson City / Roubidoux / Gasconade / Van Buren / Gunter	Prarie du Chien
Cambrian	St. Croixan		Eminence / Potosi / Derby / Davis / Bonneterre / Lamotte	

Pre-Cambrian Basement

- 35 -

East Texas—Gulf Coast—Florida Regions

System	Series	Group/Formations		
Tertiary	Plio-Pleistocene	Beaumont/Montgomery-Prairie Bently/Willis-Williana Goliad		Citronelle Tampa
	Miocene	Fleming, Catahoula, Pascagoula Anahuac Frio		Marianna
	Oligocene	Vicksburg		Ocala
	Eocene	Jackson Claiborne Wilcox	Cook Mtn. Mount Selma	Oldsmar
	Paleocene	Midway		Cedar Keyes
Cretaceous	Gulf	Navarro Taylor Austin Eagle Ford Woodbine	Eutaw Tuscaloosa	Selma Tuscaloosa
	Comanche	Washita Fredericksburg Trinity: Paluxey, Glen Rose, Hosston		(Undifferentiated limestones) Sunniland reservoir
Jurassic	Upper	Cotton Valley Haynesville-Buckner Smackover		
(age uncertain)		Eagle Mills-Louanne salt		
Paleozoic		"Ouachita facies"		

West and Central Texas

System	Series	Group/Formation		
Tertiary	Pliocene	Ogallala		
Cretaceous	Comanche	Fredericksburg Washita Trinity		
Triassic		Dockum-Santa Rosa		
Permian	Ochoa	Dewey Lake Rustler Salado Castille		
	Guadalupian: Bell Canyon, Goat Seep (Bone Spring Delaware Mt.)	Capitan Reef	Tansill Yates Seven Rivers Queen Grayburg	Whitehorse
	Leonard: Yeso, Abo	Victorio Peak Reef	San Andres El Reno Glorietta-San Angelo Clear Fork Wichita-Albany	Fullerton Tubbs Spraberry
	Wolfcamp	Hueco	Dean Wolfcamp	Admiral Pueblo
Pennsylvanian	Cisco	Reef Facies		Thrifty Graham
	Canyon			Caddo Creek Graford
	Strawn			Lone Camp Caddo
	Atoka		Bend	Smithwick Marble Falls
Mississippian	Meramec Osage Kinderhook			Barnett Chappel
Devonian	Woodford			Woodford
Silurian		Fusselman		
Ordovician		Montoya Simpson Ellenburger	McKee/Waddell	
Cambrian		Hickory		

Pre-Cambrian Basement

Oklahoma and Kansas Region

System	Series	Group/Formation	
Tertiary			Ogallala
Cretaceous	Gulf		Pierre Niobrara Carlile Greenhorn Graneros Dakota
	Comanche	Washita Fredericksburg Trinity	Kiowa Cheyenne
Jurassic			Morrison
Triassic			Dockum
Permian	Guadalupe	Chickasha Duncan Hennessey Garber	Talaga Day Creek Whitehorse
	Leonard	Wellington	Nipewalla Sumner
	Wolfcamp	Amarillo-Guymon-Hugoton gas pays	Chase Council Grove Admire
Pennsylvanian	Virgil (Cisco)	Shallow sands Pawhuska Chickasha & Hoover-Elgin Cement Tonkawa	Wabaunsee Shawnee Douglas
	Missouri (Hoxbar Canyon)	Healdton, Hewitt, Loco Marchand, Madrano Cleveland Countyline Peru Checkerboard Wheeler	Pedee Lansing Kansas City Pleasanton
	Des Moines (Deese Strawn)	Oswego Tatums, Fox Graham, Prue, Calvin Senora, Red Fork, Burbank, Skinner Tussey, Bartlesville, Earlsboro Thurman, Hartshorn, Booch	Marmaton Cherokee
	Atoka (Bend)	Dutcher, Gilcrease Upper Dornick Hills Wapanucka Union Valley, Cromwell	Atoka
	Morrow	Lower Dornick Hills, Cromwell, Springer	Morrow
Mississippian	Chester	Caney	Chester
	Meramec	Mayes	St. Genevieve St. Louis Spergen Warsaw Cowley
	Osage	Sycamore	Keokuk Reeds Springs St. Joe Gilmore City Sedalia Compton
	Kinderhook		
Silurian Devonian	Hunton	Woodford Frisco Bois d'Arc Haragan Henryhouse Chimney Hill	Chattanooga Cooper Chimney Hill
Ordovician	Cincinnatian	Sylvan	Maquoketa
	Champlainian	Viola Simpson: Bromide, Tulip Creek, Wilcox, McLish, Oil Creek, Joins	Kimmswick Simpson
	Canadian	Arbuckle	Arbuckle
Cambrian	Croixian	Reagan	Reagan

Pre-Cambrian Basement

Rocky Mountain Region

System	Series	Group/Formation		
Tertiary	Pliocene	Ogallala		
	Miocene	Arikaree		
	Oligocene	Uinta, White River		
	Eocene	Wasatch, Golden Valley, Bridger, Green River, Chuska, Tohatchi, Torrejon, Puerco, Raton		
	Paleocene	Fort Union, Tongue River, Ojo Alamo, McDermott		
Cretaceous	Montana	Hell Creek, Cannon Ball, Lance, Medicine Bow, Kirtland, Farmington, Vermejo		
		Fox Hills, Lennep, Lewis		Trinidad, Pictured Cliffs, Fruitland
			Bearpaw	Cliff House
		Pierre	Judith River	Menefee
			Claggett	Mesaverde
				Point Lookout
			Eagle	Steele
	Colorado	Niobrara	Hilliard	Mancos, Tocito, Apishapa
		Carlile, Greenhorn, Graneros	Belle Fourche	Frontier
	Cloverly	Dakota	Fall River	Aspen, Mowry
				Bear River, Thermopolis, Purgatoire
			Kootenai	Fuson, Lakota
Jurassic		Morrison		Entrada, Carmel, Navajo
				Stump, Preuss
		Sundance	Ellis	Twin Creek
			Nuggett	Keyenta, Todolito, Wingate
Triassic		Chugwater	Spearfish	Ankereh
				Thanes
		Dinwoody		Woodside, Dinwoody, Santa Rosa, Poleo
				?Shinarump, Moenkopi, Coconino?
Permian		Phosphoria, Minnekahta, Opeche, Park City, Cutler		
Pennsylvanian		Tensleep, Minnelusa, Quadrant, Wells, Weber, Casper		
Mississippian		Amsden, Big Snowy, Madison		Brazer
Devonian		Monarch	Three Forks Darby, Jefferson	
		Beaverhill Lake, Elk Point		Saskatchewan
				Manitoban-Winnipegosan
Silurian		Interlake		Stonewall
Ordovician		Stony Mountain, Red River, Winnipeg	Big Horn	Whitewood
Cambrian		Deadwood		Gallatin, Gross Ventre, Flathead
Pre-Cambrian Basement				

Northern to Southern California Region

Age		Group/Formation	
Pleistocene	Red Bluff	Tulare, Paso Robles, Santa Barbara, Saugus, San Pedro	
Pliocene	Tehama	San Joaquin, Kern River, Careaga, Etchegoin, Foxen, Pico, Jacalitos, Sisquoc, Repetto	
Miocene	San Pablo	Reef Ridge, Chanac	
		McLure, Sta. Margarita, Monterey, Modelo, Puente	
		McDonald, Fruitvale	
		Devilwater, Round Mountain	
		Gould, McVan, Temblor, Rincon, Topango	
		Button Bed	
		Media, Olcese	
		Carneros, Jewett	
		Santos	Vaqueros
		Phacoides, Vedder	
		Vaqueros, Walker	
Oligocene		Tumey, Leda, Oceanic	Kreyenhagen, Lospe, Sespe
Eocene	Markley, Domengine, Capay, Meganos	Point of Rocks	Coldwater
			Llajas
		Avenal	Cosy Dell, Tejon
		Gatchell	Famosa Matilija, Santa Susana
Paleocene	Martinez		
Cretaceous	Chico, Shasta	Moreno, Panoche	
Jurassic & Older	Knoxville	Franciscan	

Western Canada

System	Series/Group/Formation	
Cretaceous (Lower)	Viking and Blaimore ss*	Athabasca tar sands
Jurassic	Vangard ss.*, Shaunovan, Gravelburg, Watrous	
Mississippian	Charles*, Mission Canyon*, Lodgepole*, Bakken-Exshaw	
Devonian	Wabumun, Winterburn, Woodbend (Leduc)*, Cooking Lake, Beaverhill Lake (Swan Hills)*	Slave Point ls.*, Sulfur Point dol., Muskeg/Keg River reef*, Keg River ls., Chinchaga anhydrite
Lower Paleozoic		
Pre-Cambrian Belt Series and Crystalline Basement		

*Oil or gas production.

Alaska North Slope Sequence

System	Series	Formations
Tertiary		Saganavirtok
Cretaceous	Upper Cretaceous	Kogosukruk† Sentinel Hill Tuluvac† Seabee Nanushuk* Grandstand*
	Lower Cretaceous	Topagoruk† Fortress Mtn.† Oumalik†
Jurassic		Kemik Kingak†
Triassic		Shublik*
		Ivshau
Permian		Echooka†
Carboniferous	Mostly Mississippian	Lisburne*
		Kayak
Devonian		Kanayut Mt. Weller†
Basement: Low-Grade Metamorphics		

South Slope Alaska—Cook Inlet

System	Series	Formations
Tertiary	Kenai	*conglomerate sandstone, shale

*Oil or gas production.
†Probable oil or gas production.

South America

System	Series	Columbia	Venzuela
Quaternary	Pleistocene	Mesa	
Tertiary	Pliocene		Rio Yuca Parangula* Lagunillas- Quiriquire*
	Miocene	Real	
	Ohgocene	Chuspas* — Colorado / Mugrosa	
	Eocene	Choro* — Esmaraldo Cantagallo LaPaz	Attamira Pauji Massparito Gubernador
	Paleocene	Lisama	
Cretaceous	Upper	Umir La Luna*	Berguita
	Middle	Salto Simiti Tamblazo Paja*	Quevado La Morita Guyacan
	Lower	Rosa Blanca* Tamblor*	Escandalosa* Cogollo
Jurassic and Triassic		Girbon	
Basement			

*Oil or gas production

Mexico

System	Series	Group/Formation
Tertiary	Plio-Pleistocene	
	Miocene	*Anahuac (Conception-Encanto) (Amate, Zargazal)
	Oligocene	*Frio (Jackson-Wilcox), Deposito — Chicontopec
	Eocene	*Midway
	Paleocene	*Wilcox
Cretaceous	Upper	*Aqua Nueva-Escamela
	Middle	*Elabra limestone
	Lower	*Tamabra *Tamaulipas
Jurassic		*San Andres
Triassic		Red beds and evaporites
Deformed Paleozoics and Pre-Cambrian Basement		

*Oil or gas production.

North Sea

System	Series	N.W. Europe	North Sea	Britain
Tertiary	Pliocene			
	Miocene			
	Oligocene			
	Eocene & Paleocene		*Cod gas and condensate	
Cretaceous	Upper	•	†	
	Weald			
Jurassic	Malm	•	†	
	Dogger	•	†	
	Lias	•	†	
Triassic	Rhaetic			
	Keuper	•		
	Muschelkalk			
	Buntsandstein		†	
Permian	Zechstein	•	†	•
	Rotliegendes	•	major gas fields	•
Carboniferous	Stephanian			
	Westphalian			•
	Namurian			
	Lower			•
Lower Paleozoic and Pre-Cambrian Basement				

*Oil or gas production.
†Probable oil or gas production.

North Africa

System	Series	Algeria	Libya	Egypt
Tertiary	Pliocene		detrital*	limestones
	Miocene			limestone* sandstones*
	Oligocene			
	Eocene	*	caleareous* evaporites	limestone* and sandstones*
	Paleocene	*	caleareous and shales	limestone and sandstones*
Cretaceous	Upper	*	shales	limestone, sandstones*
	Lower		Nubian basal ss.*	sandstones*
Jurassic		Offshore Morocco *		
Triassic		Salt beds *		
Permian Carboniferous Siluro-Devonian Ordovician Cambrian			* * * * *	* * * * *
Pre-Cambrian Basement				

*Oil or gas production.

Niger Delta
(Successively Deposited Deltaic Sediments on Pre-Cretaceous Crystalline Basement)

System	Series	
Tertiary	Pliocene-Pleistocene	(Oil) at Okan
	Upper Miocene-Pliocene	(Oil & Gas)
	Lower Miocene	Agbada Sands (Oil & Gas) at Port Harcourt-Bamu
	Oligocene-Miocene	(Oil & Gas)
	Oligocene	(Oil & Gas) at Calabao
	Late Eocene	
	Early Eocene	
	Paleocene-Campanian	
Cretaceous	Albian-Santonian	
Basement Complex		

Persian Gulf

System	Series	Iran	Kuwait	Saudi Arabia	Oman
Tertiary	Pliocene Miocene Oligocene Eocene	Asmari Limestones*			
Cretaceous		(new)	Burgan s.s.*		*Thamama-wasia Grp.
Jurassic		(new)†		Arab Zone*	
Unexplored					

*Oil and or gas fields.

East Indies

System	Series	Sumatra	Borneo	Java
	Pliocene			Surabaja*
	Miocene	Talang Akar*	Very thick oil sand* reservoirs	Mundu* Ledoc Wonotjolo* Ngrajong* Remblang Series
	Oligocene			
Basement				

Australia

System	
Tertiary	*
Cretaceous	*
Jurassic	*
Triassic	*
Permian	*
Carboniferous	*
Devonian	*
Silurian	
Ordovician	*
Cambrian	
Basement	

*Oil and/or gas fields.

GEOLOGY—GLOSSARY

alluvial fan—a sloping, fan-shaped mass of loose rock material deposited by a stream at the place where it emerges from an upland into a broad valley or a plain.

amorphous—without form; applied to rocks and minerals without crystalline structure.

anticline—an elongate fold in which the sides or limbs slope downward away from the crest.

apparent dip—the dip of a rock layer as exposed in any section that is not at right angles to the strike. The apparent dip is a component of, and therefore always less than, the true dip.

aquifer—a geologic formation or bed that transmits water.

argillaceous—containing or consisting of clay.

artesian—pertaining to underground water that is confined by impervious material under pressure sufficient to raise it to the surface if the material is penetrated by wells or natural fissures.

asymmetrical fold—any fold in which the limbs do not dip at equal angles away from the axial plane.

barrier reef—a type of coral reef that lies some distance from a coast.

basin—an area in which the rock layers dip toward a central axis.

bed—a specific layer of earth or rock separated by visually or physically-defined boundary planes from the layers of different material lying above, below, or adjacent to it.

bedrock—the more or less solid, undisturbed rock in place either at the surface or beneath superficial deposits of gravel, sand, or soil.

bentonite—a soft, porous rock consisting largely of silica and composed essentially of clay minerals.

calcareous—containing calcium carbonate.

calcite—a common rock-forming mineral: $CaCO_3$. It is the principal constituent of limestone.

caprock—the impermeable rock overlying an oil or gas reservoir that prevents the migration of the oil or gas out of the reservoir.

carbonate rock—a rock consisting chiefly of carbonate minerals, such as limestone or dolomite.

cementation—the binding together of particles of soil by some cementing agent, such as clay or carbonates.

channel sands—sandstone deposited as a result of stream action and, therefore, having the shape of a stream or river.

chemical sediments—accumulations formed directly by precipitation from solution (as by evaporation), or occasionally by the formation of insoluble precipitates on mixing solutions of two soluble salts.

clastic rock—a consolidated sedimentary rock composed of rock fragments derived from pre-existing rocks and transported to the place of deposition.

closure—the vertical distance between the highest point of the fold and the lowest contour that "closes" around the structure.

compaction—the stage in the process of converting sediments into rocks whereby closer packing of aggregate rock particles causes a decrease in the void space between the particles; the volume reduction of sedimentary deposits due to compressive stress, usually resulting from the continual deposition of each succeeding sediment.

competent—pertaining to beds, rocks, or geologic structures that support their own weight and the overlying rock during folding.

confining bed—a body of impermeable or less permeable material that restricts or confines the migration of fluids in a reservoir.

conformable—a term applied to parallel beds of rock that have been laid down in uninterrupted succession.

conglomerate—a cemented clastic rock containing transported rock fragments of gravel or pebble size. Also known as puddingstone.

connate water—water that is contained within a rock formation, and that was originally entrapped in the interstices of the rock material (either sedimentary or extrusive igneous) at the time the material was deposited.

continental shelf—the gently sloping belt of shallowly submerged land that fringes the continents.

core—the nucleus of the earth's interior.

correlation—the process of determining the time of occurrence of one geologic phenomenon in relation to others.

crude oil—liquid petroleum in its natural state in an underground reservoir or as it emerges from a well, or after it passes through a separator, but prior to any refining or distilling process.

crust—the outermost layer of earth.

cycle of erosion—the complete series of changes or stages through which a land mass passes from the inception of erosion on a newly uplifted or exposed surface, through its dissection into mountains and valleys, to the final stage in which it is worn down to the level of the sea or other base level.

deformation—the process of folding, faulting, compression, or extension of rocks as a result of various earth forces.

degradation—the wearing down and general lowering of a land surface by weathering and erosion, especially the erosion of streams.

desiccation—the process of drying up or drying out, as in the loss of water in the pore spaces of sediments due to compaction.

detrital—pertaining to that loose material or debris that results from rock disintegration.

diapir fold—a variety of anticline, usually involving salt deposits, in which the more mobile core has been able to pierce the overlying beds.

dip—the angle which a stratum, sheet, vein, fissure, fault, or similar geological feature makes with a horizontal plane, as measured in a plane normal to the strike.

disintegration—the reduction of rock to smaller pieces, principally by mechanical means.

displacement—the relative movement or magnitude of movement of the two sides of a fault in any specified direction.

dolomite—a carbonate sedimentary rock composed mainly of the mineral dolomite.

dome—an anticlinal uplift approximately circular or elliptical in shape.

downthrow (downthrow side)—the side of a fault that has apparently moved down with respect to the other side.

drag—the bending of rock layers adjacent to a fault.

economic geology—the application of geologic principles to the discovery and procurement of materials in the earth that are useful to mankind.

effective permeability—a measure of the ability of a particular fluid to flow through a rock.

effective porosity—the portion of pore space in saturated permeable material in which movement of fluids takes place.

eolian—pertaining to the wind; a designation for rocks and soils whose constituents have been transported and deposited by atmospheric currents.

epoch—a unit of geologic time; subdivision of a period.

era—one of the major divisions of geologic time, including one or more periods.

erosion—the wearing away and removal of materials of the earth's crust by natural means.

erosion cycle—the succession of stages through which a newly uplifted land mass must pass before it is worn down to a peneplain or a surface near sea level. See *cycle of erosion*.

exposure—an unobscured outcrop of either solid rock or unconsolidated superficial material.

facies—the aspect or appearance of sedimentary rock different in one or more respects from surrounding material; a distinctive set of characteristics. The features by which facies are named and recognized are usually selected more or less arbitrarily and may be lithologic (lithofacies) or biologic (biofacies).

facies changes—lateral or vertical changes in the lithologic or paleontological characteristics of contemporaneous deposits.

failure—fracture or rupture of a rock due to stress.

fault—a break in the materials of the earth's crust in which there has been movement parallel with the surface along which the break occurs.

fault plane—an approximately planar surface along which dislocation or faulting has taken place.

fault zone—a zone in which there are a number of more or less closely spaced faults.

field—a geographical area in which a number of oil or gas wells produce from a continuous reservoir. There may be a number of separate reservoirs at various depths within a single field.

flank—see *limb*.

flood plain—a strip of relatively smooth land bordering a stream, built of sediment carried by the stream and deposited in the slack water beyond the influence of the swiftest current during flooding.

fluid—a substance that flows and yields to any force tending to change its shape.

fold—a bend or flexure in a layer or layers of rock.

footwall—the lower or underlying wall of an inclined fault, vein, ore deposit, coal bed, etc.

fossil—any evidence of the existence or nature of an organism that lived in ancient times and that has been preserved in materials of the earth's crust by natural means.

fracture—a line where a rock has broken due to folding or faulting.

friable—easily crumbled, pulverized, or reduced to powder.

gas sand—a stratum of sand or porous sandstone from which natural gas is obtained.

geologic map—a map upon which geologic information is plotted. The distribution of the formations is shown by means of symbols, patterns, or colors. The surficial deposits may or may not be mapped separately. Folds, faults, mineral deposits, etc., are indicated by appropriate symbols.

geologic time—the time dealt with by historical geology extending from the end of the formation of the earth as a separate planet to the beginning of written history.

geologist—one who is involved with the geological sciences.

geology—the science that deals with the origin, composition, structure, and history of the earth.

geomorphology—strictly defined, the science of the form of the earth. Through common usage, it has come to mean the science of land forms.

graben—a long, narrow block of the earth's crust that has been relatively depressed by normal faults along the sides.

gradient—the inclination of any slope, as of a stream bed or hillside, expressed as a fraction, percentage, or angle.

ground water—subsurface water in a zone of saturation.

group—a unit of stratigraphical classification. A local or provincial subdivision of a system, based on lithologic features.

hanging wall—an old mining term designating the side or part of an inclined geologic structure that lies above or overhangs. The terms "hanging wall" and "footwall" are used not only to describe fault structures which are inclined between 0° and 90°, but also to designate all inclined tabular structures.

hardpan—a loosely used term designating a relatively hard or impervious layer beneath the soil or in the subsoil that offers exceptionally great resistance to digging or drilling.

historical geology—the study of the history and development of the earth, including the life forms that have inhabited it; also, the sum of that knowledge.

horst—a long narrow block of the earth's crust that has been relatively uplifted between normal faults along the sides.

hydraulic action—the mechanical loosening and removal of materials by water alone.

hydrocarbons—organic compounds of hydrogen and carbon, whose densities, boiling points, and freezing points increase as their molecular weights increase. The smallest molecules of hydrocarbons are gaseous; the largest are solids.

hydrologic cycle—all movement of water vapor in the atmosphere, on the earth's surface, below the surface, and in return to the atmosphere by evaporation and transpiration.

igneous rock—rock formed by the solidification of hot, mobile, rock material (magma).

impermeable—preventing the passage of fluid.

incompetent—referring to rocks that are inherently weak or less able to transmit stress without flow or fracture than are adjacent rocks.

index fossil—a distinctive fossil that characterizes a given formation or other rock or time unit, and that makes it possible to recognize and trace such a unit from place to place. An ideal index or guide fossil is one that has a wide geographic range and a short time range.

inorganic—pertaining to a compound that contains no carbon.

in situ—in place; applied to rocks, soils, and fossils that are situated in the place where they were originally formed or deposited.

intrusive rock—igneous rock that has ascended in a hot mobile state from the depths of the earth, but that has been arrested and cooled before reaching the surface.

joint—a fracture or parting plane along which there has been little if any movement parallel with the walls.

kerogen—a fossil organic substance that is a characteristic component of oil shale. The chemical composition and source of kerogen are not established, but it is known to be essentially bituminous.

lacustrine—pertaining to, produced by, or formed in a lake or lakes.

lateral changes—horizontal changes in rock content of a specific formation.

limb—the side of a fold.

limestone—a sedimentary rock composed mainly of calcium carbonate.

lithification—the formation of sediment into solid rock involving the processes of cementation, compaction, desiccation, crystallization, and compression.

lithology—the study of stones or rocks. It is nearly synonymous with petrology but is little used except in connection with sedimentary rocks, where it designates the sum total of physical characters.

magma—hot, mobile rock material generated within the earth, from which igneous rock is formed through cooling and crystallization.

mantle—the part of the earth between the surface and the core. Mantle also refers to the loose material at or near the surface, above bedrock.

marine—pertaining to the sea or ocean.

marlstone—a consolidated rock, more correctly known as an argillaceous limestone rather than a shale.

member—a subdivision of a geologic formation that is identified by lithologic characteristics such as color, hardness, composition, and similar features, and that has considerable geographic extent.

metamorphic rock—rock that is formed from original igneous, sedimentary, or older metamorphic rock through alterations produced by pressure, heat, or the infiltration of other materials at depths below the surface zones of weathering and cementation.

migration—the movement of hydrocarbons from their source rocks, through permeable formations, into reservoir rocks.

mineral—a substance occurring in inorganic nature, though not necessarily of inorganic origin, which has (1) a definite chemical composition or, more commonly, a characteristic range of chemical composition, and (2) distinctive physical properties or molecular structure. With few exceptions, such as opal (amorphous) and mercury (liquid), minerals are crystalline solids.

mineralogy—the science that deals with the origin, composition, chemical and physical properties, occurrence, and use of minerals.

monocline—a step-like bend in otherwise horizontal or gently dipping beds. It consists of a change in the amount of dip from gentle to relatively steep and back again to gentle.

natural gas—a highly compressible, highly expansible mixture of hydrocarbons having a low specific gravity and occurring naturally in a gaseous form.

nonconformity—a type of unconformity in which an older, eroded sequence of rocks meets a younger, overlying sequence at an angle. Tilting and erosion of the lower sequence before deposition of the higher beds is implied.

normal fault—a fault in which the hanging wall has apparently moved downward in relation to the footwall.

nose—the resultant flexure when an anticline is only half developed, that is, one end left open without closure.

oblique fault—a fault that strikes obliquely or diagonally to the strike of the adjacent rock.

oil pool—the accumulation of oil in the pores of sedimentary rock that yields petroleum on drilling.

oil sand—a sandstone that yields oil.

oil shale—broadly, any of a number of sedimentary materials that have the common property of yielding oil by distillation.

oolitic—pertaining to rock consisting of small, round grains cemented together.

organic—being, containing, or relating to carbon compounds, usually derived from the remains of plant and animal life.

outcrop—a part of a body of rock that appears, bare and exposed, at the surface of the ground.

oxidation—the process whereby oxygen is added to the rocks, especially to the iron compounds.

paleontology—the study of plant and animal life of past periods.

Pangea—a hypothetical supercontinent early in the geologic past, composed of all the earth crust at the time, and from which the present continents were formed by fragmentation and displacement.

period—the fundamental unit of the standard geologic time scale; the time during which a standard system of rocks was formed.

permeability—the measure of a rock's ability to transmit fluids.

petrography—the branch of petrology concerned with the description and systematic classification of rocks. Although it may be based on observations made in the field, on hand specimens, and on thin sections, it is now generally applied to the microscopical study alone.

petroleum—oil or gas obtained from the rocks of the earth. See *hydrocarbons*.

petroleum geology—the branch of economic geology relating to the origin, occurrence, migration, accumulation, and exploration of hydrocarbons.

physical geology—a broad subdivision of the science of geology which includes such aspects as the composition and occurrence of rocks and minerals, the original and secondary structures that they may possess, and the changes that may result in their metamorphism and decay.

plate tectonics—branch of geology that affirms the existence of large blocks of continental and oceanic crust floating on viscous material in the mantle of the earth.

pore space—the volume of holes or voids in a rock.

porosity—the ratio of the voids or pores in a rock to its total volume or size.

porosity trap—a type of stratigraphic oil trap that exists because of variations in porosity, such as lack of cement in sandstone or solution cavities in limestone or dolomite.

quartzite—a quartz rock derived from sandstone, composed predominantly of quartz, and characterized by such thorough induration (either through cementation with silica or through recrystallization) that it is essentially homogenous and breaks across or through the individual grains rather than around them.

radioactive dating—determining the age of geologic materials by measuring the presence of radioactive element and the amount of its decay products.

regional dip—the more or less uniform inclination of sedimentary beds in one direction over a wide area.

relief—the variations in elevation of a land surface considered collectively.

reservoir—a subsurface, porous, permeable rock body in which oil and/or gas is stored.

rock—in a strict geological sense, any naturally formed aggregate or mass of mineral matter, whether or not coherent, constituting an essential and appreciable part of the earth's crust. A few rocks are made up of a single mineral, such as a very pure limestone. Usually, however, two or more minerals are combined together to form a rock.

rock cycle—that series of events included in the formation, alteration, destruction, and reformation of rocks, and usually involving one or more of the following processes: erosion, transportation, deposition, lithification, and metamorphism.

salt dome—a more or less circular uplift of sedimentary rocks caused by the pushing up of a body of salt.

sandstone—a compacted sedimentary rock composed of the minerals quartz or feldspar.

sedimentary bed—a layer of sedimentary rock with common physical characteristics.

sedimentary rock—rock that is composed of sediment: mechanical, chemical, or organic. It is formed through the agency of water, wind, glacial ice, or organisms, and is deposited at the surface of the earth at ordinary temperatures.

sedimentation—the act or process of forming sediment in layers.

sequence—a distinct succession of major rock units deposited under related environmental conditions.

shale—a fine-grained sedimentary rock composed of consolidated silt and clay or mud.

siliceous—pertaining to, or containing, silica.

strata—sections of sedimentary formations that consist of the same type of rock. Plural of stratum.

stratification—the natural layering or lamination characteristic of sediments and sedimentary rocks.

stratigrapher—one who studies the formation and composition of stratified rocks in the earth's crust.

stratigraphic column—a diagram that shows the sequence of stratigraphic units of a given region (oldest at the bottom, youngest at the top) arranged to indicate their relations to the subdivisions of geologic time and their relative positions to each other.

stratigraphic trap—a reservoir of oil or gas in which the chief confining element is variation in the stratum or strata concerned.

stratigraphy—the branch of geology that deals with the definition and interpretation of the stratified rocks, the conditions of their formation, their character, arrangement, sequence, age, distribution, and, especially, their correlation by the use of fossils and other means.

strike—the direction of a line formed by the intersection of a bedding plane, vein, fault, slaty cleavage, schistosity, or similar geologic structure, with a horizontal plane.

structural basin—a downward fold that is synclinal in vertical section in any direction.

structural geology—the study of the architecture of the earth insofar as it is determined by earth movements.

structural high—the crest of a dome or anticline.

structural low—the bottom of a structural basin or syncline.

structural trap—a petroleum trap that is formed because of deformation of the rock layer that contains petroleum.

structure contours—lines of equal elevation regarded as drawn on the upper or lower surface of a particular stratum.

subsidence—the gradual downward settling or sinking of the earth's surface.

subsurface contours—equivalent to structure contours; so called to distinguish them from topographic contours representing the surface of the earth.

subsurface correlation—correlation of rock units and structures that do not appear at the surface, by means of well logs, mine maps, and geophysical data.

subsurface trap—geologic structure formed in such a way that migrating oil or gas is trapped and accumulated to form a petroleum reservoir.

surface beds—divisions of rock strata which are visible on the surface and easy to identify.

syncline—a fold that is concave upward.

tectonic—pertaining to the rock structures and external forms resulting from the deformation of the earth's crust.

thrust fault—a fault with a low angle of dip along which the hanging wall has moved up with respect to the footwall, and along which the horizontal component of movement has been great. Formerly used as a synonym for reverse fault, but now used with a variety of related meanings.

tight formation—a petroleum formation of relatively low porosity and permeability.

tight sand—a sand with such little pore space, or with the pore space so filled with clay or cementing material that oil and water cannot pass through. See *tight formation*.

transportation—the movement of sediment or loose rock material from one place to another by natural agents, such as water, wind, ice, or gravity.

trap—an arrangement of rock strata or structures that halts the migration of oil or gas and causes them to accumulate.

unconformity—a surface that separates one set of rocks from another younger and bedded set, and that represents a period of non-deposition, weathering, or erosion, either subaerial or subaqueous, prior to the deposition of the younger set.

uniformitarianism—the principle that geologic processes now operating to modify the earth's crust have acted in the same way throughout geologic time, and that past geologic events can be explained by forces and phenomena observable today.

uplift—a structurally high area in the earth's crust caused by forces that raise or upthrust the rocks.

viscosity—the property of a substance offering resistance to flow.

vug—a cavity in a rock.

weathering—the erosion of earth materials by atmospheric agents at or near the earth's surface; specifically, the physical disintegration and chemical decomposition of rock.

GEOLOGY—BIBLIOGRAPHY

Amyx, J.W., D.M. Bass, Jr., and R.L. Whiting: **Petroleum Reservoir Engineering,** McGraw-Hill Book Company (New York) 1960.

Calder, Nigel: **The Restless Earth,** The Viking Press (New York) 1972.

Clark, Norman J.: **Elements of Petroleum Reservoirs,** Society of Petroleum Engineers of AIME (Dallas) 1960.

Cox, Dennis P. and Helen R. Cox: **Introductory Geology, a Programmed Text,** W.H. Freeman and Company (San Francisco) 1965.

Gary, Margaret, Robert McAfee, Jr., and Carol L. Wolf (eds.): **Glossary of Geology,** American Geological Institute (Washington, D.C.) 1972.

Hobson, G.D. (ed.): **Modern Petroleum Technology,** John Wiley and Sons (New York) 1973.

Hobson, G.D.: **Petroleum Geology,** Oxford University Press (London) 1954.

Hubbert, Marion King: **History of Petroleum Geology and Its Bearing Upon Present and Future Exploration,** U.S. Geological Survey (Washington, D.C.) 1965.

Hunt, John M.: **Petroleum Geochemistry and Geology,** W.H. Freeman and Company (San Francisco) 1979.

Kummel, Bernhard: **History of the Earth,** 2nd Ed., W.H. Freeman and Company (San Francisco) 1961.

Landes, Kenneth: **Petroleum Geology,** John Wiley and Sons, Inc. (New York) 1963.

Landes, Kenneth K.: **Petroleum Geology of the United States,** John Wiley and Sons (New York) 1970.

Leet, L. Don and Sheldon Judson: **Physical Geology,** Prentice-Hall, Inc. (Englewood Cliffs, N.J.) 1970.

Levorsen, A.I.: **Geology of Petroleum,** 2nd Ed., W.H. Freeman and Company (San Francisco) 1967.

Moore, Carl A.: **Handbook of Subsurface Geology,** Harper and Row (New York) 1963.

Scientific American, Inc.: **Continents Adrift: Readings from Scientific American,** W.H. Freeman and Company (San Francisco) 1972.

Shelton, John S.: **Geology Illustrated,** W.H. Freeman and Company (San Francisco) 1966.

Wheeler, Robert R. and Maurine Whited: **Oil: Prospect to Pipeline,** 3rd Ed., Gulf Publishing Company (Houston) 1975.

ACKNOWLEDGEMENTS

Figures 2.30 through 2.37 are taken from **Elements of Petroleum Reservoirs,** by Norman J. Clark, and are used with the permission of the Society of Petroleum Engineers of AIME. The Society of Petroleum Engineers of AIME owns the copyright to these drawings.

CHAPTER 3:
EXPLORATION

I. INTRODUCTION

Before any production activity can be initiated, oil and gas must first be found. Without the continual discoveries of new oil and gas fields, progress and growth would be impossible; consequently, the industry relies heavily on exploration ventures. The geologists and geophysicists who perform this function have made possible the growth of the petroleum industry, and without their efforts, the petroleum industry as we know it today in the United States could not exist.

Because exploration is so vital, it is included in the activities of most oil and gas producing companies. The role of the geologist and geophysicist is very often the first and most important one in many of these companies. The exploration function determines the long-range stability of the company and its growth potential.

A term which is often used to express this long-term stability is called the **R/P ratio**. This is the ratio of a company's total reserves of oil and gas in the ground, divided by its yearly production. This ratio indicates the company's ability to continue in business over a given period of time. The following equation defines this R/P ratio:

$$R/P = \text{hydrocarbon reserves} \div \text{production per year} = \text{years of producing capacity}.$$

The amounts of recoverable oil or gas contained underground on a company's property constitute its hydrocarbon reserves.

To illustrate the use of this equation, let us look at two examples:

Example One:

An oil company has an oil field with 10 million barrels of recoverable oil in place under the ground. This field is developed and is producing at a rate of 1,000 barrels of oil per day, or 365,000 barrels of oil per year. The R/P ratio can then be calculated:

$$R/P = 10 \text{ million BO} \div 365,000 \text{ BOPY} = 27.4 \text{ years}.$$

The R/P ratio indicates that this company has a producing life for its reserves of approximately 27 years. If exploration efforts do not develop additional reserves by the end of that time, the company will no longer be in the oil producing business.

Example Two:

Another oil company has recoverable gas reserves amounting to 50 billion cubic feet. These reserves are being produced at the rate of 10 million cubic feet per day, or 3.65 billion cubic feet per year. The R/P ratio is determined using these figures:

$$R/P = 50 \text{ billion cubic feet} \div 3.65 \text{ billion cubic feet per year} = 13.7 \text{ years}.$$

The R/P ratio projects that this company has a producing life of approximately 14 years. New gas reserves must be developed during this time if the company is to continue in the gas producing business beyond these 14 years.

The search for oil and gas beneath the earth's surface is one of the most risky undertakings in any industry. Indeed, it is quite a gamble to invest money in an exploration wildcat well, because there are so many unknowns involved in the analysis and study of the earth. As noted in the previous chapter on Geology, a number of variables must be present in the proper configuration to produce the right conditions for an oil or gas field. Several scientific and technical procedures are now available to the geologist to assist him or her in trying to predict whether these variables may have combined to create an oil or gas field, but the results are not guaranteed.

Few people realize the extent of this risk, and how few exploration wells drilled in the United States today actually result in the discovery of oil and gas. Approximately one out of seven new wells drilled ever produces oil or gas, and only one out of 50 exploratory wells ever finds significant new reserves of petroleum. The risk is further enhanced by the expense involved in drilling new wells. In 1979, the average drilling cost for new onshore wells in the United States was approximately $275,000, an increase of 126% over the previous five years. For offshore wells, the average cost was approximately $2,500,000, or three times as much as it was five years before. With inflation driving up the cost of materials and related expenses, these figures have continued to climb. Today, the cost of an exploratory well in the North Sea can exceed 6.5 million dollars.

The high cost of drilling for new oil and gas fields demands that proper exploration work be done by any company involved in searching for new reserves. This has become increasingly more evident during the last few years in the United States, since the discovery of new oil and gas fields is declining. The present success ratio in drilling for new oil and gas fields is only a little over half of what it was in the late 1940s, despite today's improved exploration techniques. While the success ratio has gone down, the costs for drilling exploratory wells has continued to escalate. Consequently, new and better scientific methods to search for oil and gas are constantly being developed. With the increasing scarcity of petroleum resources, the rising cost to develop these resources, and our nation's demand for petroleum products of all types, utilization of the best possible exploration methods is vital to the modern oil company.

II. EXPLORATION METHODS

A. History

The use of exploration techniques to find oil and gas began over 100 years ago, when Drake drilled the first successful oil well in Pennsylvania in 1859. Drake's well touched off a drilling boom that eventually spread to other parts of the country. Since then, many advances have been made in the science of oil exploration, and the history of exploration techniques has produced an interesting and colorful chapter in the development of the oil industry in the United States.

Early oil explorers originally looked for oil in gas seeps and slicks along low places in creek valleys. Water wells, with their occasional oil shows, provided additional clues to the existence of petroleum beneath the surface.

It was during this early period, too, that the term "wildcat well" was first used. In those days, the woods of Pennsylvania were full of wildcats, or bobcats, and mountain lions. At night, while oil drillers were working on their rigs, the cats often could be heard screaming in the woods. As a result, the early exploratory wells drilled in this part of the country came to be called "wildcat wells." This term has been used throughout the history of oil exploration, and still refers to those wells that are drilled to find oil and gas in previously unexplored areas.

The early booms in the oil industry created an entirely new group of people who developed unorthodox methods to search for oil and gas. The techniques were generally crude, even primitive, and relied for the most part on superstition or unscientific paraphernalia. Oil explorers sometimes employed the service of clairvoyants, fortune tellers, spiritualists, and oil "smellers" to aid them in the search for oil. Divining rods or "wiggle sticks," so often used in water exploration, and "doodle bugs," small chemical or electrical devices, also were considered useful.

The discovery of oil initially along streams led to the search for petroleum along other streams. Oil explorers also examined surface rocks and cuttings from producing wells, compiling all this information into a primitive "science" they called "creekology." Because of the abundance of shallow and easily found oil deposits, these techniques were sometimes quite successful.

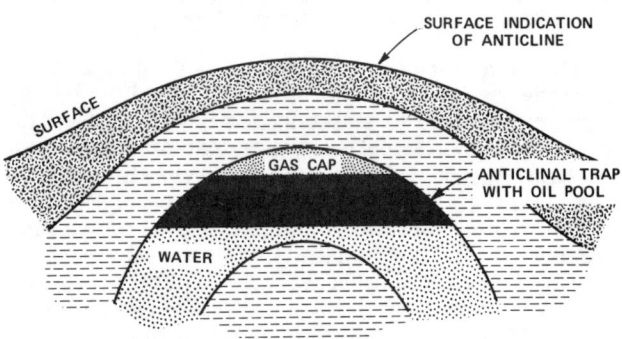

Figure 3.1. ANTICLINAL OIL TRAP.

Not long after oil became a commercial commodity, the science of geology began to be applied to the petroleum exploration effort. Geologists realized that oil was often found in subsurface traps known as **anticlines**. These traps were first recognized by their sometimes distinctive surface features (Figure 3.1). The frequent association of oil with this type of structure led to the development of the "anticlinal theory." Early oil explorers looked for indications of such anticlinal traps on the earth's surface and drilled in these locations in the hope of finding oil.

As geologists gathered more information from the growing number of new oil field discoveries, they learned that oil was often found in other types of geological structures, such as salt domes and faulted zones (see the previous chapter on Geology for an explanation of these). The anticlinal theory was combined with the study of salt domes and fault zones, forming the "structural theory," which indicated that oil could be found in types of geologic structures other than anticlines. The theory enabled exploration geologists to consider many new areas for oil drilling operations—areas that had been previously disregarded because no anticlines were present. This led to the development of modern geology as it applied to petroleum exploration. As the amount of information available to geologists increased, the theories and techniques they used advanced as well.

The modern science of exploring for petroleum deposits has developed far beyond the original geological methods used in the late 1800s and early 1900s. Today, explorationists use many different approaches in their attempts to "see into the earth." The science of geology is aided in this effort by another modern scientific discipline—**geophysics.**

In the following sections we will look at some of the major categories of exploratory methods used by the modern explorationist. These may be generally described as **surface** and **subsurface geological** techniques, and **geophysical** techniques.

B. Surface Geology

Surface geology has long played an important role in exploring for petroleum reservoirs. The character of surface outcrops, the dip of surface rocks, the presence of faults and other geological phenomena on the surface—all of these are clues that lead to the identification of reservoir traps. The geologist usually compiles this data on various types of maps, and uses this information when exploring for oil and gas.

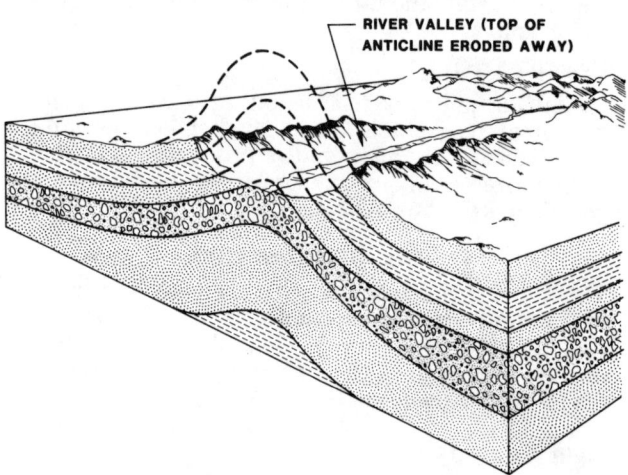

Figure 3.2. ERODED ANTICLINAL STRUCTURE.

Surface geological maps can be helpful only where some significant subsurface geological feature, such as a fold or fault in the sedimentary beds, extends to the surface to provide an indication of what lies underground. This technique is useful where the surface topography has sufficient character to determine these features. Figure 3.2 illustrates how these surface features help identify potential underground reservoir traps. A geologist can obtain a good indication of the presence of a subsurface anticlinal trap by observing the location and dip of the surface beds, which may provide an image of the subsurface feature. In Figure 3.2, the river has eroded away the highest portion of the anticline, so the surface expression of the anticline is quite evident.

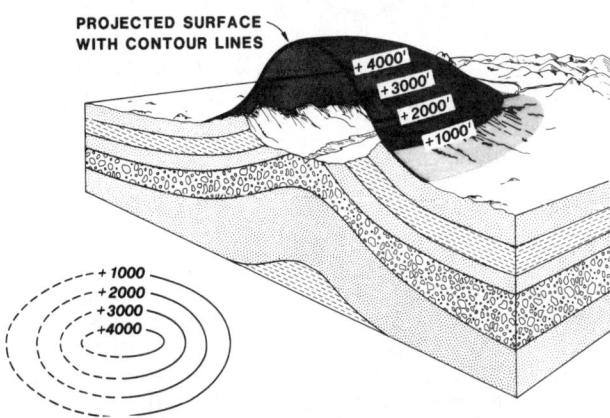

Figure 3.3. STRUCTURE CONTOUR MAP OF PROJECTED SURFACE. Dotted lines indicate that portion not shown on three-dimensional drawing.

A **surface structural map** could be made of this example (Figure 3.3). The anticlinal structure is indicated by **contour lines** which have been superimposed over the topographical features. These contour lines were developed by observing and measuring the elevation, dip, and orientation of selected reference beds on the surface rock outcropping. By making these measurements at many points along the river valley, a complete picture of the anticlinal bulge or structure can be mapped. In this illustration, the surface anticline which has been mapped reflects an identical feature underground where oil may be trapped.

In many cases, surface topographical features do not expose sufficient geological information to permit the preparation of satisfactory surface structural maps. When this happens, shallow stratigraphic test holes may be drilled to supplement surface information in the process of developing an adequate geologic interpretation.

Another common type of surface geologic map is the **photogeology map.** This map is made from aerial photographs and, generally, documents the type of rock exposed on the surface. Other important geological features such as faults, bed orientations, and dips are also shown. Figure 3.4 is an example of a photogeology map of a small mountain.

Many other types of surface geological maps are prepared, and may be used for particular exploration projects. (Surface maps are particularly valuable in exploration projects for such minerals as coal and uranium.) At the turn of the century, surface geology did lead to the discovery of many new fields, but today most of the obvious surface features have already been explored.

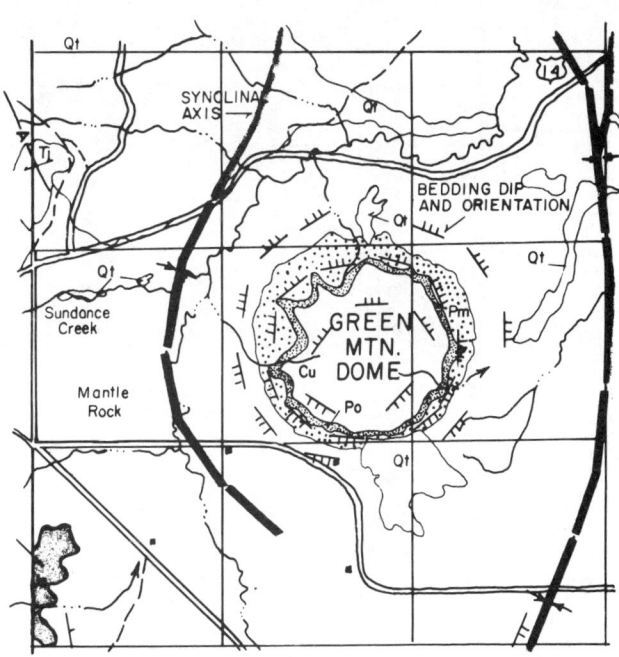

Figure 3.4. PHOTOGEOLOGY MAP.

Surface geologic maps are of limited value where the geologic data is insufficient or where the surface geology does not reflect the subsurface structures. The oil industry has had to develop new methods, since many of the newly discovered fields give little surface indication of their presence. Procedures which combine the use of subsurface geologic studies with geophysical techniques have lead to the discovery of such fields.

C. Subsurface Geology

Subsurface geology became a more sophisticated extension of surface geological techniques. Originally, only surface information was available to the geologist, but as more wells were drilled, more subsurface information was gathered, leading eventually to the widespread use of subsurface geology in oil and gas exploration. Geologists can now use several different subsurface geological techniques to help identify the possible presence of oil and gas. The methods used to acquire the necessary subsurface geologic information are summarized below.

1. Subsurface Geological Procedures

a. Well Samples

The most direct source of information available to the subsurface geologist is the examination of **rock cuttings** taken from the wellbore during drilling. Rock cuttings consist of small fragments of rock that have been broken off by the bit during the drilling process. They provide a lithologic record of the sedimentary deposits encountered in the well. After they have been collected at the well site during the drilling, these samples are cleaned and stored to provide an historical record of the lithology of the well. The rock cuttings also provide clues to the presence of hydrocarbons and are examined closely for shows of oil and gas. When examined by a stratigrapher and properly described, well samples provide a good record of the sedimentary strata penetrated in the well. These sample logs may be used for regional geological studies, as well as detailed field studies.

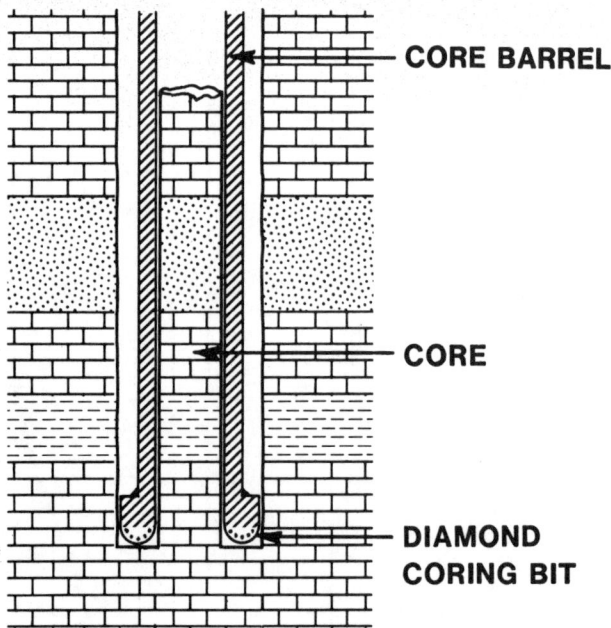

Figure 3.5. TAKING A CORE SAMPLE.

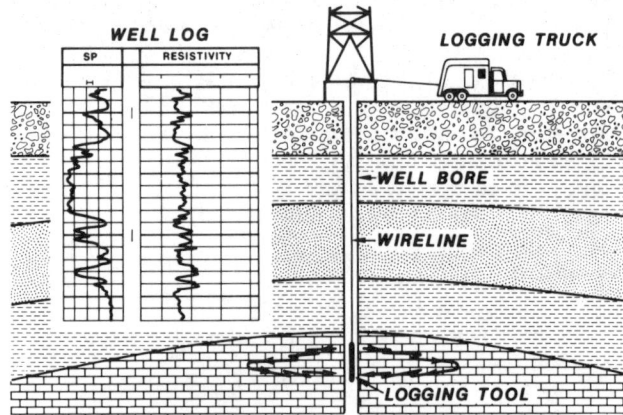

Figure 3.6. WELL LOGGING.

b. Cores

The subsurface geologist may also take cores from the rock in a zone of specific interest in the well. A **core barrel,** which has a donut-shaped diamond bit at the end of the drill pipe, cuts out a long, cylindrical piece of rock, or **core,** while the well is being drilled (Figure 3.5). Because these cores provide such an accurate sample of the zone in question, they are used for numerous geological and engineering studies. In addition to providing information on the lithology of the cored zone, the cores are particularly valuable in identifying the locations of fractures in the rock, zones of porosity and permeability, and the type and amount of fluid which occupies the pore space in the rock.

In cases where full-size cores cannot be taken, **sidewall core samples** may be used to collect rock data. In this method, a small, cylindrical cutter is fired by explosives into the side of the wellbore to retrieve a small plug of rock. The plug can be tested to determine such rock properties as lithology, porosity, permeability, etc. This method is generally used after a well has been drilled, and the geologist wants to take rock samples from zones which were not previously cored.

c. Geophysical or Wireline Logs

Geophysical logs are widely used by subsurface geologists in their search for oil and gas. A number of different types of logs are available, corresponding to the information required. In general, all of the logging tools consist of an electrical or radioactive device, which is lowered into the wellbore on a cable to take certain measurements of the rock. The electrical signals from this device are transmitted through the cable to the surface, where they are recorded on a film. The film is used to make a log showing the recorded measurements at all points throughout the depth of the well (Figure 3.6). These measurements are used to analyze the various rock properties of the formations, such as porosity, fluid saturation, and lithology. Some of the major types of logs used today are described below.

(1) Electrical Surveys

Resistivity logging is generally used to measure the electrical resistivity of the formation surrounding the wellbore. Resistivity is that property of the rock which relates to its ability to conduct electricity. The mineral framework of the rock itself generally has an infinite resistance to the flow of electric currents; that is, it will not conduct electricity. However, the fluid-filled pores of the rocks found in oil fields have varying degrees of resistivity, depending on the type and amount of fluid they contain. For example, oil has a very high resistivity reading, whereas salt water is very conductive, and will give a rock filled with this saline fluid a low resistivity reading.

Resistivity logs help determine the type of fluid which occupies a rock's pore space, and within certain limitations, the relative saturations of oil and water in formations that contain both of these fluids. Most oil reservoirs contain some water with the oil, and this can be detected when resistivity logging techniques are used.

Some of the common resistivity logging devices are described briefly in the following paragraphs.

(a) Induction Electric Log

The induction electrical log is a fundamental resistivity device, which is commonly used both for resistivity determination and for correlating one well with another. Well logs are matched or correlated with each other to determine where certain formations are present in each well. The induction electric log is most effective in formations with medium to high porosity.

(b) Dual Induction Focused Log

This is a more sophisticated resistivity log, which not only measures formation resistivity, but also permits analysis of the effect of drilling fluid invasion on the formation. Fluid invasion occurs when drilling fluids force their way into the formation and displace the natural formation fluids. This log is also used to correlate wells, and works best in zones with medium to low porosity.

(c) Laterolog

The laterolog is a focused resistivity log which is designed to be run into wells drilled with drilling mud made from salt water. The focusing system is particularly useful in identifying those thin formation beds which might be missed by conventional resistivity devices.

(d) Microlog

This resistivity device is used to determine the porous and permeable zones in the rock by measuring the resistivity of the flushed or invaded zone close to the wellbore.

(e) Micro-Laterolog

The micro-laterolog uses a focused resistivity system to measure a small amount of rock near the well, and provides information on the zone which has been flushed or invaded by wellbore fluids. By comparing this with the resistivity in the undisturbed zone, the geologist or engineer can determine the amount of moveable oil present in the formation, as well as the effective pay thickness, or that part of the oil-bearing rock which may produce oil.

(f) Spontaneous Potential Log (SP)

This log measures the electrical potential of the rock and is used to define lithology. It also is used for correlation purposes. Under certain circumstances, this device can help determine the natural resistivity of formation water encountered in the well.

(2) Radioactive Logging

Radioactive logging may be divided into two types: **gamma ray logging,** which measures the natural radiation of the rocks that have been penetrated; and **induced radioactive logging,** in which the rocks are bombarded from a radioactive source, and a measurement is made of the induced radiation.

(a) Gamma Ray Log (GR)

The gamma ray log measures the natural radioactivity of the formation, and is primarily used to define lithology and to correlate wells. It is normally run simultaneously with neutron and sonic logs.

(b) Density Log

This porosity device uses a gamma ray source which sends radiation into the rock. The rock's radioactive response is measured and used to determine porosity. When used with the sonic or neutron log, this log aids in identifying lithology.

(c) Neutron Log

A neutron log detects neutron radiation from the formation. It not only is useful for porosity and lithology determination, but also is helpful in locating gas zones.

(3) Auxiliary Logs

In addition to resistivity, porosity, and lithology information, different wireline logs can also provide other data.

(a) Sonic Log

The sonic log utilizes sound pulses, which are sent into the formation. The time required for these sound waves to travel through the rock is measured and used to determine porosity. When used with other porosity devices, it can be helpful in determining lithology.

(b) Dipmeter

A dipmeter is used to determine the dip and strike of different beds in the stratigraphic section encountered in the wellbore. This information is valuable when structure maps are made.

(c) Directional Survey

This is a mechanical device which measures the deviation from vertical in the well at all depths, and is used to accurately determine the physical location of a well at any point along its depth. Wells are seldom straight holes, and directional information is necessary for the geologist or engineer to keep an accurate record of the location of different events which occur during the drilling of the well. This information is particularly important in offshore operations where a number of wells are directionally drilled from one surface location.

(d) Caliper Log

The caliper log measures the diameter of a hole. It aids in locating washouts or restrictions in the wellbore, and generally is used with one of the resistivity or porosity logs.

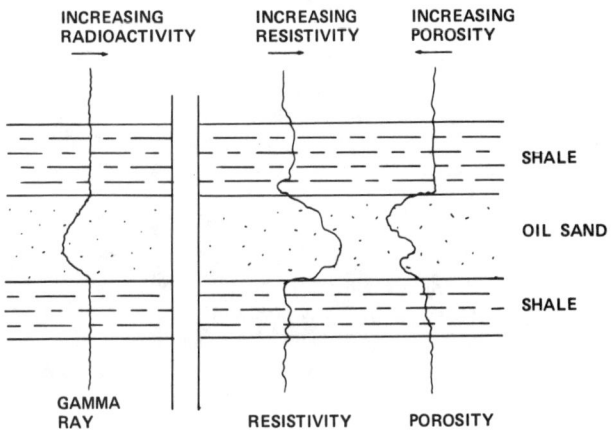

Figure 3.7. DIFFERENT LOGS OF THE SAME WELL.

Figure 3.7 illustrates how some of these logs would look if they were used to log a well with a typical shale-sand sequence. Note that the oil-bearing sand is characterized by a decrease in radioactivity on the GR Log and an increase in resistivity and porosity. The increased resistivity indicates the presence of oil, while the increased porosity identifies reservoir rock.

d. Drill-Stem Testing

The drill-stem test is used to test a well's potential pay zones for the presence of oil or gas. It is one of the few methods available that gives an estimate of what a newly discovered pay zone might produce once the well has been completed. In the drill-stem test, the **drill stem** or **drill pipe** used during drilling is used to flow-test the well for a short period of time. By means of the drill pipe and a **packer** assembly, the well is temporarily "completed" only in the zone to be tested, and its flow performance is observed. After the test has been satisfactorily concluded, drilling operations may be resumed if additional strata are to be penetrated.

In contrast to drill-stem testing, **production testing** involves completion of the well before the flow-test begins. Steel casing is run in the well and cemented in place; perforations are made in the casing opposite potential pay zones, and flow from the zone is evaluated. (A more detailed explanation of well completion can be found in the chapter on Drilling.) The disadvantage of production testing is that it usually occurs only at the

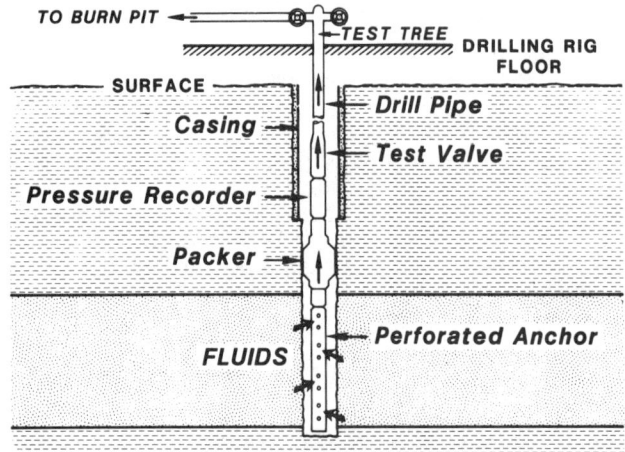

Figure 3.8. DRILL-STEM TEST.

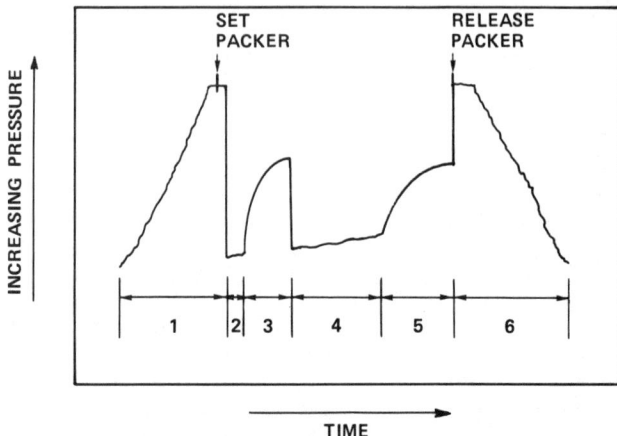

Figure 3.9. DRILL-STEM TEST PRESSURE CARD. 1) Drill string and test tools run in hole. 2) Initial flow period. 3) Initial shut-in period. 4) Final flow period. 5) Final shut-in period. 6) Drill string and test tools removed from hole.

end of drilling operations. There is no further penetration of new strata after the well has been completed. The drill-stem test, therefore, is a more convenient method of evaluating pay zones during the drilling process, by temporarily completing only parts of the well.

The following procedure is used during a drill-stem test, as illustrated in Figure 3.8. The drill string with the test tool attached is run into the hole, which is usually filled with drilling mud. The **test valve** is closed prior to this to keep the inside of the pipe empty.

When the string of tools hits bottom, its weight on the rubber packer forces the packer to expand against the sides of the hole. This seals off the portion of the hole below the packer from the hydrostatic pressure of the mud above it.

Mechanical movement of the drill string opens the test valve, exposing the formation to the testing tool attached to the drill pipe. If the formation has any significant pressure and permeability, flow of fluids into the drill pipe will result, since formation pressure is greater than the atmospheric pressure in the drill pipe.

Pressure gauges, or "bombs," are used to record flowing pressures and pressure build-up during the test. After sufficient flow data is accumulated, the test valve is shut again to allow pressure under the packer to build back up to the pressure of the formation. Weight is taken off the packer, causing it to contract, and the entire assembly is brought up out of the hole.

The sample of the reservoir fluid contained in the drill stem is retrieved when the test tools are lifted from the hole. This sample provides information on the type of fluid present, as well as the pressure in the formation.

In actual practice, most drill-stem tests involve several flow periods and several shut-in periods. This allows the geologist and engineer to collect the maximum amount of information from the test. During the procedure, the pressure information is recorded on a card which forms a permanent record of the data. Figure 3.9 illustrates a card showing a test with two flow periods and two shut-in periods.

Information obtained from a drill-stem test includes: 1) reservoir pressure, 2) average formation permeability, 3) an estimate of formation damage from drilling fluids, 4) an indication of barriers or depletion (if the reservoir is small), and 5) a description of the physical characteristics of formation fluids.

2. Subsurface Geological Mapping Techniques

From the information gathered through the use of the above procedures, a variety of maps can be developed to aid in identifying possible new hydrocarbon reservoirs. The quality of these maps depends on the amount of data available. In an area where many wells have been drilled, very detailed and accurate maps can be made. In areas where very few wells have been drilled, subsurface maps are difficult to construct and involve a great deal of interpretation on the part of the geologist. In these areas, geophysical techniques are extremely valuable.

It is worth noting that these maps are only approximations of what lies beneath the earth's surface, and are never really completed. New information is continually being added as more wells are drilled, and the maps require constant revision. A complete representation of all the subsurface features is impossible, since the necessary data is never totally acquired.

There are several types of subsurface maps, and generally, a geologist uses more than one type in his or her exploratory studies.

a. Structure Maps and Cross Sections

Oil is frequently present in folded or faulted rock formations which form traps for the oil. These traps are identified and defined through the use of **structure maps**. A **reference** or **datum horizon** is chosen in the stratigraphic section, such as a boundary between two different sedimentary formations or the top of an oil producing zone, and the elevation of this point above or below sea level is calculated. By plotting these data points from several wells on a map, the geologist gets a picture of the size, shape, and elevation of the structure.

The structure map in Figure 3.10 shows an anticlinal reservoir with an asymmetrical axis. In this example, the number of wells present in the structure make a detailed map possible. Note that the elevations shown on the contour lines indicate that the structure is located below sea level. The highest point is in the south center portion of the structure, at an elevation of approximately −2,000 feet, or 2,000 feet below sea level. Note, also, that the gas wells are located at the highest point, oil

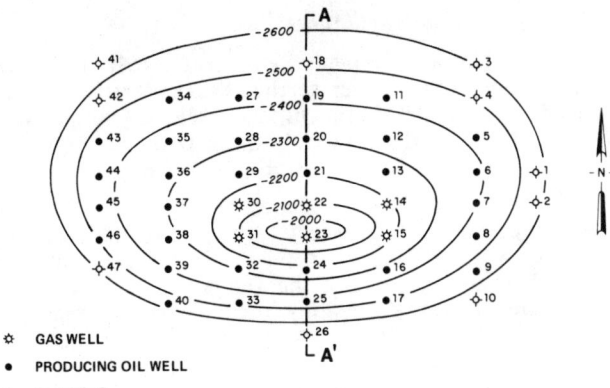

Figure 3.10. STRUCTURE MAP.

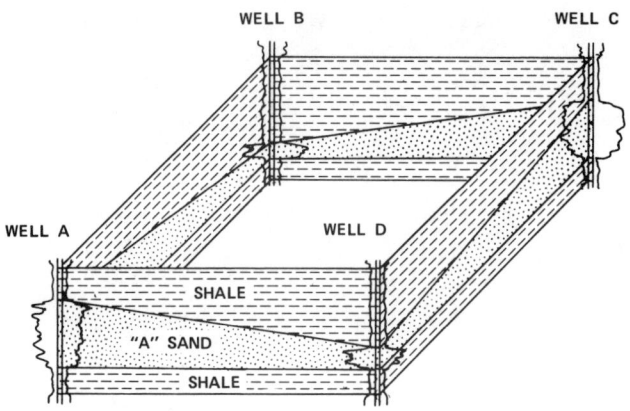

Figure 3.12. FENCE DIAGRAM.

wells are in the middle, and the "dry holes" or water wells are on the flank of the structure.

Cross sections also can illustrate information concerning the structural features of an oil reservoir. Cross sections are made by plotting logged sections of formations for each well, and then connecting equivalent points in various wells.

A cross section for the structure mapped in Figure 3.10 is shown in Figure 3.11. This cross section illustrates the asymmetrical nature of the structure. It has a much steeper dip on the south side than on the north. The cross section also clearly indicates the locations of gas, oil, and water interfaces. This provides a valuable means of predicting where to drill additional wells in the field, and where to avoid the water zone or gas zone.

b. Isopach Maps

Isopach maps illustrate the thickness of a formation when measured between two reference planes (normally the formation boundaries). If sufficient well data is available, isopach lines can be contoured to show areas of uniform formation thickness, etc. Isopach maps may be used to plot net pay areas, or those portions of a formation that have sufficient oil saturation to produce oil in commercial quantities.

Figure 3.13 illustrates an isopach map for a buried river channel sand in a sandstone formation. The thickest portions of the sand are in the center of the ancient, buried river channel. If the sand were full of oil, this should be where the most productive wells would be located.

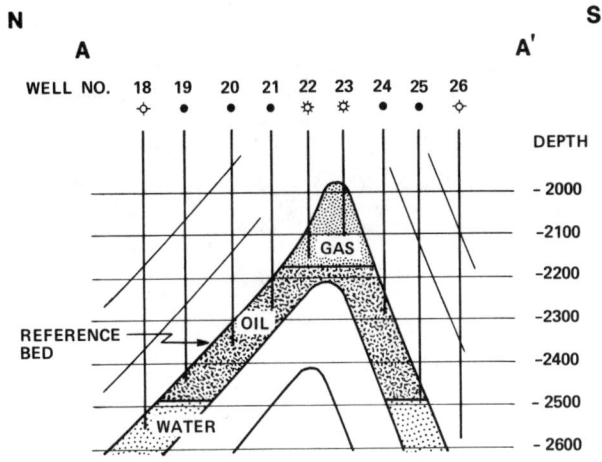

Figure 3.11. CROSS SECTION.

Cross sections are often used to construct **fence** or **panel diagrams.** These consist of several connected cross sections to illustrate the subsurface geological information in a three-dimensional manner (Figure 3.12).

Fence diagrams are especially useful in illustrating complicated subsurface features such as channel sands. The diagram's three-dimensional character permits rapid and easy visualization of complex geological features. Fence diagrams are frequently used in special reports, or when it is necessary to explain information to groups of people.

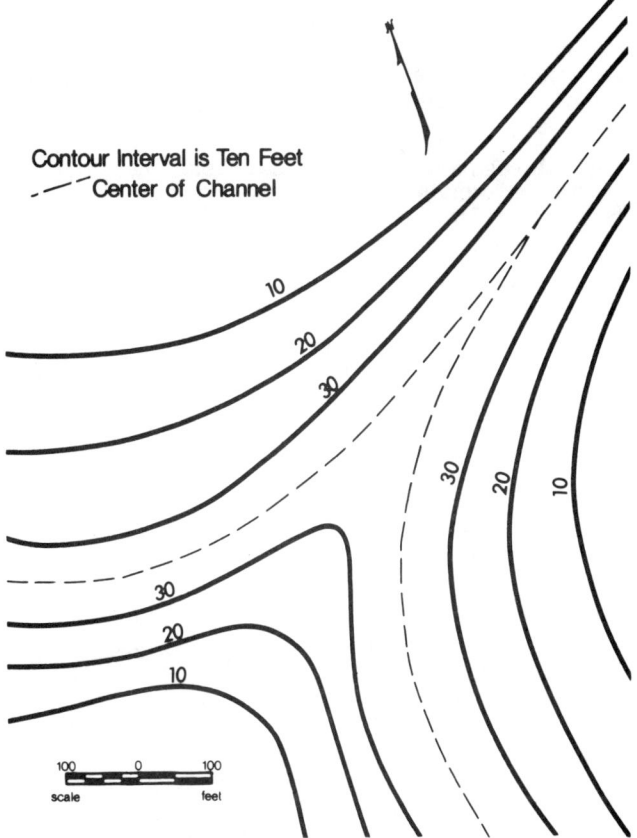

Figure 3.13. SUBSURFACE SAND ISOPACH MAP.
River channel sand.

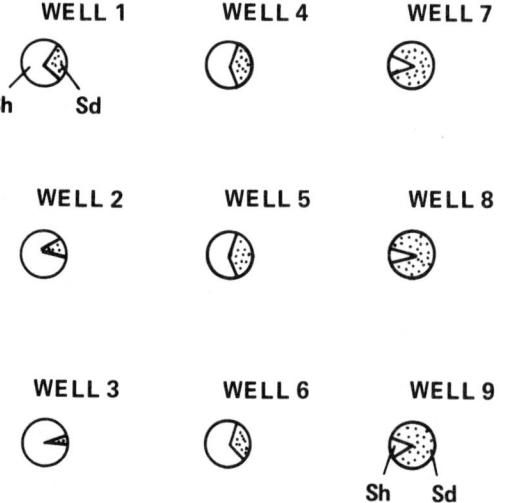

Figure 3.14. LITHOFACIES MAP.

c. Facies Maps

Facies maps show the lateral (or horizontal) variation of rock type in a given formation. A good example of this variation is sandstone, which often has different amounts of shale dispersed throughout the formation. Although a sandstone formation may be spread horizontally over a large area, the quality of the rock may vary considerably. Some portions will contain large amounts of shale, while others will be much cleaner. This range can be represented on a facies map, or more precisely in this case, on a **lithofacies map,** since it deals with changes in lithology (Figure 3.14). As the map in Figure 3.14 shows, the wells to the right contain the greatest amount of clean sand, while the amount of shale in the formation increases significantly in the wells to the left.

Facies maps can be constructed from many different types of information besides lithology, such as porosity facies maps. However, the lithofacies map is one of the more common ones used in petroleum geology.

D. Geophysical Techniques

The geophysical branch of the exploration science is primarily concerned with defining subsurface geological features through the use of seismic techniques, or the study of energy wave transmissions through rock. Seismic techniques can be used to make subsurface maps similar to those developed by standard geological methods. Geophysicists, who are specialists in seismic, magnetic, and gravity interpretation, often work side by side with geologists to determine the best prospects to drill for new petroleum reservoirs. Some of the best exploration is accomplished when good geophysical and geological techniques are combined.

Geophysics is directly concerned with measuring the physical properties of rocks. In exploration work, the three main rock properties that the geophysicist studies are 1) elastic characteristics, 2) magnetic properties, and 3) density. Although studies of density and magnetic properties provide useful information, elastic characteristics are considerably more important since they govern the transmission of energy waves through rock. It is this elastic characteristic that is studied in seismic surveys.

1. Seismic Techniques

The word **seismic** pertains to earth vibrations which result from either earthquakes or artificially induced disturbances. Earthquake seismographs were invented to record ground vibrations from earthquakes, and are used world-wide to study these events.

In the mid-1800s, evidence revealed that seismographs could be employed to study geological characteristics near the surface of the earth. Scientists discovered they could measure the velocity of shock waves from accidental explosions that early seismographs recorded. During World War I, both the Allied and German forces developed seismic techniques to pinpoint the location of large artillery guns on the front. The recoil from the guns created ground disturbances, or shock waves, which could be recorded. After the war, some of these methods were modified and refined for petroleum exploration purposes in Europe and, later, in the United States. The discovery of several salt domes along the Gulf Coast by a German seismic company in 1924 marked the beginning of serious seismic exploration efforts in the United States.

a. Reflection Seismic

Initially, seismic exploration utilized the refraction method, in which refracted waves resulting from a seismic disturbance were recorded. The method had limitations, however, especially when deeper exploration became necessary. Today, the geophysicist's primary resource is the reflection seismic method.

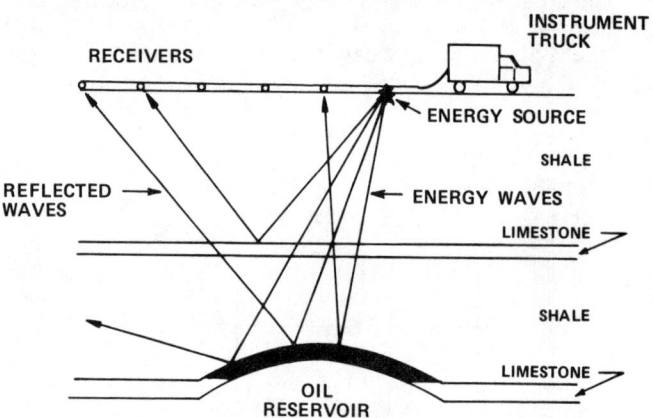

Figure 3.15. REFLECTION SEISMIC: LAND OPERATION.

Reflection seismic surveys record the seismic waves that return or reflect from subsurface formation interfaces after a seismic shock wave has been created on the surface. By measuring the time required for different waves to be reflected from different formations, the geophysicist can identify structural variations of the formations. Figure 3.15 illustrates this process in a typical land survey operation.

A similar technique is used for marine operations. A ship carries the necessary recording instruments and tows the energy source and receivers behind it in the water.

The objective of seismic work is to develop maps that indicate structures which might form traps for oil or gas. Several steps are necessary to convert seismic data into useful structural and stratigraphic information.

(1) Field Data Acquisition

The equipment for acquisition of field data in land operations, such as illustrated in Figure 3.15, includes three basic components: an **energy source, receivers,** and **recorders.**

(a) Energy Source

Early seismic work in geophysical exploration relied almost exclusively on the use of dynamite to generate seismic energy waves. Shallow holes were drilled in the earth to contain the energy released by the dynamite. This method has become relatively obsolete because it takes too much time to prepare the explosive charges. When digital computers were introduced to handle field data, a faster, more controllable method of producing the shock energy became necessary, which led to the development of higher quality, more convenient energy sources.

Today, a number of different types of energy sources are used in seismic exploration. One of the most common land devices is the **thumper.** A truck carries a large, suspended weight, which is dropped periodically on the ground at predetermined locations to create the necessary seismic waves. Other energy sources include hydraulically-actuated **vibration devices** and pneumatically-operated **air guns.** Away from inhabited areas, small charges of explosive, called **mini-charges,** are still used for certain types of reflection seismic work.

A number of devices specifically designed for marine work have also been invented. These include **gas exploders,** in which an explosive gas mixture is detonated in a contained chamber; and "sparkers," which utilize the discharge of stored electrical energy to create energy waves in the water.

(b) Receivers

Following the release of seismic energy, the energy waves travel through the rock and are reflected from formation interfaces back to the surface. The receiver's primary function is to detect the returning seismic waves, and convert them into electrical signals to be recorded for further processing.

For land operations, receiving devices called **geophones** (also known as **seismometers** or "jugs") pick up signals from a coil suspended in a magnetic field, which is sensitive to earth motion. The geophones are carefully arranged on the surface in various geometric patterns which are designed to eliminate as much "noise" as possible during the recording process.

In marine work, **hydrophones** are used for the same purposes. These consist of pressure-sensitive crystals which pick up small pressure changes from passing seismic waves in the water, and convert them to an electrical response that can be recorded on magnetic tapes. The hydrophones are strung out along a cable behind the recording ship.

(c) Recorders

The recording system amplifies the signals transmitted by the receivers, filters out "noise," and stores it in a form which allows for more efficient processing of the information. The data usually is recorded on a nine-track, digitally-recorded, **magnetic tape,** which is ideal for computer processing.

(2) Processing

High-speed, digital computers process the seismic field data from the recorded magnetic tapes. In recent years, advances in data processing have progressively enabled more information to be handled with far greater efficiency.

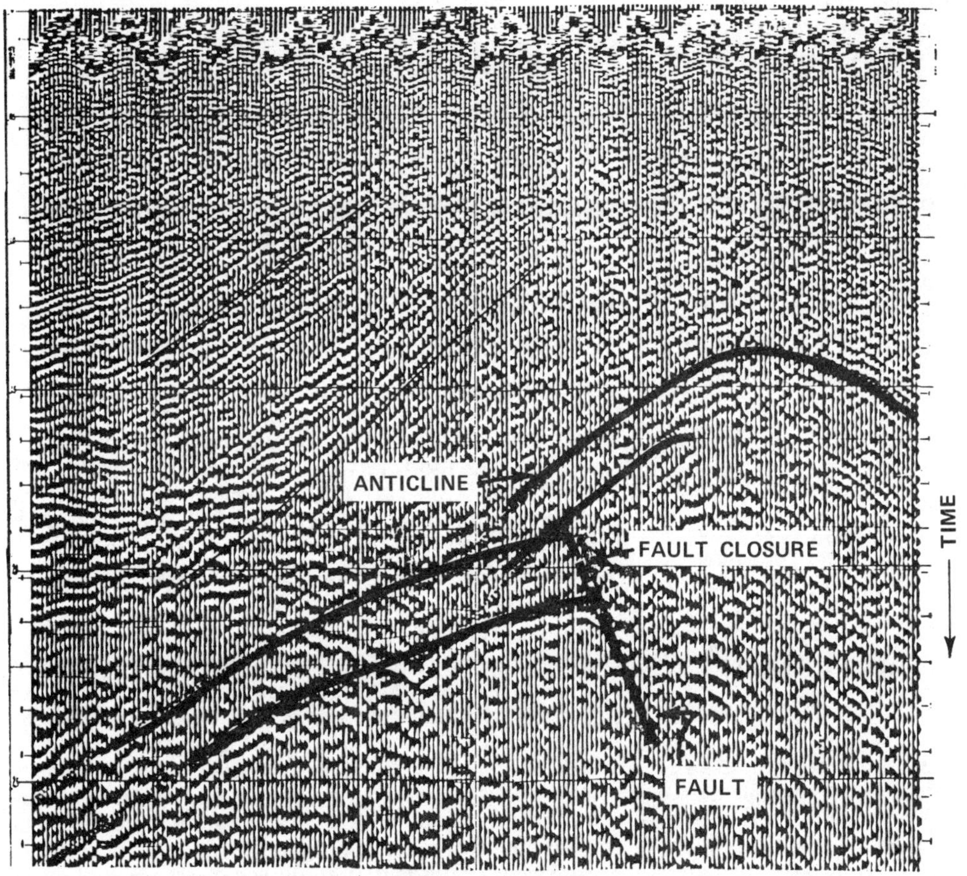

Data processing involves analysis of the entire field data acquisition system. Because the geophones are placed in various geometrical patterns on the surface, the same seismic wave may reach different receivers at different times. Computers can break this information down and analyze the proper sequence of responses. Since an energy source creates unwanted "noise" and disturbance, a primary purpose of data processing is to filter out all this "noise" and obtain an accurate measurement of the true seismic response. Signal distortion caused by waves moving through different subsurface environments with different velocities poses another problem. Complex data processing techniques can eliminate or reduce these effects.

The result of high-speed data processing is that **a record section,** which pro-

Figure 3.16. RECORD SECTION INTERPRETATION.

vides a visual display of the processed seismic signals, can be seen during the field operation (Figure 3.16).

(3) Interpretation

The object of geophysical interpretation work is to develop geological maps from the data provided on the record cross sections. The geophysicist makes these maps by "picking" changes in dips, possible faults, and other geological features that are indicated in the record sections. Even the best data processing techniques cannot completely filter out all the unwanted "noise" and distortion, so making good, reliable "picks" requires judgment and experience. The record section in Figure 3.16 has been marked to show one interpreter's "pick," which indicates a high structural feature and a large fault.

Record sections are taken from a number of **shot points** (the point at which a detonation is made) and combined with known geological information to make structure and isopach maps similar to those used by subsurface geologists. The geophysicist must convert the record sections, which are recordings of seismic waves on a time frame, into depths in feet, by making surveys of rock velocities.

The most reliable method of velocity survey involves lowering recording devices into deep wells and measuring the travel times for waves generated by surface energy sources. Another method takes information from sonic logs used to evaluate formation porosity. The information may also be determined from computer analysis of the seismic data itself, but this method is generally not as reliable as the others.

After the structural and stratigraphic data on the record sections have been "picked" and correlated with available geological data, the information is plotted on a map according to the location of each shot point. Structural cross sections and isopach maps are then constructed, as illustrated in Figures 3.17 and 3.18.

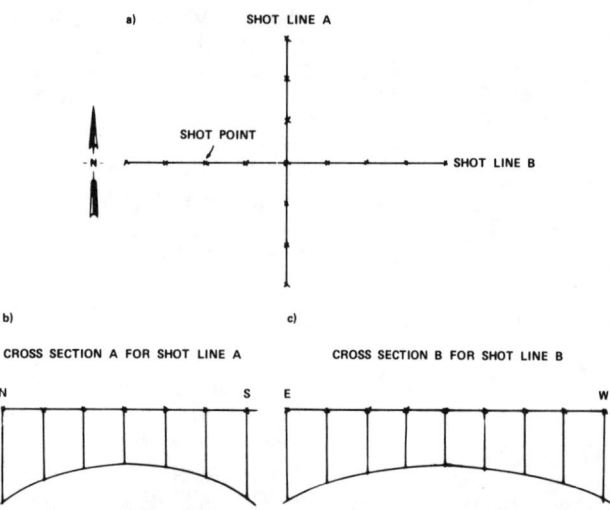

Figure 3.17. CONSTRUCTION OF CROSS SECTIONS.

During the past few years, interpretation techniques have progressed beyond identification of subsurface geological features to include methods which can determine zones of abnormal subsurface pressure and gross lithology of stratigraphic sections. A recent development is the "bright spot" technique, which was

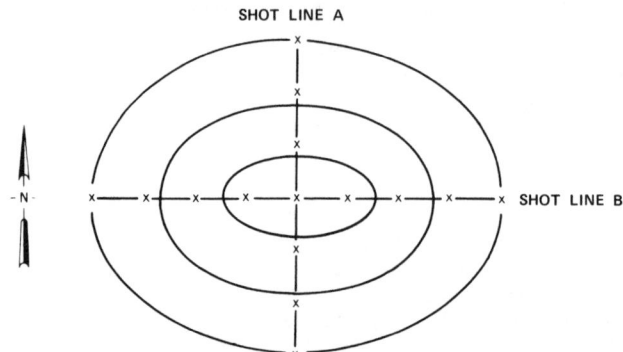

Figure 3.18. STRUCTURAL MAP. Made from a composite of information from shot lines A and B.

designed to detect hydrocarbons, particularly gas zones. This is possible because sands filled with hydrocarbons create stronger seismic waves than water-filled rocks. With the proper analysis of field data, horizontal surfaces formed by gas-oil or gas-water contacts may also be detected. The technique is useful not only in locating potential hydrocarbon reservoirs, but also in determining their probable size. In large lease sales, in which an oil company must pay a significant sum of money for a lease before drilling even begins, any information about the location and size of a potential reservoir helps the company to decide what to bid for the lease.

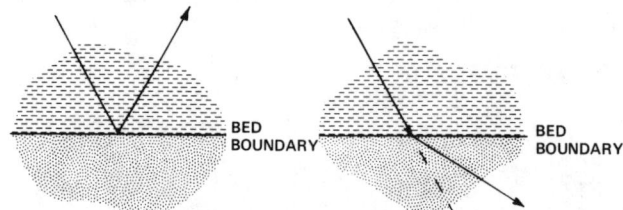

Figure 3.19. LEFT: REFLECTED WAVE. RIGHT: REFRACTED WAVE.

b. Refraction Seismic

Although they were used more extensively in the early days of geophysical exploration than they are today, refraction seismic techniques are still employed when only shallow investigation is sufficient. Reflection seismic methods rely on the detection of waves that are reflected from subsurface physical discontinuities, such as boundaries between beds of different rock formations. In the refraction method, the refraction or deflection of seismic waves at bed boundaries is studied (Figure 3.19).

In reality, seismic waves are continually reflected **and** refracted as they travel through the earth; some return to the surface as reflected waves and some continue on as refracted waves. Geophysicists have discovered that at a certain "critical angle," refracted waves can be made to travel parallel to a bed boundary (Figure 3.20). These angles vary depending on the types of rock forming the boundaries. By studying the effects of refracted waves traveling parallel to bed boundaries, geophysicists can identify shallow subsurface geological features.

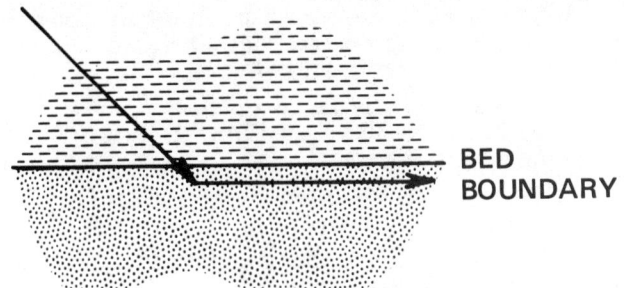

Figure 3.20. REFRACTED WAVE AT THE CRITICAL ANGLE.

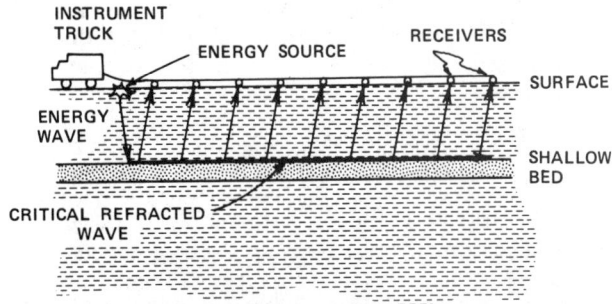

Figure 3.21. REFRACTION SEISMIC METHOD. Determination of depth of shallow bed.

An operation for determining the depth of a shallow horizontal bed is illustrated in Figure 3.21. The receivers are laid out in a straight line, and the distance between them is carefully measured. By recording the time required for the seismic waves to reach each receiver, the depth of the horizontal bed at each point can be determined.

This method was used when many of the early salt domes along the Gulf Coast were discovered. While the salt domes and their associated oil traps did not extend to the surface, they did cause deformation of the shallow horizontal beds, and could be detected by refraction seismic techniques.

The identification of a salt dome by the refraction seismic method is illustrated in Figure 3.22. The deflection of seismic waves from the horizontal reference bed indicates the presence of an underlying salt dome. The location and shape of the dome can be determined by shooting successive lines across the dome in different directions.

2. Magnetic Surveys

Magnetic surveys, while not extensively used in petroleum exploration, do provide useful information to the explorationist. These surveys determine the strength of the earth's magnetic field in a specific area. The earth's magnetic field, which is centered at the north and south poles, is affected in certain areas by concentrations of magnetic minerals in the earth's crust. These minerals (the most common is magnetite) are primarily present in igneous and metamorphic rocks. Normally, sedimentary rocks do not contain appreciable amounts of magnetic minerals, and do not influence the earth's magnetic field. However, sedimentary rocks generally lie above igneous or metamorphic rocks, and the study of magnetics does provide information on these "basement" rocks, thereby providing indirect information about the overlying sedimentary rocks, which are of concern to the petroleum explorationist.

The earth's magnetic field must be carefully measured at many points over large areas of interest. An instrument called a **magnetometer,** which is usually towed by an airplane, measures the field and records detailed magnetic information. The information can be plotted on a map which shows variations in the strength of the magnetic fields. In this way, the general distribution and depth of basement rocks in the area can be determined. When used in combination with other more detailed subsurface geological and geophysical maps, magnetic maps can aid in locating possible hydrocarbon accumulations.

3. Gravity Methods

Gravity methods are seldom considered reliable enough by themselves to justify a drilling project, but they are helpful in petroleum exploration when used in conjunction with subsurface geological, magnetic, or seismic information.

Gravity methods employ an instrument known as a **gravity meter,** which is an extremely sensitive device capable of measuring small differences in the gravitational pull of the earth. The gravity meter might be considered a weighing device which takes accurate weight measurements on the earth's surface. The amount of gravitational pull at any particular point is influenced by the type of rock at or near the surface. Heavy, dense material, such as igneous rock, has a higher gravity reading than light rock, such as sandstone.

When conducting gravity surveys, gravity readings at the precise locations of many points are plotted. Areas of higher or lower than average readings on gravity maps are called **anomalies,** and these aid the explorationist in interpreting subsurface features. A feature such as a large anticlinal structure will have a higher gravity reading than the surrounding area because the heavier "basement" rocks have been pushed up, and provide a denser rock mass closer to the earth's surface. This type of structural feature is called a **maximum gravity anomaly** (Figure 3.23).

A feature such as a salt dome would be recorded as a **minimum gravity anomaly** since the salt has less density and a lower gravitational effect than the surrounding sedimentary rocks (Figure 3.24).

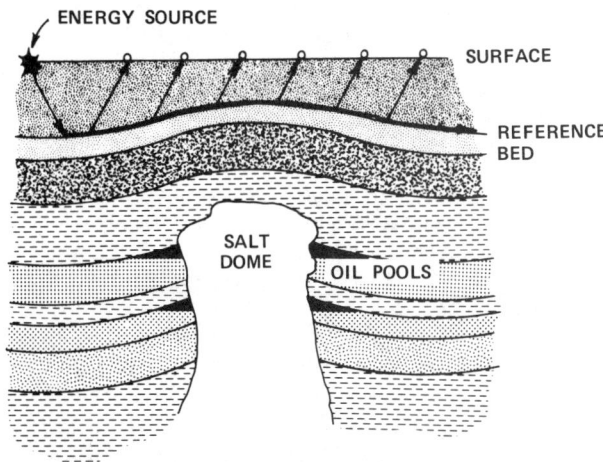

Figure 3.22. REFRACTION SEISMIC METHOD. Detection of a salt dome.

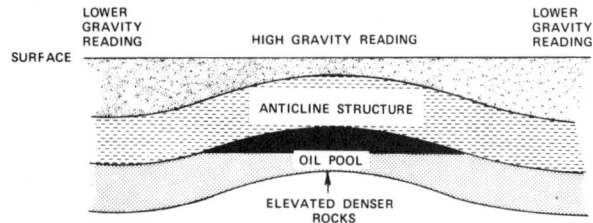

Figure 3.23. MAXIMUM GRAVITY ANOMALY.

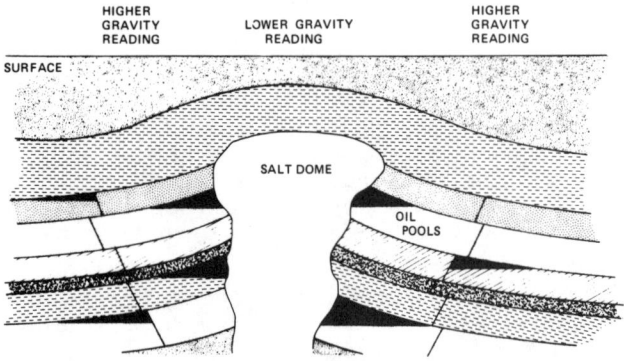

Figure 3.24. MINIMUM GRAVITY ANOMALY.

III. THE EXPLORATION CYCLE

Besides the particular techniques and methods that are used to search for oil and gas, there also exists a general exploration process. Most exploration goes through a typical cycle or pattern, although more activity is required in searching for new fields than in locating extensions of old or existing ones.

The exploration geologist's first task is to choose a general area for exploration. In the United States, there are numerous locations, generally known as sedimentary basins, where major exploration activities are conducted. The Rocky Mountains have many large basins of sedimentary deposits which contain hydrocarbons. Geologists must choose a particular area or basin which they consider promising or suitable for their company's purposes. They must further isolate or define their target by making regional studies of this basin. They may choose to follow previous trends in existing oil fields, or they may try to identify new areas which have not been drilled. Throughout the entire process, they will attempt to identify the hydrocarbon traps or potential reservoirs using all of the methods discussed previously.

In making a regional study of a large area, the explorationist must answer many questions. Each area requires an individual approach and may require certain specialized types of information. Whatever the individual circumstances might dictate, the following general questions must be considered in most exploratory programs:

1. What is the characteristic rock environment being explored?
2. What are the stratigraphic relations of these rocks?
3. What are the potential reservoir rocks in this specific environment?
4. What type of geological unconformities might be expected in the area, and where would they most likely be encountered?
5. Does folding and faulting occur, and how and where is this likely to affect the presence of petroleum reservoirs?
6. Do lateral changes in rock type occur in the potential reservoir rocks?
7. What rocks could be considered as potential source rocks for oil and gas?
8. What types of reservoir energy will be available in potential new fields to drive oil and gas out of the rock?
9. What are the directions and rates of subsurface water movement?
10. What is the effect of reservoir fluids on logging tools?
11. What are the sizes and types of petroleum reservoirs which might be found in this area?
12. What type of recovery factors might be expected from such fields?

To answer these questions, as much information as possible must be gathered. Once this has been accomplished, the explorationist attempts to go from the known to the unknown in order to predict the most likely places to drill for oil and gas.

One of the critical functions of the exploration cycle is to insure that prospects are defined well enough in advance to permit the procurement of oil and gas leases for the drilling of a wildcat well. Once the mineral leases are obtained, the oil company must also secure permission from the landowner for surface rights-of-way. Acquiring leases for oil and gas exploration will be discussed in greater detail in the next chapter.

After the necessary land work has been completed, the next step is to drill test wells. Despite all of the refined exploration techniques designed to locate petroleum reservoirs, no method has been devised that can positively identify such occurrences without actually drilling a test well. The initial exploratory program may include the drilling of "strat holes," minimum-cost holes drilled strictly for stratigraphic information purposes. After detailed geological information from these "strat holes" has been evaluated, actual test holes may be drilled to determine whether producible hydrocarbons are present. Exploration programs generally consist of more than one test well, since the chances of defining a prospect with the first well are slim.

Dry holes are a common occurrence in any exploration drilling program. There are many possible explanations for failure to find oil or gas in a test well:

1. The predicted reservoir trap did not exist.
2. Although a trap may have been present as anticipated, no reservoir condition was found; that is, porosity and permeability were so poor that it was impossible for oil to be present or to flow into the wellbore.
3. The anticipated trap may have been encountered at a shallower or deeper position than expected, and as a result, no hydrocarbons were present.
4. The test well may have been unexpectedly deflected enough to miss the potential reservoir trap.
5. Although the anticipated reservoir trap may have been present with satisfactory reservoir rock, no

oil or gas was present in the trap.

6. The test well may have encountered undetected or unrecognized hydrocarbons and was mistakenly considered a dry hole. This was more common in the early days of oil exploration because of poor detection and evaluation techniques, but occasionally this happens even today.

Once a successful test well or series of wells has been drilled, the geologic information and the nature and quality of the petroleum reservoir must be evaluated. It is at this juncture that many exploratory wells which look encouraging at first are eliminated because they produce only limited amounts of hydrocarbons. However, if the test well looks promising and all the other geological data point to a major discovery, the next step is to determine economic potential.

Several factors are considered in determining a new reservoir's economic potential. These considerations include an estimate of possible oil and gas reserves, the probable selling price, the cost of further exploration efforts, the cost to develop the field once exploration has been completed, and the taxes and other expenses associated with producing the oil. (The chapter on Economics examines these factors in more detail.)

If the venture looks promising, the final step is taken —development of the new field. At this point, the exploration geologist may turn the project over to production geologists, who handle the drilling of additional wells to define the limits of the field, and to fully develop it.

Often, there is only a fine line of distinction between the exploration geologist and the production geologist; they both utilize the same methods and engage in similar activities. Although the production geologist works in areas where oil or gas production has already been established, the geologic problems are no less challenging. New subsurface geological and production information requires constant detailed study and evaluation to insure that oil and gas fields are efficiently and completely developed.

The production geologist must also answer certain questions in the course of his or her work:

1. What is the extent and size of the new field?
2. What are the total reserves of oil or gas in place in the reservoir?

    What percentage of these reserves might be recovered?

3. What is the detailed distribution of oil, gas, and water throughout the reservoir?
4. What variations in rock porosity and permeability occur in the reservoir? Are faults or fractures present?
5. How many different pay zones are there within the reservoir?
6. How many wells will be required to completely develop the new field, and where should they be located?
7. Are there possible extensions to the field which should be explored?

The production geologist usually works closely with many other people when trying to find answers to these questions. New information about the reservoir is constantly being accumulated from field production data and engineering studies, and is used to revise and refine the knowledge at hand.

EXPLORATION—GLOSSARY

anomalies—differences in gravity readings in a particular area.

anticlinal theory—the theory that water, oil, and gas accumulate in anticlines.

basement rock—igneous and metamorphic rock which lies below sedimentary rock beneath the surface of the earth.

bright spot technique—a method of processing and interpreting record sections of seismic wave responses to locate hydrocarbons, by detecting the interfaces between oil and gas zones or water and oil zones.

contour—an imaginary line on a surface, all points of which are at the same elevation in relation to a specified datum surface; an imaginary line along which a certain quantity has the same value.

core—a cylindrical sample taken from a formation for geological analysis.

core analysis—laboratory examination and analysis of core samples taken from the wellbore. This analysis is made to determine the capacity of the formation to contain oil or gas, the ability of fluids to flow through the formation, and the degree of saturation of the formation with oil, gas, and water. Other determinations of the behavior of various fluids in the core sample under different conditions may also be made.

core barrel—a tubular device from 25 to 60 feet long run at the bottom of the drill pipe in place of a bit to cut a core sample.

core catcher—the part of the core barrel that holds the formation sample.

core hole—a hole drilled into the earth by a diamond core drill for the purpose of obtaining rock samples.

creekology—an early "science" used to find oil that was based on looking for oil seeps along creek banks and streams.

critical angle—the point at which refracted waves can be made to travel parallel to a bed boundary.

cross section—an illustration of vertical sections of rock formations in adjacent wells.

doodle bug—a small chemical or electrical device used by early oil explorers in trying to locate oil.

drilling boom—sudden intense burst of exploratory drilling activity, usually initiated by the discovery of a new field or area with high potential for crude oil recovery.

drill-stem test—a method of gathering data on the potential productivity of a formation in which the drill string is used to flow-test a well for the presence of oil or gas.

DST—see *drill-stem test*.

electric log—a survey or record of certain electrical characteristics of the formations encountered in a wellbore from signals transmitted to the surface by an electrical device inserted into the well. The purpose is to identify the formations, determine the nature and amount of fluids they contain, and estimate their depth.

explorationist—a person engaged in the search for natural accumulations of petroleum by geological, geophysical, or other methods.

exploration well—a well drilled to test for the presence of oil or gas in a previously undeveloped area.

facies map—a map showing the lateral variation of one or more rock types within a given formation.

fence diagram—a drawing which consists of several connected cross sections to illustrate subsurface geological information in a three-dimensional manner.

fluid invasion—the effect of drilling fluids which force their way into a formation when the well is drilled; this action causes the natural formation fluids to be displaced.

formation—a bed or deposit composed throughout of substantially the same kinds of rock.

gas exploder—a chamber used in marine seismic operations that contains an explosive gas mixture which is detonated to create seismic waves.

gas seep—an area where gas escapes from the ground into the atmosphere.

geophone—a receiving device used in seismic operations that picks up signals from a suspended coil in a magnetic field; also called "jugs" or seismometer.

geophysicist—a person engaged in the study of the earth's structure, composition, and development through observation of its physical properties.

geophysics—broadly, the physics of the earth, including the fields of meteorology, hydrology, oceanography, seismology, volcanology, magnetism, and geodesy. In the more popular and practical sense, the term implies the application of electrical, thermal, magnetic, gravimetric, and seismic methods to the exploration for petroleum, metals, and underground supplies of water.

gravity meter—an instrument designed to measure differences in the gravitational pull of the earth in a specific area.

gravity methods—techniques which measure the gravitational pull of the earth.

hydrophone—a pressure-sensitive detector used in marine seismic operations to pick up small pressure changes from passing seismic waves in the water.

isopach map—a map which illustrates the thickness of a formation measured between two reference planes or formation boundaries.

lithofacies map—a map showing the lithology of a stratigraphic unit.

logging—the recording of information about subsurface geological formations.

magnetic survey—a procedure which measures and determines the strength of the earth's magnetic field at points in a designated area.

magnetite—a common magnetic mineral present in igneous and metamorphic rock; an iron oxide.

magnetometer—an instrument that measures the earth's magnetic field.

mini-charges—small, explosive charges used in seismic exploration designed to induce vibrations in the earth.

oil shows—the detectable presence of hydrocarbons in a wellbore, as determined by examination of cores or cuttings.

oil smeller—a person common to the early oil industry who professed to have the ability to smell, from the surface, oil that was beneath the surface of the earth.

panel diagram—see *fence diagram*.

pay thickness—the part of a hydrocarbon-bearing rock which is capable of producing petroleum.

photogeology map—a map constructed from geological interpretations of aerial photographs of a certain area.

pneumatic air gun—a device used in seismic exploration, powered with compressed air, that is used to generate vibratory seismic waves.

radioactive well log—the record of the natural or induced radioactive characteristics of subsurface formations.

receiver—an electrical device designed to pick up transmission of seismic waves.

reflection seismic—a seismic technique which records seismic waves that reflect up from subsurface formations after a seismic disturbance has been created on the surface.

refraction seismic—a seismic technique which records seismic waves which refract, or continue downward, from bed boundaries after a seismic disturbance has been created on the surface.

resistivity—that property of a rock denoting its resistance to conducting electricity.

rock cuttings—rock fragments chipped from a formation by the bit during drilling.

seismic—pertaining to earth vibrations which result from earthquakes or artificially induced disturbances.

seismograph—an instrument designed to record and measure earthquakes and other vibrations in the earth.

seismology—the science concerned with observing and recording the generation and propagation of elastic waves in the earth, including waves both of artificial and natural origin.

shot point—the point where the detonation of a charge is made in seismic work.

sidewall core sample—a rock sample taken from the wall of a wellbore by a coring device.

sonic log—a record of the time required for a sound wave to travel a specific distance through a formation.

sparker—a discharge of stored electrical energy used in marine seismic operations to create energy waves in the water.

strat hole—a hole drilled strictly for the purpose of obtaining information on the stratigraphy of sedimentary rocks in a specific location.

structural theory—the geological theory that oil can be found in many types of geological structures, such as salt domes, faulted zones, anticlines, etc.

structure map—a map constructed by plotting points of elevation on a reference stratum in a stratigraphic section in order to obtain a representation of the size, shape, and elevation of a structure.

subsurface geology—the branch of geology that deals with rock formations and other features beneath the earth's surface.

surface geology—the branch of geology that is concerned with rock formations and other features on the earth's surface.

test hole—a hole drilled to determine if oil or gas is present.

thumper truck—a truck which carries a suspended weight that is dropped periodically on the ground to create seismic waves.

topography—the physical features, particularly the relief and surface contours of an area, as illustrated on maps.

well samples—rock cuttings taken from the well during drilling. An examination of these samples can provide information on the type of rock and formations being drilled and estimates of oil and gas content.

wildcat well—see *exploration well*.

wireline log—an electric log obtained by running logging instruments into the wellbore on a wireline or twisted steel cable.

zone—a specific formation or rock bed separate from those surrounding it.

EXPLORATION—BIBLIOGRAPHY

Dobrin, Milton B.: **Introduction to Geophysical Prospecting,** McGraw-Hill Book Company, Inc. (New York) 1960.

Dresser Atlas Division: **Log Review 1,** Dresser Industries, Inc. (Houston) 1974.

Haun, John D. and L.W. LeRoy (eds.): **Subsurface Geology in Petroleum Exploration,** Colorado School of Mines (Golden, Colo.) 1958.

Heroy, William B. (ed.): **Unconventional Methods in Exploration for Petroleum and Natural Gas,** Southern Methodist University (Dallas) 1969.

Hobson, G.D. (ed.): **Modern Petroleum Technology,** John Wiley and Sons (New York) 1973.

Hubbert, Marion King: **History of Petroleum Geology and Its Bearing Upon Present and Future Exploration,** U.S. Geological Survey (Washington, D.C.) 1965.

Moore, Carl A.: **Handbook of Subsurface Geology,** Harper and Row (New York) 1963.

Newendorp, Paul D.: **Decision Analysis for Petroleum Exploration,** PPC Books (Tulsa) 1975.

Sheriff, R.E.: **Encyclopedic Dictionary of Exploration Geophysics,** Society of Exploration Geophysicists (Tulsa) 1973.

Ver Wiebe, W.A.: **How Oil is Found,** Edward Brothers, Inc. (Ann Arbor, Mich.) 1951.

CHAPTER 4:
LAND

I. INTRODUCTION

Up to this point in our discussion of the oil industry, we have concentrated on technical subjects: basic geology, the nature and geology of petroleum, and the methods used to explore for oil and gas. Before moving on to explanations of how wells are drilled and oil is produced, we will examine in this chapter the fundamental principles of the petroleum "land business."

In this segment of the industry, men and women, commonly known as "landmen," work to obtain and maintain leases and other property rights, as an essential part of a company's effort to find and produce oil and gas fields. An additional aspect of land work involves the negotiation of various business arrangements with other companies or individuals.

Land work usually begins once the decision has been made to acquire mineral (or other) rights in a particular area. First, the landman determines who owns the land. Once this has been done, the landman contacts the owner to obtain a lease or other agreement. The type of agreement depends on the nature of the work to be done. For example, permits are needed for geophysical surveys, while rights-of-way are necessary for pipelines. The most common agreement is an oil and gas (or mineral) lease for drilling and producing petroleum.

Even after a lease or agreement has been signed, much work still remains to be done. In the case of leases, ownership often must be verified by an attorney; accurate maps and records are needed; required payments must be made correctly and on time; and various provisions, as well as numerous governmental regulations, require careful compliance. Thus, land work demands people with many different skills to perform these tasks.

This chapter will discuss several aspects of the petroleum land business, including land description methods, ownership, leases and other agreements, regulations, and common types of business arrangements. Additional information is available from the references listed at the end of the chapter.

II. LAND DESCRIPTION

A. Introduction

In most land work, an adequate description of the location of each tract of land is essential. An "adequate description" is one which enables anyone reasonably familiar with the area to identify a particular tract of land apart from any others in the same general area.

In the United States, there are four methods used to describe the location of tracts of land:
1. U.S. system of rectangular surveys
2. Metes and bounds
3. Town lots
4. Offshore rectangular grid systems

The method used to describe the location of a particular tract of land depends on where the tract is located and its shape.

B. U.S. System of Rectangular Surveys

The rectangular survey system of land measurement and description is used in thirty-one states, and covers a large portion of the land in this country. The system was proposed by Thomas Jefferson in 1784, and adopted the following year by the Continental Congress, to provide an orderly system for future measurements and descriptions of public land.

The rectangular system is based on two reference lines, which intersect at right angles. The first line, called the **base line,** runs east and west parallel to the equator. The second line, called the **principal meridian,** runs north and south parallel to the prime meridian, which runs north-south through Greenwich, England.

A series of east-west lines, called **township lines,** are established parallel to the base line, and at six mile intervals north and south of that line. Similarly, north-south lines, called **range lines,** are established at six mile intervals east and west of the principal meridian. Together, these east-west township lines and north-south range lines establish a grid, which divides the land into rectangles (actually squares) which are six miles on each side (Figure 4.1).

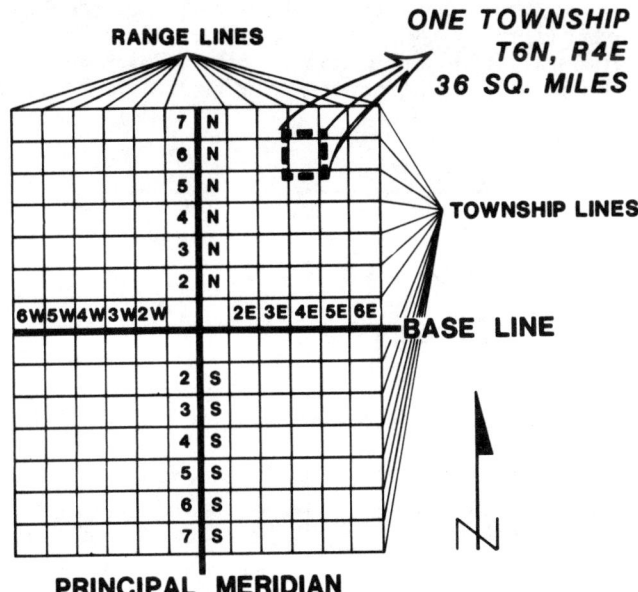

Figure 4.1. RECTANGULAR SURVEY SYSTEM.

Each square can be identified according to its distance north or south from the base line, and east or west from the principal meridian. For example, the six mile square identified as Township 6 North, Range 4 East, is in the sixth row of squares north of the base line, and in the fourth row of squares east of the principal meridian. For convenience, a description such as Township 6 North, Range 4 East is often shortened to T6N, R4E.

Each six mile square, called a **township,** is divided into 36 **sections.** A standard section is one mile square and contains 640 acres. The sections are numbered from 1 to 36, beginning with number 1 in the northeast corner of the township and ending with number 36 in the southeast corner of the township (Figure 4.2).

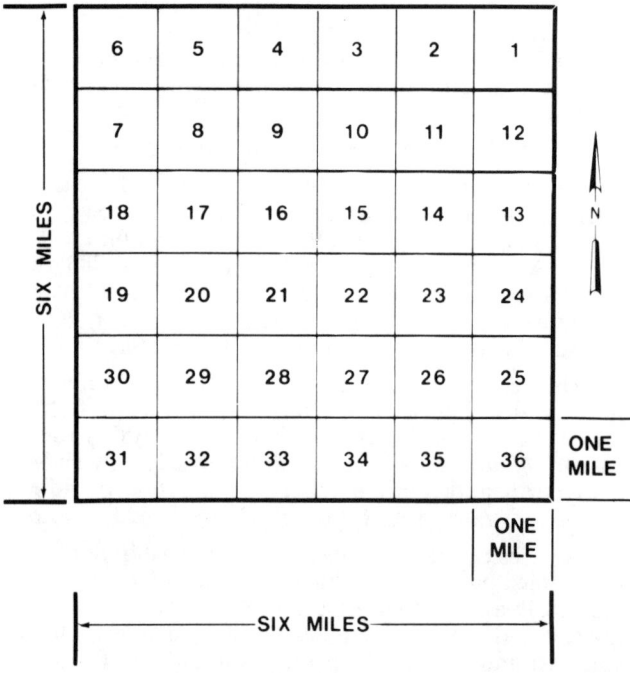

Figure 4.2. TYPICAL TOWNSHIP WITH SECTION NUMBERS INDICATED.

In order to describe tracts of land smaller than 640 acres, sections may be divided into halves, or into four parts identified as the northeast (NE), northwest (NW), southeast (SE), and southwest (SW) **quarter sections** (Figure 4.3). Each quarter section contains 160 acres.

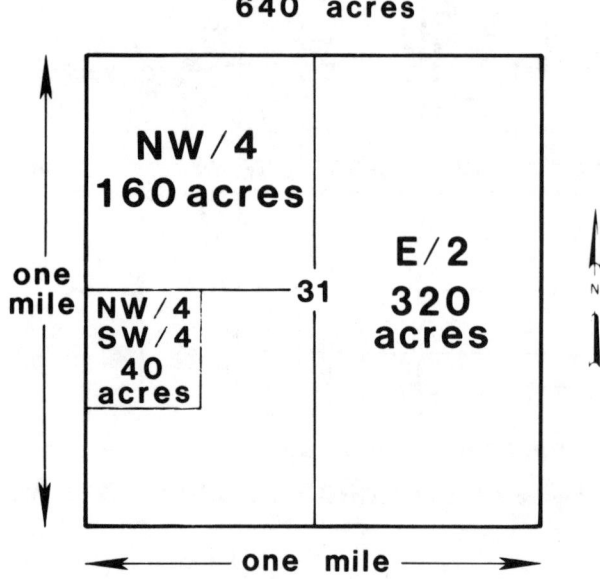

Figure 4.3. DIVISION OF A SECTION.

Similarly, quarter sections are divided into four parts containing 40 acres each, or one-sixteenth of the total area of the section. These **quarter-quarters** are also identified as northeast, northwest, southeast, and southwest. The description for a 40-acre tract which is located in the northwest quarter of the southwest quarter of a section is abbreviated as NW/4 SW/4, or NW SW.

Great care must be taken with the placement of commas in describing land with the rectangular survey system. For example, NW/4, SE/4 Section 8 describes two 160-acre tracts, or 320 acres total, while NW/4 SE/4 Section 8 means one 40-acre tract.

Using the rectangular survey system, the description of a particular 40-acre tract of land is written as follows: NW/4 NW/4 Sec. 20, T2N, R4W, MDB&M, Contra Costa Co., California (Figure 4.4). The 40-acre tract in the example is the northwest quarter of the northwest quarter of Section 20, Township 2 North, Range 4 West, Mount Diablo Base and Meridian, in Contra Costa County, California. Note that it is necessary to identify the base line and principal meridian only in states where more than one set of these reference lines have been established.

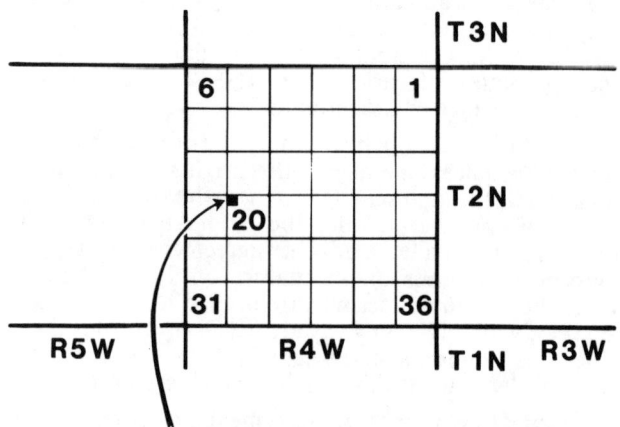

Figure 4.4. NW/4 NW/4 Sec. 20, T2N, R4W, Mount Diablo Base and Meridian, Contra Costa Co., California.

Due to variations in the land surface, and to variations in human accuracy in measuring, sections as surveyed on the ground may not be perfect squares containing 640 acres. Non-standard quarter-quarters ("lots") are usually located along the northern and western borders of a section. In the same manner, non-standard sections are usually located along the northern and western borders of a township.

C. Metes and Bounds

The metes and bounds system of land description is used in the original thirteen states plus Maine, Vermont, Kentucky, West Virginia, Tennessee, and Texas. Ohio is a hybrid state where both systems are used in different parts of the state. Certain large tracts of land in southern and southwestern states, which were originally grants from the French or Spanish governments, are also described using the metes and bounds system. In any state, irregularly shaped tracts of land, which do not fit the rectangular grid, must be described using metes and bounds.

Before the rectangular system was adopted in 1785, early settlers in the eastern states had divided much of the land into parcels of widely varying sizes and shapes, using convenient features such as ridges, streams, trees, and roads to mark the boundaries. Of course, these irregular tracts could not be transferred to the rectangular system. Also, French and Spanish land grant boundaries were established without the rectangular system. When Texas joined the union, land not privately owned was retained by the state rather than being turned over to the federal government. As a result, most of the land in Texas is also divided into irregular parcels, which must be described using the metes and bounds system.

What is the metes and bounds system? "Metes" means **measurements,** and "bounds" means **boundaries.** The description of a tract of land with this system requires an actual survey, which begins at a convenient point on the tract boundary. The selected beginning point should be one which can be located again, even after considerable time has elapsed; and its position should be carefully described.

The surveyor than measures, in turn, the direction and length of each straight-line segment which forms a part of the perimeter of the tract. If the surveyor's work is accurate, a map from the measured distances and directions will show that the last line segment ends where the survey began. If the last line segment does end at the beginning point, the survey is said to "close." A "bust" occurs when the first and last points do not coincide, indicating an error in the measurements.

Each line segment of the boundary is described by a direction or bearing in degrees, minutes, and seconds, and by a length, usually in feet or yards. The **vara,** a Spanish unit of length equalling approximately 2.8 feet, is often found in surveys in Texas and California.

Figure 4.5 is a map of a tract of land to be described using the metes and bounds system. The description is written:

Hood County, Texas: Beginning at a rock marked with an X in the EBL of the J.D. Anderson Surv. and 1,000 ft. N of the SEC of same, thence N 45°30' E 800 ft., thence W 400 ft., thence S 45°30' W 800 ft., thence E 400 ft. to the point of beginning. Said tract contains 5.15 acres more or less.

The description is read: Hood County, Texas: Beginning at a rock marked with an "X," which rock is located in the east boundary line of the J. D. Anderson Survey, at a distance of 1,000 feet due north of the southeast corner of the J. D. Anderson Survey, then go 800 feet along a line which bears 45 degrees and 30 minutes east of due north, then turn and go due west a distance of 400 feet, then turn and go 800 feet along a line which bears 45 degrees and 30 minutes west of due south, then turn and go due east a distance of 400 feet to the point of beginning. The tract is calculated to contain approximately 5.15 acres.

Since the tract shown in Figure 4.5 has only four sides, the description above is relatively simple. Many metes and bounds descriptions, even for small tracts, contain a large number of line segments, particularly when the boundary is an irregular line such as the center of a creek. The description of a 40-acre tract can occupy half a page or more, instead of the few lines required by the rectangular system.

D. Town Lots

A third land description method in the United States is used in urban areas. Small tracts **(town lots)** for

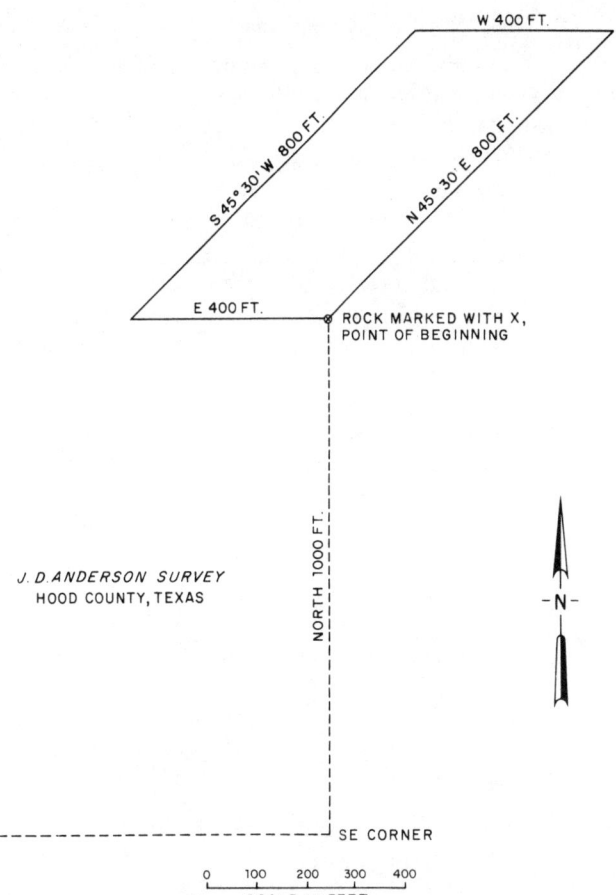

Figure 4.5. METES AND BOUNDS SYSTEM.

homes and businesses are numbered and located by reference to an official map of a subdivision. The map is part of the public records of the county in which the subdivision is located. This land description system is not often used in the petroleum industry, because oil and gas exploration and development activities are not usually carried out in urban areas.

E. Offshore Rectangular Grid Systems

The federal government and coastal state governments have adopted various forms of rectangular grid systems, whereby offshore lands can be divided into readily identifiable and locatable lease blocks. The most common grid is the **universal transverse mercator (UTM)** system. In most grids, square blocks are established which are approximately nine square miles in area.

In land work, accurate land descriptions are absolutely essential to avoid lawsuits and costly mistakes, such as drilling on the wrong property.

III. LAND OWNERSHIP

A. Introduction

In addition to land description systems, an understanding of the basic principles of land ownership is important in land work.

One fundamental principle of land ownership is that: **all land belongs to someone.** There are no ownerless tracts. **Ownership** is defined as: **the right to use the property and to transfer all or part of the ownership to someone else.**

B. Classification of Landowners

Land and landowners can be classified into two major groups: **public** and **private**.

Public lands include:

1. Lands owned or controlled by the federal government or one of its branches. Federally owned lands are divided into four categories:
   a. **public domain lands,** which have never been transferred, or patented, to another owner;
   b. **acquired lands,** which were previously patented to another owner, but subsequently reacquired by the United States;
   c. **Indian lands,** which are divided into two categories:
      (1) **tribal lands,** which lie within the boundaries of an Indian reservation and are controlled by a tribal council;
      (2) **allotted lands,** which have been allotted to the use of individual Indians;
   d. **other lands,** such as naval petroleum and oil shale reserves, national parks and monuments, and other special tracts.
2. Lands belonging to one of the following:
   a. states
   b. counties
   c. cities
   d. school systems
   e. boards, commissions, etc.

Private lands, often called **fee** lands, belong to:

1. individuals
2. corporations
3. institutions
4. all other non-public owners

C. Ownership Transfers (Title Transactions)

Land ownership, or **title,** may be transferred in a variety of ways. In order for a transfer to be valid it must be: **a legal act under applicable laws, in writing, and irrevocable.**

Some of the ways in which ownership may be transferred are: **conveyances** such as deeds, **inheritance,** and **court actions.**

D. Divided (Severed) Ownership

In its simplest form, ownership of a tract of land is undivided: all rights are held by one owner in what is known as the **fee simple estate.** However, ownership can also include the right to transfer only a part of what is owned. As a result, a single tract of land may have multiple owners, with each owner controlling only specific rights to the property, and none having total control.

Ownership can be divided in five principal ways: **surface, minerals, water, time,** and **depth.** Examples are given below of some of the ways each type can be further subdivided:

1. **Surface**
   a. rights-of-way (strips of land set aside for pipelines and roads)
   b. hunting and fishing rights
2. **Minerals**
   a. oil and gas only, or coal only
   b. royalty only (receives share of minerals produced, but not of other payments which may be made under a lease)
3. **Water**
   a. surface
   b. subsurface
4. **Time**
   a. life of owner only
   b. specified number of years
5. **Depth**
   a. surface to 5,000 feet below
   b. everything deeper than 5,000 feet (or any other selected depth)

Divided, or severed, ownership can result from many types of title transactions. For example, a fee simple owner sells the surface and half the underlying minerals, but reserves or keeps the other half of the minerals; or, divisions can be made among the heirs when an estate is settled after a death.

When the mineral estate has been severed from the surface estate, the surface owner will not receive any revenue from production of the underlying minerals. The surface owner is entitled only to payment of reasonable "damages" to compensate for necessary use of the surface to extract the minerals.

Occasionally, a surface owner tries to prevent surface access entirely. In these cases, the conflict may end up in court. Through many such cases, the courts have established that the mineral estate is dominant: mineral ownership includes the right to use the surface to the extent necessary to enjoy ownership of the minerals.

E. Ownership Information and Records

The sources of ownership information on land vary according to the type of owner involved.

1. Land Maps

A preliminary determination of ownership is usually made using a land map (Figure 4.6). These maps provide a compact display of pertinent information, such as surface or mineral ownership, existing leases with their holders and expiration dates, and in many cases, wells which have been drilled in the area.

Land maps are available from commercial companies which specialize in providing this information for the petroleum industry. The land departments of oil companies may also maintain their own maps for areas of interest to the company.

2. Private Land

If land is privately owned (fee) land, there are two sources of ownership information. The primary source is the records maintained by the county in which the land is located. In addition, the files of an abstract company may be accessible. These companies specialize in researching and compiling the land ownership documents filed in the county records.

a. County Records

The county records are usually kept at the courthouse in the county seat. In these records, the originals or copies of various documents relating to land ownership are filed as public information.

The instruments (deeds, wills, liens, mortgages, oil and gas leases, etc.) are often filed separately by type. For example, all deeds may be filed in one set of books, while mortgages and liens on the same property are filed in another set of books. Court proceedings (probate, tax sales) and tax records may be kept in still other loca-

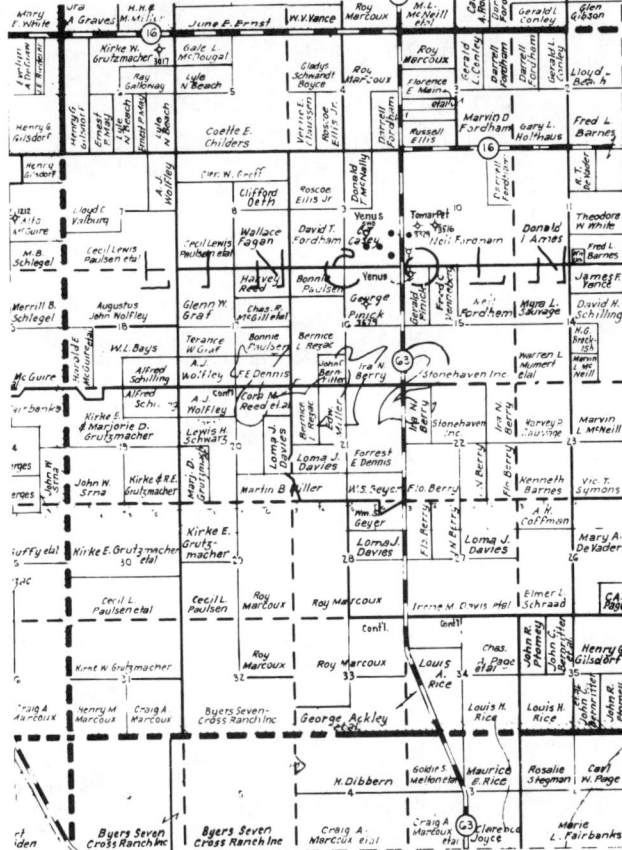

Figure 4.6. LAND MAP.

tions. Thus, a check of county records for ownership information on a tract of land not only requires knowledge of how a particular county's records are organized, but also may involve considerable time in researching the information.

b. Abstract Company Records

A second source of information on private land ownership is the records of an abstract company. These companies usually have complete indices, or lists, of documents affecting title to each tract of land in their county. Depending on the policy of the particular abstract company, the lists may be available for inspection without charge, for a fee, or not at all.

3. Federal Land

The Bureau of Land Management (BLM) is the agency responsible for managing most of the land owned by the federal government, both onshore and offshore. Records concerning Indian lands are maintained separately.

a. Public Domain and Acquired Lands

Federal records relating to public domain and acquired lands are located in the BLM offices in Washington, D.C., and in state land offices. Ordinarily, the state office has all the necessary information under the present system, which is referred to as the "new records." The new records consist of the following parts:

- The **Master Title Plat** is a map based on the official survey of the area, with differently weighted letters and lines added to indicate areas included in national forests, wildlife refuges, and other special locations.

- **Use Plats,** which include several kinds of maps, are prepared to indicate the different uses being made of federal lands, except for grazing leases. For example, oil and gas leases are shown on maps called "O G Plats."

- The **Historical Index** is a chronological record of all past and present actions which affect the ownership of public lands and resources. The same information appears on the Master Title and Use Plats.

- The **Serial Register** is an index of all the transactions related to a particular offer to lease federal lands. Such transactions include the date a lease was issued, approval of any assignments or transfers, and similar information.

- The **Case Files** contain the original documents related to oil and gas leases, plus any correspondence. The case files are identified by the number in the Serial Register, and may be examined only by persons giving a satisfactory reason for being interested in the contents.

b. Indian Lands

When land is owned either by an Indian tribe or by individual Indians, records are maintained in the following places: the office of the Commissioner of Indian Affairs in Washington, D.C., the offices of area directors, and in the records of the realty and finance officers in the agency office for each reservation. The information in the agency office is most readily available.

c. Offshore Lands

Beyond the limits of state territorial waters, offshore lands of the United States are under the jurisdiction of the Bureau of Land Management's division for oil and gas related matters.

4. State Lands

For state lands, ownership records are usually kept in the office of the State Land Commissioner, or whatever agency handles most of the land matters for that particular state. Other state agencies may also own lands, including park departments, highway departments, schools, etc. For all state-owned land, the first step is to determine the appropriate agency to contact for ownership information; then, that agency's records can be researched.

F. Recording Ownership Documents

Each state has a law establishing a system for filing in the public records all instruments affecting property. Documents must be recorded in the county where the land is located, and must be properly signed and notarized, or "acknowledged."

Documents received by the county clerk's office are placed in the appropriate book. Each page of the document is marked with a volume and page number for reference.

All documents affecting the title, or ownership, of land are usually recorded to protect the interests of the parties involved. The importance of recording is illustrated in the following example. Mr. Jones sells his land to Mr. Smith, but Mr. Smith is busy and does not bother to have the deed recorded. Mr. Smith knows that the deed is binding on himself and Mr. Jones without recording. Later Mr. Jones, being absent-minded, sells the same land to Mr. Harris, who promptly has his deed recorded. Now, who owns the land? Mr. Harris does, not Mr. Smith. Mr. Harris bought the land in good faith: there was nothing in the public records to inform him of the earlier sale to Mr. Smith. Mr. Smith's deed does not protect him against the sale to Mr. Harris because it was not recorded. The laws require that title instruments be recorded to be completely effective.

G. Verification of Ownership (Clear Title)

1. The Sovereign (Original) Owner

The ownership history of a tract of land begins with an original owner, known as the **sovereign.** The identity of the sovereign varies with the early history of the area. In the western United States, the federal government is usually the sovereign. In the east, and in some southern and southwestern states, the sovereign is an English, French, or Spanish king. (The Indians have somehow been ignored.)

2. Chain of Title

Under ideal conditions, ownership records go as far back as the sovereign, and include all documents which have affected title to the land down to the present time. When the records are complete, there is an unbroken "chain of title" from the sovereign to the present owner.

3. Identification of Present Owners (Record Checks)

Since oil companies rarely own the land outright on which they work, present owners must be identified and contacted to obtain the appropriate rights to the property.

To find the present owner, a landman begins with the sovereign, and searches forward to the current owner, by following the chain of title through various documents in the records. This "record check" may begin at some more recent point, if the older records are no longer available.

When running a records check, the landman searches out not only the most recent ownership information, but also any court proceedings, unreleased leases or mortgages, unpaid taxes, or other circumstances which could affect ownership. From the records, a list of pertinent documents, called a "take-off," is compiled.

At this point, "record title" has been determined, and the owners can be contacted to obtain the necessary rights—most often an oil and gas lease.

4. Title Examination (Establishing Clear Title)

The record check does not confirm that the present owner has "clear title," or full legal ownership. After the record check has been conducted, further steps are taken to establish clear title, before large amounts of money are spent, such as drilling a well.

If title is not checked and a lease obtained from all mineral interest owners prior to drilling, the result can be costly should production be established. For example, notice the differences in the distribution of one hundred dollars of revenue between mineral rights fully leased and those only half leased.

Fully Leased (1/8 royalty)	**Half Leased Only** (1/8 royalty)
$ 12.50 mineral owner (1/8)	$ 6.25 half mineral owner (leased) (1/2 x 1/8)
87.50 company (7/8)	50.00 half mineral owner (unleased)
$100.00	43.75 company
	$100.00

a. Private Land

(1) Abstracts

Further determination of ownership of private lands begins with the preparation or updating of an abstract covering the property. An abstract consists of copies of all the available instruments which have affected title from sovereignty (or some early point) to the present. For a fee, the documents are assembled and certified correct by an abstractor. Only one copy of an abstract is made.

(2) Title Opinions

The documents in the abstract are examined by an attorney who knows the applicable laws and court decisions of the state in which the land is located. The attorney must decide whether the various instruments are valid, and what their effect is on ownership of the property.

The attorney also checks for defects or gaps in the chain of title. Such gaps may result from an unrecorded deed, an unsettled estate, or other circumstances. The attorney must evaluate each problem and decide whether any corrective action is required to "cure" the defect.

The attorney puts the results of the examination in a report called a **title opinion.** The opinion describes the land, lists the abstract(s) and other documents examined, and comments on any problems discovered. When defects are serious enough to require corrective action, the opinion includes an **objection,** which states the problem, and a **requirement,** which explains what action should be taken to cure the defect.

The attorney's title opinion also states the present ownership of the land surface, the underlying minerals, and other rights, based on the documents examined and any requirements being satisfactorily cured. Information is also included on mortgages, leases, unpaid taxes, and other relevant items. If the land is already leased, an analysis of the terms and conditions of the lease will be included.

b. Public Land

(1) Status Report

For publicly-owned land, the process of verifying ownership usually begins with a status report. An attorney examines the public records, and prepares a factual summary of all the pertinent information in the records. If the land has been acquired from private owners, an abstract from the county records may also be required.

(2) Title Opinions

Generally, the chain of title for public lands is simpler, and fewer defects are found, than for private land. Nevertheless, many companies have a title opinion prepared by an attorney prior to drilling on public land. The reason is to verify title, which is not established by the factual status report.

c. Curative Work

When a title opinion indicates a problem with the title serious enough to necessitate corrective action, the landman is the person usually responsible for the work which must be done to "cure" the title.

Curative work may include obtaining copies of missing documents, taking affidavits from knowledgeable persons if the requested documents are not available, securing releases of old leases, and other similar work.

All curative documents are submitted to an attorney for examination. If they are satisfactory, the title has been "cured," and "good" title has been established.

IV. ACQUIRING LAND RIGHTS

A. Introduction

The first step in acquiring land rights is to determine the specific rights which are needed. The owner then must be identified and contacted to obtain an appropriate agreement. Unless such an agreement is reached and the owner's permission obtained, work on the land cannot proceed.

B. Surface Rights

The type of surface rights required depends on the nature of the work to be done:

1. **Rights-of-way** are agreements authorizing the use of a strip of land to install a pipeline, construct a road, or put in a power line.
2. **Surface leases** authorize the use of the property for the installation of field production equipment or for other facilities.
3. **Permits** are issued to give access to the property, usually for geophysical exploration.
4. **Purchases** of the surface may be made when a large facility, such as a refinery, is to be built.
5. **Surface damage payments** are required when the land surface and the underlying minerals are separately owned. The courts have decided that the mineral estate is dominant, and that the mineral owner has the right to use the surface to the extent necessary for extracting the minerals. Since the surface owner will not receive any share of the mineral owner's compensation, reasonable "damages" are paid to the surface owner for the use of the property.

The procedures for acquiring these and other surface rights vary, depending on the owner and the type of rights required.

C. Mineral Rights

The mineral rights needed for drilling and production are most commonly acquired by means of an **oil and gas lease.** If the minerals have already been leased, these might still be acquired by a transfer, or **assignment.** Occasionally, the minerals are acquired by a direct purchase.

1. Leases

An oil and gas lease is an agreement which gives one party (the **lessee**) the right to explore, drill, and produce oil and gas on a particular tract of land. In return, the mineral owner (the **lessor**) receives various financial and other types of compensation.

a. An Early Lease

The first oil and gas lease may have been signed as early as 1853, six years before the Drake well. The lease was a single paragraph containing 95 words:

> "Agreed, this fourth day of July 1853, with J. D. Angier, of Cherrytree township, in the county of Venango, Pa., that he shall repair up and keep in order, the old oil spring on land in said Cherrytree township, or dig and make new springs, and the expense to be deducted out of the proceeds of the oil, and the balance if any, to be equally divided, the one half to J. D. Angier and the other half to Brewer, Watson & Co., for the full term of five years from this date. If profitable.
> (signed) Brewer, Watson & Co.
> J. D. Angier"

(Source: **AAPL Guide for Landmen**)

Over the years, leases have evolved into documents two or three pages long, written in very small print.

b. Typical Lease Provisions

No single form of oil and gas lease is acceptable in every state, or under all circumstances. Instead, a wide variety of forms have been developed to meet the requirements of different federal and state laws, court decisions, mineral owners, and specific circumstances.

In spite of the variety of forms in use, there are a number of provisions which are found in most leases, although the exact language may vary. Common provisions of leases on private land are reviewed in the section below. A discussion of federal and state leases is presented in following sections.

(1) Fee Leases

This section examines the typical clauses contained in fee leases and the preparation that must be done for these leases. The first page of a typical oil and gas lease on fee land is shown in Figure 4.7.

(a) Clauses

Effective Date—Usually the date signed.
Parties—The names and addresses of lessors and lessees; each must be a legal entity.
Consideration—The money and/or promises contained in the lease given by the lessee to the lessor. Even though the total dollar amount may not be stated in the lease, the consideration usually includes a lump sum cash payment at the time of signing—the **signature bonus.**
Words of Grant—The lease is given or granted for specific purposes (generally exploration, drilling, and production) and for stated substances (commonly oil and gas).
Land Description—The county and state are given, followed by a detailed description of the covered tract. The number of acres is qualified "more or less" in case the measurement of the tract is not totally accurate.
Term—The length of time the lease will be in effect. A specific number of years will be given, but the term may be extended past the **primary term** (original limit) by any one of several circumstances. The most common ways a lease may be extended beyond its primary term are:
1. production of oil and/or gas—often a minimum quantity is required to hold the lease in effect after the end of the primary term;
2. continuous operations—if production stops after

FORM C 88 EASTERN   OIL AND GAS LEASE   RA-346 (REV. 6/67) PRINTED IN U.S.A

THIS AGREEMENT, made and entered into this _____ day of _____, 19_____,
between _____
_____
_____
_____, hereinafter called lessor,
and _____, hereinafter called lessee,

WITNESSETH: 1. That lessor, for and in consideration of the sum of Ten Dollars ($10.00) and other good and valuable consideration, in hand paid, and of the covenants and agreements hereinafter contained to be performed by lessee, has this day granted and leased and hereby grants, leases and lets unto lessee for the purpose and with the exclusive right to explore and operate for and produce oil and gas, including casinghead gas and casinghead gasoline, condensate and all related hydrocarbons and all products produced therewith; lay pipe lines, build tanks, store oil, build powers, stations, telephone lines and other structures thereon to produce, save, take care of and manufacture all of such substances; and to store and remove gas as provided in paragraph 13 hereof; the following described tract of land situate in Warrant-Lot No. _____ in the Township or District of _____, County of _____, State of _____, bounded substantially as follows:

containing _____ acres of land, more or less, being the same land described in deed from _____
_____
dated _____ and recorded in Book _____ at Page _____ in the Recorder's office of the said County, such deed being referred to for a more particular description of such land but not to limit the interest or estate covered hereby, and also, in addition to the above described land, any and all strips or parcels of land adjoining or contiguous to the above described land and owned or claimed by lessor.

2. This lease shall remain in force for a primary term of ten (10) years and as long thereafter as oil, gas or any other mineral covered by this lease is produced or this lease is extended by any subsequent provision hereof.

3. (a) Lessee shall deliver to the credit of lessor as royalty, free of cost, in the pipe line to which lessee may connect its wells the equal one-eighth part of all oil produced and saved from the leased premises, or at the lessee's option, may pay to the lessor for such one-eighth royalty the market price for oil of like grade and gravity prevailing on the day such oil is run into the pipe line, or into storage tanks.

(b) Lessee shall pay lessor, as royalty, for gas produced from any well on leased premises and used by lessee off leased premises or in the manufacture of gasoline or any other product, one-eighth of the market value of said gas, as such, at the mouth of the well. If such gas is sold by lessee, then lessee shall pay lessor, as royalty, one-eighth of the net amount realized by lessee, computed at the mouth of the well. Lessor shall have gas free of charge from any well on leased premises producing gas only for stoves and inside lights in the principal dwelling house on said land by making his own connections with the well, the use of said gas to be at lessor's sole risk and expense.

(c) The royalties herein provided shall not be payable on any oil, gas or other mineral covered by this lease which is produced from any strata being utilized for storage purposes pursuant to paragraph 13 hereof.

(d) This lease shall continue in full force for so long as there is a well or wells on leased premises capable of producing oil or gas, but in the event all such wells are shut-in and not produced by reason of the lack of a market acceptable to lessee, by reason of Federal or State laws, executive orders, rules or regulations (whether or not subsequently determined to be invalid), or for any other reasons beyond the reasonable control of lessee, then on or before each succeeding anniversary of the date hereof occurring ninety (90) or more days after all such wells are so shut-in and prior to the date production is commenced or resumed, or this lease surrendered by lessee, lessee shall pay or tender to the lessor, or to the credit of lessor in the depository bank hereinafter designated, as royalty, an amount equal to the delay rental hereinafter provided for, which payment or tender may be made by lessee's check or draft.

4. If operations for the drilling of a well for oil or gas on leased premises are not commenced within one year from the date hereof, thereupon this lease shall terminate unless, within such period, lessee shall pay or tender, or in good faith attempt to pay or tender, to lessor or to the credit of lessor in the _____
_____ Bank at _____
or its successors, which bank and its successors are lessor's agent and shall continue as depository for all payments hereunder regardless of changes of ownership of leased premises or of the right to receive rentals hereunder, the sum of
_____ Dollars ($_____) as rental which shall extend for one year the time within which such operations for drilling may be commenced. Upon like payments or tenders annually, the time within which such operations for drilling may be commenced shall be further extended for successive periods of one year each during the primary term hereof. Such payments may be made by check or draft, and the deposit thereof in the mail within the times above specified for payment

Figure 4.7. FIRST PAGE OF AN OIL AND GAS LEASE.

the end of the primary term, the lease will still remain in effect, if work to drill another well or to restore production of an existing well begins within a certain number of days.

Royalty—A portion (1/8 or more) of the production reserved for the lessor, at no expense. In almost all cases, the oil or gas is sold by the lessee, and the royalty paid with money. Royalty payments take precedence over all other payments made from lease revenues.

Delay Rental—An annual payment of a specified amount made by the lessee to the lessor to keep the lease in effect, when the lessee does not want to proceed with drilling. Delay rental payments cannot, however, extend the lease beyond the end of the primary term.

Shut-in Royalty—An annual payment, often equivalent to the delay rental, which is made when there is a well (usually a gas well) on the leased land capable of producing, but not in actual production. The purpose of this payment is to hold the lease until actual production can begin.

Pooling—This clause gives the lessee the right to combine all or part of the leased land with other nearby or adjoining tracts. The reason for such an action would be to assemble a tract of land large enough to meet state spacing regulations for drilling a single well. (Further information on spacing is contained in the section on regulations in this chapter.) After leases are pooled, there is in effect, one lease for all purposes, except for the payment of royalty. Once pooling has been accomplished, royalty payments are apportioned among the lessors based on the number of acres each contributed to the total (Figure 4.8).

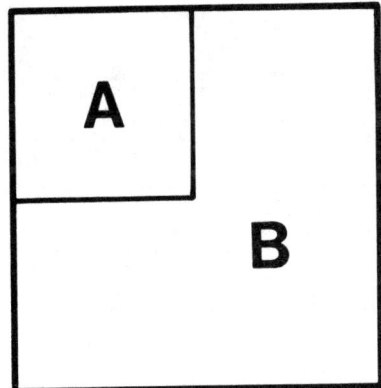

Figure 4.8. POOLING. Lease A (40 acres) and Lease B (120 acres) pool leases to meet 160-acre spacing requirement.

Right of Assignment—An assignment clause gives both the lessee and the lessor the right to transfer, or **assign,** all or part of their interests in the land and the lease to other parties. All rights and obligations contained in the original lease are transferred from the **assignor** (the original lessee) to the **assignee.**

Lesser Interest—If the lessor owns less than 100 percent of the oil and gas under the leased land, payments will be reduced proportionately.

Free Use—The lessee has the right to use free of cost any oil and gas produced and used **on the same lease.** The lessor also may have free use of produced gas for domestic purposes at the principal dwelling on the property.

Surrender—The lessee may surrender all or part of the leased acreage at any time. Delay rentals would then be reduced proportionately.

Warranty—The lessor agrees to defend title to the lands covered by the lease.

Force Majeure—The lease will not be lost if certain conditions beyond the control of the lessee prevent compliance with its terms.

Special Provisions—Many other provisions may be added as agreeable to all parties. Any provisions added to a standard lease form take precedence over the printed text, in case there is any conflict.

Other Provisions—A number of additional provisions may be included in a particular lease as they are needed to fully state the agreement between the parties.

Implied Provisions—Even though the lease does not specifically say so, the laws and court decisions have established that the lessee has the obligation to perform as a "prudent operator," doing a workmanlike job of developing the acreage and marketing the oil and gas produced.

(b) Preparation

When oil and gas rights are being acquired by a lease, great care must be taken that the lease is properly prepared and signed. A lease form must be used whose language is not only acceptable to all parties, but also is legal in the state where the land is located. All blanks on the form should be filled. The lease should be signed by the legal owners, and their signatures witnessed, or "acknowledged," by a notary public. Both the signatures and the acknowledgement must be in the correct legal form. Finally, the lease should be "recorded" in the county records, and the recording data noted on the lease.

(2) Federal Leases

This section reviews the clauses and preparations that are involved in federal leases. The first page of a federal lease form is shown in Figure 4.9.

(a) Clauses

Federal leases contain many clauses similar to those in fee leases, plus others which are included to conform to the various laws and regulations. Federal leases also differ depending on whether the lease is issued through competitive or non-competitive procedures.

Effective Date—Specified on the lease.

Parties—The names and addresses of the lessees, each of whom must meet certain qualifications. For example, individuals must be U.S. citizens, but cannot be minors, even if otherwise qualified. The maximum number of acres in a state which can be held directly or indirectly by one individual, corporation, or association is 246,080; except for Alaska, which allows 300,000 acres in each of two districts.

Consideration—Money and/or promises, as with fee leases. For competitive leases offered to the highest bidder, one-fifth of the amount must accompany the bid, with the rest usually to be paid within 30 days if the bid is successful. For other leases, only a ten dollar filing fee is charged.

Words of Grant—The purposes of the lease are stated, and the covered substances are identified. The lease terms, however, may be modified by other agreements approved by the Secretary of the Interior.

Land Description—As with fee leases, an accurate description is required. The acreage requested initially

Form 3110—1
Eleventh Edition
(March 1977)
(formerly 3120—3)

UNITED STATES
DEPARTMENT OF THE INTERIOR
BUREAU OF LAND MANAGEMENT

Form approved
Budget Bureau No. 42–R0990

Office _____

Serial No. _____

Fill in on typewriter
or print plainly in ink
and sign in ink.

OFFER TO LEASE AND LEASE FOR OIL AND GAS
(Sec. 17 Noncompetitive Public Domain Lease)

The undersigned hereby offers to lease all or any of the lands described in item 2 that are available for lease, pursuant and subject to the terms and provisions of the Act of February 25, 1920 (41 Stat. 437, 30 U. S. C. sec. 181), as amended, hereinafter referred to as the Act and to all reasonable regulations of the Secretary of the Interior now or hereafter in force, when not inconsistent with any express and specific provisions herein, which are made a part hereof.

Mr.
Mrs.
1. Miss _____
(First Name, Middle Initial, Last Name)

Please notify the signing officer of any change of address.

_____
(Number and Street)

_____
(City, State, ZIP Code)

2. Land requested: State   County   T. : R. : Meridian

3. Land included in lease: State   County   T. : R. : Meridian

Total Area ____ Acres ____

(Offeror does not fill in this block)   Total Area ____ Acres ____ Rental retained $ ____

4. **Amount remitted:** Filing fee $10, Rental $ _____, Total $ _____
5. **Undersigned certifies as follows:**
   (a) Offeror is a citizen of the United States. Native born _____ Naturalized _____ Corporation or other legal entity (specify what kind): _____

   (b) Offeror's interests, direct and indirect, do not exceed 200,000 acres in oil and gas options or 246,080 chargeable acres in options, offers to lease and leases in the same State, or 300,000 chargeable acres in leases, offers to lease and options in each leasing district in Alaska. (c) Offeror accepts as a part of this lease, to the extent applicable, the stipulations provided for in 43 CFR 3103.2. (d) Offeror is 21 years of age or over (or if a corporation or other legal entity, is duly qualified as shown by statements made or referred to herein). (e) Offeror has described all surveyed lands by legal subdivisions, all lands covered by protracted surveys by appropriate subdivisions thereof, or all unsurveyed lands not covered by protracted surveys by metes and bounds, and further states that there are no settlers on unsurveyed lands described herein.
6. Offeror ☐ is ☐ is not the sole party in interest in this offer and lease, if issued. (If not the sole party in interest, statements should be filed as prescribed in Item 6 of the Special Instructions.)
7. Offeror's signature to this offer shall also constitute offeror's signature to, and acceptance of, this lease and any amendment thereto that may cover any land described in this offer open to lease application at the time the offer was filed but omitted from this lease for any reason, or signature to, or acceptance of, any separate lease for such land. The offeror further agrees that (a) this offer cannot be withdrawn, either in whole or in part, unless the withdrawal is received by the land office before this lease, an amendment to this lease, or a separate lease, whichever covers the land described in the withdrawal, has been signed in behalf of the United States, and (b) this offer and lease shall apply only to lands not within a known geologic structure of a producing oil or gas field.
8. If this lease form does not contain all of the terms and conditions of the lease form in effect at the date of filing, the offeror further agrees to be bound by the terms and conditions contained in that form.
9. It is hereby certified that the statements made herein are complete and correct to the best of offeror's knowledge and belief and are made in good faith.

Offeror duly executed this instrument this _____ day of _____, 19____

_____   _____
(Lessee signature)   (Lessee signature)

                           _____
                           (Attorney-in-fact)

This lease for the lands described in item 3 above is hereby issued, subject to the provisions of the offer and on the reverse side hereof.

THE UNITED STATES OF AMERICA

By _____
(Signing officer)

Effective date of lease _____   _____   _____
                                                  (Title)   (Date)

THIS OFFER MAY BE REJECTED AND RETURNED TO THE OFFEROR AND WILL AFFORD THE OFFEROR NO PRIORITY IF IT IS NOT PROPERLY FILLED IN AND EXECUTED OR IF IT IS NOT ACCOMPANIED BY THE REQUIRED DOCUMENTS OR PAYMENTS.

18 U. S. C. sec. 1001 makes it a crime for any person knowingly and willfully to make to any Department or agency of the United States any false, fictitious or fraudulent statements or representations as to any matter within its jurisdiction.

This form may be reproduced provided that the copies are exact reproductions on one sheet of both sides of this official form, in accordance with the provisions of 43 CFR 3123.1(a).

Figure 4.9. FIRST PAGE OF A FEDERAL LEASE.

by the lessee may be changed by the BLM before the lease is issued. Land leased under competitive procedures is subject to a maximum of 640 acres per lease. For non-competitive leases, 640 acres is usually the minimum, with a maximum of 2,560 acres.

Term—The initial term is five years for almost all federal leases. For non-competitive leases, a second five years may be applied for, with the extension usually being approved. When the leased land has been included within the limits of a producing area as determined by the U.S. Geological Survey, the extension period will be two years.

Royalty—For non-competitive leases, the basic rate is 12½ percent of production, although the rate may increase as production increases, to a maximum of 25 percent for oil and 16-2/3 percent for gas. For competitive leases, the royalty is stated in the published notice of the sale.

Rentals—On non-competitive leases, the rental is fifty cents or one dollar per acre per year, depending on the leasing procedure used. For competitive leases, the rate is two dollars per acre per year.

Minimum Royalty—After oil or gas is discovered on the leased land, the minimum annual royalty payment is one dollar per acre per year, in lieu of rentals.

Communitization—This is similar to pooling, in that the objective is to meet well spacing requirements.

Right of Assignment—Leased lands may be assigned wholly or in part, provided that the assignee meets the same qualifications required for lessees. The assignment must also be approved by the BLM.

Lesser Interest—If the United States owns less than 100 percent of the minerals, royalty payments will be proportionately reduced, but rentals and minimum royalties will not.

Surrender—All or part of a lease may be relinquished at any time.

Equal Opportunity—The lessee will not discriminate against employees or applicants for employment on the basis of race, creed, color, or national origin.

Bonds—The lessee will furnish a bond in an amount double the annual rental for a competitive lease (minimum $1,000, maximum $10,000). For non-competitive leases, the usual amount is $1,000. For drilling, the required bond is $10,000.

Reports and Confidentiality—The lessee must furnish the BLM with various reports, maps, well logs, and other data. If the lessee so requests, the Bureau will not release the information to the public until the lease expires.

Other Provisions—Federal leases include a number of other provisions dealing with the powers of the Secretary of the Interior in certain areas, the protection of the land surface and natural resources, and other matters.

(b) Preparation

Federal leases must be prepared on the designated form. Leases must be accompanied by the required payments, such as filing fees and advance rentals, and supporting documents. For example, corporations must furnish a statement listing the state of incorporation, stockholders holding over ten percent of the stock, stock owned by aliens, and other relevant information.

(3) Indian Leases

Leases on Indian lands are administered by the Department of the Interior, according to federal laws and regulations, and as specified by the constitution or charter of the tribe.

While there are many similarities to federal leases, there also are significant differences. For example, twenty-five percent of the bid amount must accompany the bid, not twenty. More detailed information is required from corporations who are prospective lessees of Indian lands. Assignments covering Indian tribal lands must cover the whole lease; there is no provision for geographic subdivision.

(4) State Leases

Each producing state has laws and regulations establishing the form and terms of leases covering state-owned land. Many provisions are similar to those in fee and federal leases, with others added or modified to conform with the special requirements of the particular state. For information on a particular state, the agency in charge of oil and gas leasing for that state should be consulted.

2. Assignments

When a particular tract of land is already leased to another party, all or part of the mineral rights may be acquired by means of a transfer, or **assignment**. When a lease is assigned, all the rights and obligations are transferred from the assignor (the original lessee) to the new holder (the assignee). A typical assignment form is shown in Figure 4.10.

a. Restrictions

While most leases include the right of assignment, some restrictions may be imposed. In private leases, the number of assignments which can be made without the lessor's permission may be limited. In federal leases, the assignees must meet the same qualifications as the original lessees. The assignment also must be submitted to the BLM within 90 days after signing, and approved by the Bureau.

b. Consideration

When an assignment is made, some consideration is usually included for the assignor. The consideration commonly takes the form of a cash payment at the time the assignment is made, plus a share of production, known as an **overriding royalty**. This royalty is in addition to the royalty share of the mineral owner. Both forms of royalty take precedence over all other payments made from lease revenues. When an assignment is part of some type of business agreement among several parties, the consideration may take other forms.

c. Clauses

The assignment will include:

- effective date
- consideration (see above)
- words of grant stating what interest is assigned
- parties whom the assignment is in favor of
- lands and leases included
- other appropriate provisions

d. Preparation

To be fully effective, the assignment must be in the proper form, with legal signatures correctly acknowledged. In most cases, the assignment is recorded.

3. Purchases

Occasionally, mineral rights are acquired by outright purchase. In those instances, the transfer is permanent; whereas, the rights acquired through a lease

ASSIGNMENT
(Reserving Overriding Royalty)

KNOW ALL MEN BY THESE PRESENTS:

That the undersigned _____,
hereinafter designated "ASSIGNOR", for and in consideration of One Dollar ($1.00) and other good and valuable considerations, and subject to the reservation of an overriding royalty interest hereinafter set forth, hereby does transfer, assign, set over, and convey unto _____
hereinafter designated "ASSIGNEE", its successors and assigns, that certain oil and gas lease bearing No. _____
dated _____, 19____, from _____,
as Lessor, to _____, as Lessee,
recorded in Book _____ at page _____ of the Records of _____ County,
_____, in so far as it covers and includes the following described land situate in said County and State, to-wit:

together with all rights incident thereto and all personal property thereon, appurtenant thereto, or used in connection therewith.

As an additional consideration for this assignment, ASSIGNOR shall be entitled to receive and ASSIGNEE, by its acceptance hereof, agrees to pay or deliver to ASSIGNOR overriding royalties as follows:

(a) ASSIGNEE shall deliver to the credit of ASSIGNOR, free of cost, in the pipe line to which ASSIGNEE may connect its wells _____ of _____
of all oil produced and saved by ASSIGNEE from the above described lands or from time to time, at ASSIGNEE'S option, pay ASSIGNOR for such overriding royalty oil at the current market value of such oil at the well; and

(b) ASSIGNEE shall pay ASSIGNOR for gas produced from any well on the above described lands and used or sold by ASSIGNEE off said premises, or used by ASSIGNEE in the manufacture of gasoline or other products, _____
_____ of _____ of the current market value of said gas, as such, at the mouth of the well. If such gas is sold by ASSIGNEE at the well, then ASSIGNEE shall pay to ASSIGNOR _____ of _____ of the net proceeds derived from such sale.

The overriding royalties above provided shall be applicable only to the oil and gaseous substances produced and saved by ASSIGNEE from the above described land under the terms of the lease hereby assigned, after deducting that part used for lease operations, and if the leasehold estate hereby assigned covers less than the full oil and gas rights in any portion of the land above described, then the overriding royalties above provided, as to the production from such portion, shall be reduced and paid or delivered to ASSIGNOR only in the proportion which the leasehold estate thereon and assigned hereby bears to the full oil and gas leasehold estate thereon. Nothing herein contained shall impose or be held to imply any obligation on the part of ASSIGNEE as to the exploration, development or operation of the above described lands or to require ASSIGNEE to continue in force the lease hereby assigned. No transfer of this overriding royalty, however accomplished, shall be binding upon ASSIGNEE until after it has been furnished with evidence of such transfer satisfactory to it.

For the same considerations, ASSIGNOR covenants with ASSIGNEE, its successors and assigns, that ASSIGNOR is the lawful owner of and has good title to said lease, free and clear of all liens, encumbrances, and adverse claims; that said lease is a valid and subsisting lease on the land above described, and all rentals and royalties due hereunder have been paid, and that all conditions necessary to keep the same in full force have been duly performed, and that ASSIGNOR will warrant and forever defend the same against all persons whomsoever lawfully claiming or to claim the same.

IN WITNESS WHEREOF, this instrument is duly executed this _____ day of _____, 19___.

WITNESSES:

_____   _____ (SEAL)
                                                                Assignor

_____   _____ (SEAL)
                                                                Assignor

Figure 4.10. ASSIGNMENT FORM.

expire when the lease is no longer in effect.

D. Acquisition Procedures for Mineral Rights

1. Fee Lands

To acquire mineral rights to private lands, either by lease, assignment, or purchase, the procedure is direct negotiation with the owner or lessee. As long as the instrument used is legally valid, the actual terms are a matter of mutual agreement between the parties involved, although going rates for signature bonuses, royalties, rentals, and other terms will be established in areas of active leasing.

The terms of a particular lease, assignment, or purchase may vary considerably from the regional norm. For example, the dollar amount of the cash signature bonus for a lease is influenced by a number of variables, such as the size of the tract, the current price per acre in the area, the nearness to production, the negotiating skill of the landman, the size of the royalty and other payments included in the lease, and the competition.

The old standard amount for royalties was one-eighth of production, but leases with one-sixth or one-fourth are now common in many areas.

2. Federal Lands

a. Competitive Leases

Federal leases are issued through competitive procedures (either sealed bids or public auction) for offshore tracts beyond a state's territorial limits, and for onshore tracts known as **KGS** land. A **Known Geologic Structure** (KGS) is technically the trap in which an accumulation of oil or gas has been discovered by drilling, and has been determined to be productive. The limits of this structure include all the acreage that is presumptively productive. The U.S. Geological Survey decides which lands should be included in the KGS classification.

The Bureau of Land Management holds periodic lease sales. Anyone interested in a certain KGS or offshore tract may request that it be included in a future sale. The public notice of each sale will list the lands offered, and will specify the royalty and rental which will be provided for in the lease.

At the sale, the BLM officer will award the lease on each tract to the qualified party offering the highest bid. The Bureau reserves the right to reject all bids.

The winner usually has thirty days to sign or execute the lease on the proper form, pay the balance due of the bonus amount (four-fifths), pay the first year's rental in advance, and furnish any bond which may be required. When the lease covers acquired lands, consent to the lease must also be obtained from the agency under whose jurisdiction the lands fall.

b. Non-Competitive Leases

(1) Simultaneous Filings

On the third Monday of each month, the BLM land offices publish lists of public domain and acquired lands (other than KGS) on which the leases are no longer in effect. Offers to lease these tracts are submitted on a "Simultaneous Oil and Gas Entry" card, accompanied by a ten dollar filing fee. The cards may be submitted to the BLM up to 10:00 A.M. on the fifth working day after the third Monday. A few days after the filings close, three cards are drawn from all those submitted on each parcel. The party having submitted the card first drawn has the right, if qualified, to pay the first year's rental and acquire the lease. If the first party does not acquire the lease, then the second party has the right to pay the rental, and so on.

(2) Offers to Lease

Any tract of land not leased when offered through the simultaneous filing procedure may be acquired later by filing an "Offer to Lease" on the appropriate form. The offer must be accompanied by a ten dollar filing fee and full payment of the first year's rentals. If the offer is made by a qualified party and accepted by the BLM, the lease will be issued. Leases of acquired lands are also subject to approval by the agency having jurisdiction over these lands.

The procedures mentioned in the preceding paragraph are those followed when tracts of federal non-KGS land are to be leased which have never previously been under lease.

c. Assignments

Mineral rights to federal lands may be acquired through assignment, if the assignee meets the qualifications of a lessee, and if approved by the BLM.

3. Indian Leases

Leasing procedures for Indian lands are generally similar to those for federal lands, with the most common method being competitive bidding. The exact procedures to be followed are contained in the federal laws and regulations, and in tribal documents. One notable difference between Indian leases and other leases is that the bidders are required to share in the costs of advertising the sale.

4. State Leases

The laws, regulations, and procedures governing leasing and assignments of leases on state land vary considerably from state to state. Anyone desiring to lease land from a particular state, or wanting to assign a state lease, must first determine the requirements of that state. This information is available from the state agency responsible for oil and gas leasing.

V. SEQUENCE OF LAND WORK

This section briefly outlines the typical sequence of events that is followed when land work is done. Note that many of the concepts discussed in the previous sections are incorporated into this outline.

1. **Initial interest.** Land work typically begins when someone decides that a particular area or tract of land has potential for oil and gas production. The initial interest may come from an explorationist who assembles and interprets geological and/or geophysical information. A "prospect" area is identified, and land work begins.

Alternatively, land work may precede exploration. A landman (often a lease broker) may notice that leasing activity is picking up in a certain area, or that an exploratory well is to be drilled in the vicinity. The broker then may begin to acquire leases, with the objective of putting together a group, or package, of leases. When this package has been assembled, the block of leases will be assigned, if possible, to another party for exploration and/or drilling.

2. **Type of rights.** Once an area has been selected, the next step is to determine what rights are specifically required. Since oil and gas leases are most common, the balance of this section will trace the sequence related to these leases.

3. **Determination of ownership.** The landman determines the type of ownership from land maps. A further check of county, federal, or state records is made to establish the current owner, and whether the land is available for lease.

4. **Lease acquisition.** As noted in the preceding section, the type of ownership will determine the procedure used by the landman to acquire a lease. If the land is already leased, an assignment may be possible, or a **top lease,** which will become effective upon expiration of the current lease.

5. **Title verification.** Usually prior to drilling, ownership is checked to be sure that all mineral owners have signed leases. An up-to-date abstract will be required for private (and some public) lands. An abstract is a collection of all the documents pertaining to the ownership of a particular tract of land. The abstractor who assembles these documents is a specialist in searching through county records.

For public lands, a status report listing all the pertinent documents and facts found in the public files is prepared by an attorney.

An attorney then examines the documents in the abstract or in the status report, and issues a title opinion stating current ownership and any title defects which should be corrected or "cured."

6. **Curative work.** If the title opinion indicates some important defect, the landman must try to obtain the appropriate documents to cure the defect. Such material will be submitted to the attorney for examination, and another opinion will be written.

7. **Recording.** At some point in time, usually just prior to drilling (if not before), the lease is recorded in order to make it fully effective.

8. **Analysis and records.** Even after the title has been verified and the lease signed and recorded, much still remains to be done. A lease analyst must carefully read the lease to ensure that its obligations are understood and can be met. The lease is read as a complete entity rather than as individual clauses, since the provisions in the lease often are interrelated. The appropriate files and records are then established.

9. **Compliance.** This is the phase of the overall sequence in which the payments are made at the proper time to the proper party. For example, annual rentals are usually paid well in advance to ensure that the lease is maintained. If a payment is not made on time or is made incorrectly, a ratification of the lease by the mineral owners is usually necessary for the lease to be retained.

In the event of production of oil and/or gas, royalty payments will be required. The first step normally involves the preparation of a **division order title opinion** by an attorney, stating the proper parties to whom royalty should be paid. After each party has signed the division order verifying their correct share of the royalty, payments begin.

10. **Release.** When a lease has terminated for any reason, a release should be placed in the public records to notify anyone who may be interested that the lease is no longer effective.

VI. REGULATIONS

A. Well Spacing and Permits

Before a well can be drilled, it is necessary to obtain a permit from the state. Drill sites are restricted by state regulations, which are designed to promote the most efficient development of the reservoir (**spacing**), and to prevent drainage of adjacent tracts (**footage location**).

1. Spacing

Producing states have enacted regulations that specify a minimum number of acres to be allotted for each well drilled. If the area has not been previously drilled, the minimum is often 40 acres. Once oil and/or gas has been discovered, the number of acres can be adjusted up or down, depending on the nature of the production.

Since gas can move more easily than oil through reservoir rock, spacing for gas reservoirs is likely to be greater. Because fewer wells are required to drain a gas reservoir efficiently, spacing of 160 and 320 acres (or more) is common.

If the oil found in the reservoir is thick, or viscous, in nature and does not move readily through the reservoir rock, the spacing will be adjusted downward, since more wells will be required to produce this oil. Sometimes the spacing units will be as small as two acres.

Spacing may also be related to depth, with deeper wells being assigned to drain a larger section of a reservoir.

When a lease is too small to drill on, it may be combined, or **pooled,** with other adjoining leases, provided there is a pooling clause in all of the leases involved. If the situation should arise in which a mineral owner does not want a lease to be pooled, the state may require that pooling be done to promote efficient development of a reservoir. This particular process is called **forced pooling,** and is usually done only after a hearing by the responsible state agency.

2. Footage Location

In order to minimize the possibility of a well draining oil or gas from formations beneath adjoining tracts, state regulations require that wells be located a certain distance inside lease lines. In addition, wells must be situated a minimum distance from each other. In some cases, a specified well location cannot be drilled due to existing structures, geographical barriers (lakes, rivers, etc.), or some other restrictions. In these instances, non-standard locations may be approved by the state agency, with the permission of the surrounding mineral owners and lessees.

Many leases have provisions that address the problem of possible drainage. For example, a lease may provide that, if a well is drilled on an adjacent tract within a specified distance from the boundary line, the lessee will drill an offset well to prevent drainage. **Compensatory royalty** may be paid to the mineral owner if the lessee does not want to actually drill such a well. This royalty payment serves as a substitute for the revenue the mineral owner would have received if a productive well were actually drilled.

B. Well Location Description

The location of a well site must be described using references to established property boundaries. For rectangular survey areas, the location is usually given by its distance from two section lines (Figure 4.11). For areas using the metes and bounds system, distance and directions may be given from established property lines.

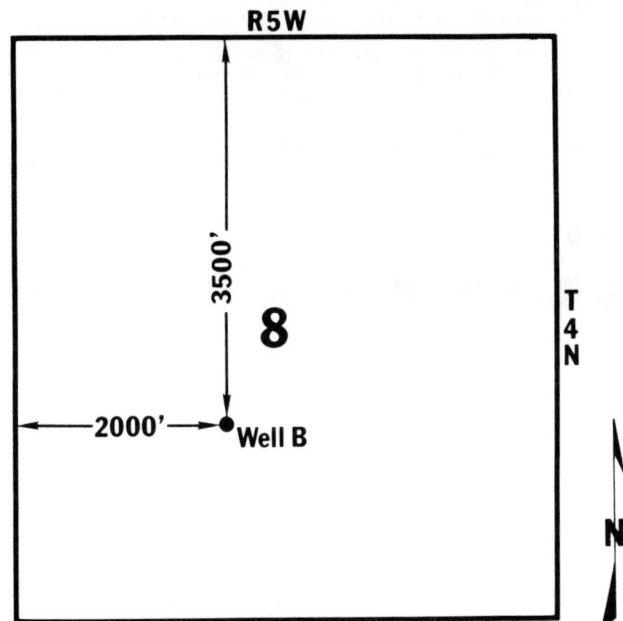

Figure 4.11. WELL SITE. Well B: 3500 ft. FNL, 2000 ft. FWL, Section 8, Township 4N, Range 5W, Indian Base and Meridian.

C. Legislative Acts

Numerous legislative acts affect and restrict the land work and other activities of the oil industry. Some of the more important pieces of federal legislation are:

- National Environmental Policy Act (1969)
- Clean Air Act of 1970
- Federal Water Pollution Control Act of 1972

State and local governments also have laws and ordinances controlling and restricting land use. The preparation of environmental impact statements and the compliance with various legislative requirements often represent a considerable investment in time and money for the petroleum industry.

VII. BUSINESS ARRANGEMENTS

A. Introduction

Numerous types of agreements have been developed in the petroleum industry to cover the various cooperative efforts between companies and/or individuals. Some of the most common business arrangements are described briefly in this section.

B. Farmins/Farmouts

In a typical farmout agreement, one company holds land or leases in which another company wants to earn an interest by drilling a well (or wells). (The first page of a farmout agreement is shown in Figure 4.12.) The terms of any farmout agreement are negotiable between the parties, and are highly variable. In some cases, the interest in the land will be assigned only if the well is productive; other agreements may state that the interest is earned merely by drilling the well, regardless of the results.

C. Joint Ventures

A joint venture is an arrangement in which two or more parties agree to do something together. Each party pays a portion of the expense, and receives a portion of the revenue, if there is any. Joint ventures make many projects feasible that would otherwise be impossible, because two companies are sharing a risk that would be too great for one company to shoulder alone.

Typically, one company serves as the **operator** to conduct routine business on behalf of the group. The operator is usually compensated for this service, and is required to report to the other parties, or the nonoperators. Major decisions are normally made by a committee representing all of the participants.

D. Units

In most cases, a unit is created by combining leases, or parts of leases, covering the surface above an oil reservoir. There are two reasons for establishing such a unit: 1) increased oil recovery from the reservoir through the installation of an enhanced oil recovery project, such as a waterflood; and 2) efficient operation. Both objectives require that the reservoir be managed as a single tract, rather than as fragmented leases.

All of the parties having an interest in any of the tracts of land to be unitized should consent to the formation of the unit by signing a "Unit Agreement." This basic document will be supplemented by an "Operating Agreement" delineating the operational procedures of the unit, and the responsibilities and authority of the various participants. An "Accounting Procedure" that establishes how the funds will be handled and accounted for will usually be attached.

For federal lands, leases are also combined, or **communitized,** for exploratory purposes. Acreage included in a federal unit can no longer be charged against the total acreage held by an individual or corporation.

E. Investment Arrangements

To attract capital for industry operations, a number of investment vehicles have been devised. Among these are **limited partnerships,** in which investors (limited partners) provide a certain amount of money that the general partner uses for drilling a well (or wells).

Funds are also raised by selling company stock, either by **private placement** or **public offerings.** Public offerings are usually under the strict control of the Securities and Exchange Commission regulations.

F. Types of Interests

Many different types of interests in a tract of land can be created by the mineral owner, by lessees, and through various business arrangements. Some common interests are described below:

Royalty—The mineral owner's share of production, at no expense.

Overriding Royalty—An additional share of production payable to someone other than the mineral owner, at no expense. "Overrides" terminate when the lease ends. Under certain conditions, federal leases limit overrides.

Participating Royalty—A share of the production, but includes a participation in some of the expenses.

Term Royalty—This royalty ends at a specific point in time.

Working Interest—Working interest owners pay all the expenses, and receive whatever revenue remains after royalty payments are made.

Production Payments—A share of the production, up to a specified amount.

Net Profits Interest—A share of the funds remaining after royalties and operating expenses have been

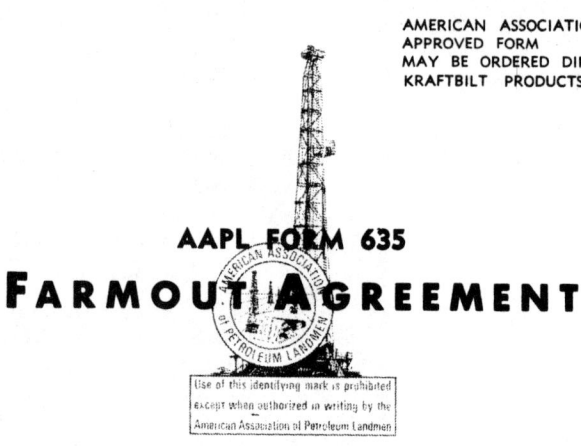

AMERICAN ASSOCIATION OF PETROLEUM LANDMEN
APPROVED FORM   A.A.P.L. NO. 635
MAY BE ORDERED DIRECTLY FROM THE PUBLISHER
KRAFTBILT PRODUCTS,   BOX 800 TULSA 74101

AAPL FORM 635
FARMOUT AGREEMENT

DATE:

TO: RE:

In consideration of the benefits to accrue to the parties hereto and the covenants and obligations to be kept by you, it is hereby mutually agreed as follows:

I ACREAGE:

We represent without Warranty of Title of any kind or character that we hold Oil and Gas Leases or Mineral Interests described as follows:

We agree to deliver to you such abstracts and other title papers as we have in our files at this time, and at your sole cost, risk and expense you agree to conduct such Title Examinations and secure such curative matter as is necessary to satisfy yourselves that Title is acceptable to you.

II OBLIGATIONS:

(A) TEST WELL: On or before the ____ day of _____, 19____, you agree to commence, or cause to be commenced the actual drilling of a well for oil and/or gas at the following location:

and you further agree to drill said Test Well with due diligence in a workmanlike manner to a depth sufficient to thoroughly test the following:

Figure 4.12. FIRST PAGE OF A FARMOUT AGREEMENT.

paid. Net profits also may be computed only after investment is recovered.

Carried Interests—In some business arrangements, one party pays all the expenses on behalf of the other parties, up to an established limit, thus "carrying" the other parties' shares for them. For example, in a group of four, three may each pay 1/3 of the expenses, but revenue will be divided four ways.

Back-ins—A carried interest owner may change to a working interest owner after the parties who have paid the expenses have recovered their investment, or their investment plus an additional amount.

LAND—BIBLIOGRAPHY

Matthew Bender and Company: **The Law of Federal Oil and Gas Leases** (New York).

Brown, Earl A.: **The Law of the Oil and Gas Leases,** 2 Vols., Matthew Bender and Company (New York).

Bureau of Land Management: **Regulations Pertaining to Oil and Gas Leasing,** Circular No. 2357, Title 43, Code of Federal Regulations (Washington, D.C.) 1980.

Hoffman, Leo J.: **Voluntary Pooling and Unitization,** Matthew Bender and Company (New York).

Kulp, Victor H.: **Oil and Gas Rights,** Little, Brown and Company (Boston).

McLane, A.E.: **Oil and Gas Leasing on Indian Lands,** Matthew Bender and Company (New York).

Moses, Leslie: **AAPL Guide for Landmen: From Lease to Release,** AAPL (Fort Worth) 1970.

Myers, Raymond M.: **The Law of Pooling and Unitization,** 2 Vols., Matthew Bender and Company (New York).

Office of the Federal Register: **Code of Federal Regulations: Title 25, Subchapter P, Part 171** (Indian Tribal Lands), U.S. Government Printing Office (Washington, D.C.) 1980.

Office of the Federal Register: **Code of Federal Regulations: Title 25, Parts 172, 172.1, et. seq.** (Indian Allotted Lands), U.S. Government Printing Office (Washington, D.C.) 1980.

Rocky Mountain Mineral Law Foundation: **Landman's Legal Handbook,** 3rd Ed. (Boulder, Colo.) 1977.

Sullivan, Robert E.: **Handbook on Oil and Gas Law,** Prentice-Hall, Inc. (Englewood Cliffs, N.J.).

Williams, Howard R. and Charles J. Meyers: **The Law of Oil and Gas,** 7 Vols., Matthew Bender and Company (New York) 1959-1976.

CHAPTER 5:

DRILLING AND COMPLETION

I. INTRODUCTION

The basic techniques discussed in the chapter on Exploration have become indispensable to the petroleum industry's search for new oil and gas reservoirs. Despite the continuing refinement of these techniques, however, they can only provide indications of where petroleum "might" be found. The one sure method of "proving" the existence of oil or gas in underground formations is by putting a test hole into the ground; in short, by drilling a well.

The development of drilling methods constitutes a rich history, dating back several thousand years to early China. Over the course of time, many mechanical improvements and operational adjustments have been made in adapting general drilling techniques to the search for oil and gas. Today, new exploration frontiers, such as the Arctic polar regions and the ocean depths, call for continued research and advancement in drilling technology and methodology. The developments made in these areas will help determine where we can look for oil and gas in the future.

After a brief review of some of the historical aspects of drilling, this chapter will examine modern rotary drilling techniques, various types of onshore and offshore rigs, special drilling operations, and methods of well completion and stimulation. A glossary is included at the end of the chapter as a review of terms used in the chapter and to provide definitions of additional terms.

II. HISTORY OF DRILLING

A. Historical Summary

The process of drilling wells originates early in human history. In an effort to find water or salt, prehistoric peoples dug the first wells by hand, using implements made of bone, wood, or stone. Although they still used a manually operated system, the Chinese were the first to reach a level of drilling sophistication with their spring pole drilling technique (see Chapter One). The science and methods of drilling wells evolved through the centuries from these primitive techniques to the highly technical and mechanized processes we have today.

The following excerpt from an article by J. E. Brantley, in **The History of Petroleum Engineering,** published by the American Petroleum Institute, provides a summary of historical events which have contributed to the evolution of modern drilling techniques.

"1. Peoples of prehistoric cultures dug holes and wells by hand, without the use of tools other than those in natural form; and the same procedures can be observed in present-day Paleolithic cultures.

"2. From prehistoric times to about 600 B.C. was the period of hand-dug wells using crudely manufactured tools of the bar, pick-and-shovel type, with skins or baskets for removing waste, and rope of leather or natural fibers. With the invention of the wheel by the Sumerians some 4000 B.C., windlasses and pulleys doubtless came into use.

"3. The period from about 600 B.C. to 1800 A.D. was made notable by the work of the Chinese of the Chou Dynasty in drilling brine wells with percussion tools, and by deep hand-dug holes in Japan, Burma, Europe, and probably Africa. It seems quite reasonable to assume that the Greeks and Romans, with their broad technical knowledge and mechanical skills, practiced the art of water-well digging and possibly drilling, although the contemporary writers recorded little data on such activities. By 1500 A.D., the Chinese seem to have developed the art of drilling "deep" wells, possibly as much as 2000 feet.

"4. From 1800 to 1859, in both America and Europe, there was rapid improvement in drilling equipment and operating methods—so much so that by the end of the 1850s many wells had been drilled to depths of around 2000 ft.; and some of the European water wells of the period were as large as 2 ft. in diameter at depths in excess of 1500 ft.

"5. Succeeding the completion of the Drake well in 1859 and continuing to about 1880, the percussion well-drilling industry made great advances in mechanical equipment and technical skill. Drilling machines and standard rigs were developed into types, forms, and principles of operation that have carried down to modern times without basic changes except in small details of machinery and in the sources of power.

"6. From about 1880 to 1930 the development of standard drilling rigs and tools was completed, but relatively few additions were made to the tools except those used on the derrick floors for handling pipe. Many portable and semi-portable machines were likewise developed, built and used.

"7. Beginning in 1930 or thereabouts, the need for faster drilling, rig portability, and greater overall efficiency brought about the development of the one-package internal-combustion-engine-driven cable-tool drilling machine of modern manufacture. These machines were not basically new, for many of similar type had been built directly after the turn of the century. However, the older rigs did not have the depth capacity or the portability of the modern machines because of the lack, at that time, of suitable steels and of a power plant of reasonable size and weight.

" In 1901, came the first important oil well to be

completed with rotary tools—the Spindletop gusher—and although cable tools continued to drill the great majority of all wells for some years to come, rotary equipment soon replaced percussion tools in the areas of younger and softer formations. By 1920, rotary rigs were drilling practically all wells in such areas as the Coastal Plains and had begun to make inroads on all cable-tool rigs in the softer areas of Oklahoma, Kansas, and North Texas. The first rotary reached these areas about 1914 or 1915.

" With the perfection of the rolling-cutter rock bits for rotary drilling in the late 1920s, cable tools commenced to lose ground rapidly in all of the oil-producing areas. By 1950 practically all of the oil wells of the United States and of the world at large were being drilled by rotary rigs, except in a few regions such as the oil fields of the northeastern Paleozoics of the United States, parts of the Permian Basin in Texas, and a few minor areas abroad. "

B. Spring Pole and Cable Tool Drilling

The innovative technique of spring pole drilling was first used by the Chinese, then much later by the 19th century Europeans and Americans. An example of an early European spring pole drilling machine is shown in Figure 5.1.

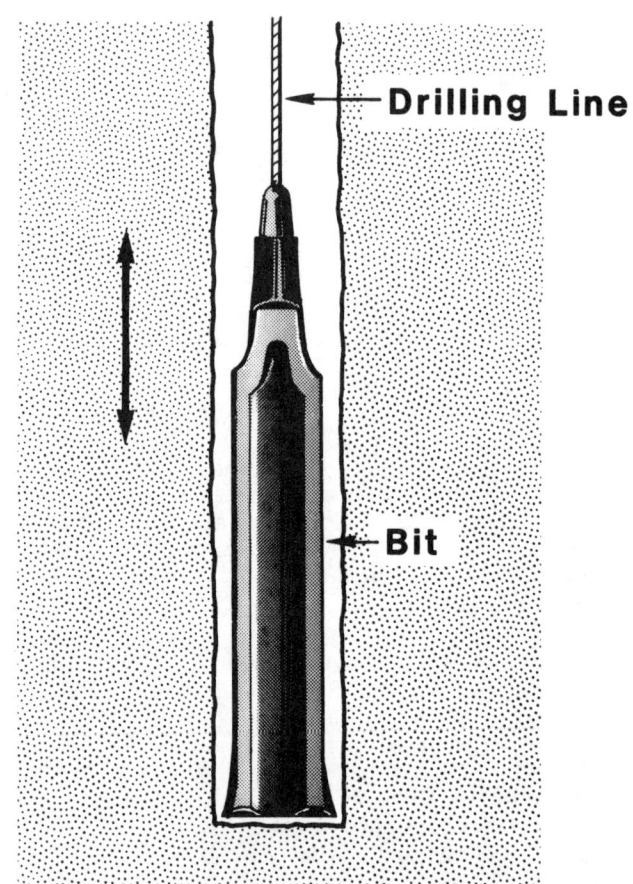

Figure 5.2. **PERCUSSION BIT.** The constant raising and dropping of the bit drills the hole.

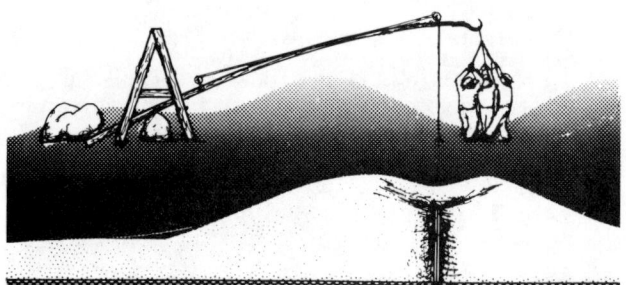

Figure 5.1. **SPRING POLE DRILLING.** A Jobard drilling machine used in Brussels in 1828.

The mechanics involved were fairly simple and are known as **percussion drilling.** A hardened bit suspended on a cable was dropped against the bottom of the hole (Figure 5.2). This constant pounding succeeded in increasing the depth of the hole. The spring pole device was an aid in lifting the bit from the bottom of the hole, and in the process, eliminated some of the manual labor.

From the spring pole came the **cable tool rig,** shown in Figure 5.3. Cable tool drilling was used in the United States in the latter half of the 1800s and early 1900s. Wells were often drilled to 3,000- or 4,000-foot depths with this type of rig. Essential to cable tool drilling was a beam, known as the **walking beam,** which was mounted on a post and was used to raise and drop the bit in the hole. Rock ground up at the bottom of the hole was removed at periodic intervals during the drilling process by a **bailer.** The cable tool rig was usually driven by a steam engine, which added speed and efficiency to the

1. Boiler
2. Steam engine
3. Sand reel
4. Band wheel
5. Pitman
6. Walking beam
7. Sampson post
8. Sand line
9. Crown block
10. Drilling cable
11. Derrick
12. Bull wheel

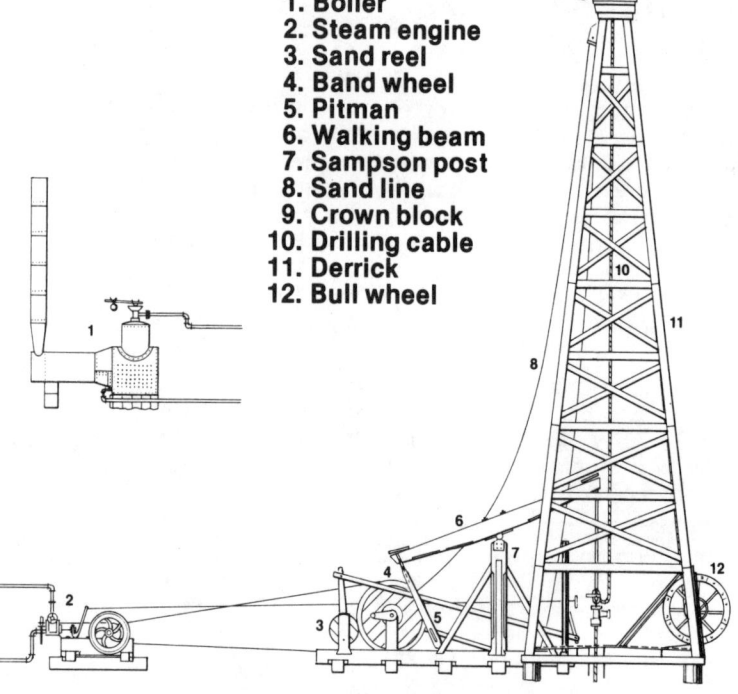

Figure 5.3. **CABLE TOOL RIG.**

drilling process. In contrast, the spring pole method depended entirely on manual labor.

C. Modern Drilling Methods

The majority of well drilling in today's petroleum industry is accomplished by the rotary drilling method, which is discussed in detail in the next section of this chapter. It is now possible to drill wells to much greater depths (sometimes beyond 30,000 feet) than could be drilled using cable tool or spring pole equipment. Cable tool drilling was very slow and provided no means of controlling high pressures often encountered in deeper wells. Rotary drilling has proven to be much faster and provides for control of high pressure oil, gas, or water flows. A fascinating new area of technology allows for offshore drilling operations in waters several thousand feet deep. In the United States today, approximately ten percent of all drilling units are located offshore.

III. CONTEMPORARY ROTARY DRILLING

A. Introduction

Tremendous improvements in contemporary drilling technology have been accomplished through mechanical ingenuity. Nowhere is this more evident than in modern rotary drilling equipment, essential to today's petroleum industry. Years of practical drilling experience and developing technology have produced a highly efficient mechanical system, adaptable to land or water operations, and capable of drilling to great depths in the search for oil and gas.

In examining rotary drilling, we will emphasize the major components of a rotary rig and their relationship to the basic purpose of the system, known in the vernacular as "makin' hole." For a more detailed discussion of the equipment, the interested reader is encouraged to research some of the excellent references at the end of the chapter. Of course, the best way to learn more about drilling and to gain some firsthand experience is to visit a drilling site where a rotary rig is in operation. Such a field trip can increase one's understanding of the drilling process and of the machinery being used.

A rotary drilling rig with some of its major components identified is illustrated in Figure 5.4. There are five major systems essential to the operation of a rotary rig: (1) the drill string and bit, (2) the fluid circulating system, (3) the hoisting system, (4) the power plant, and (5) the blowout prevention system. A discussion of each of these systems and their interrelationship follows.

B. The Drill String and Bit

The most crucial part of the rotary drilling system is that portion which physically creates the hole: the drill string and bit. Figure 5.5 illustrates this portion of the system in detail. The term "rotary" drilling refers to the rotation of the drill string and the drill bit, thereby applying a cutting action against the rock at the bottom of the hole. This rotating action is achieved through the operation of the drill string and bit system.

The **swivel** is latched onto the **hook** at the bottom of the traveling block, and serves the dual function of supporting the weight of the drill string and allowing the string to rotate. Attached to the swivel is a four or six-

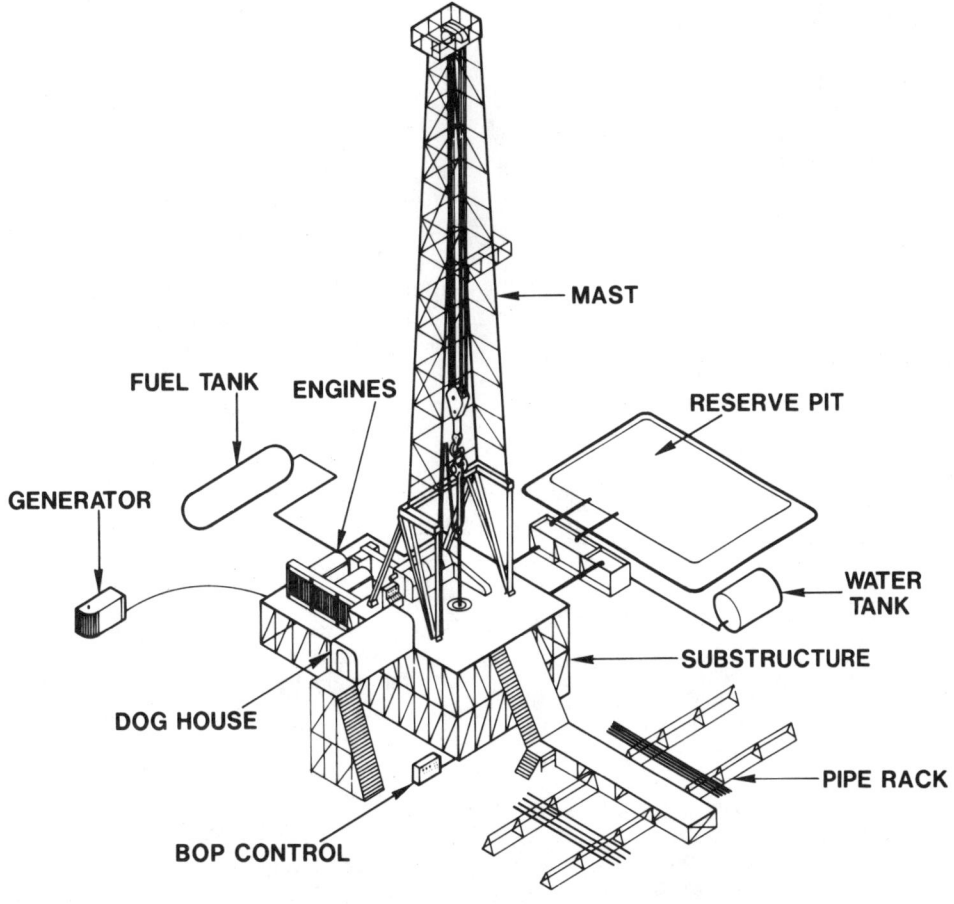

Figure 5.4. ROTARY DRILLING RIG.

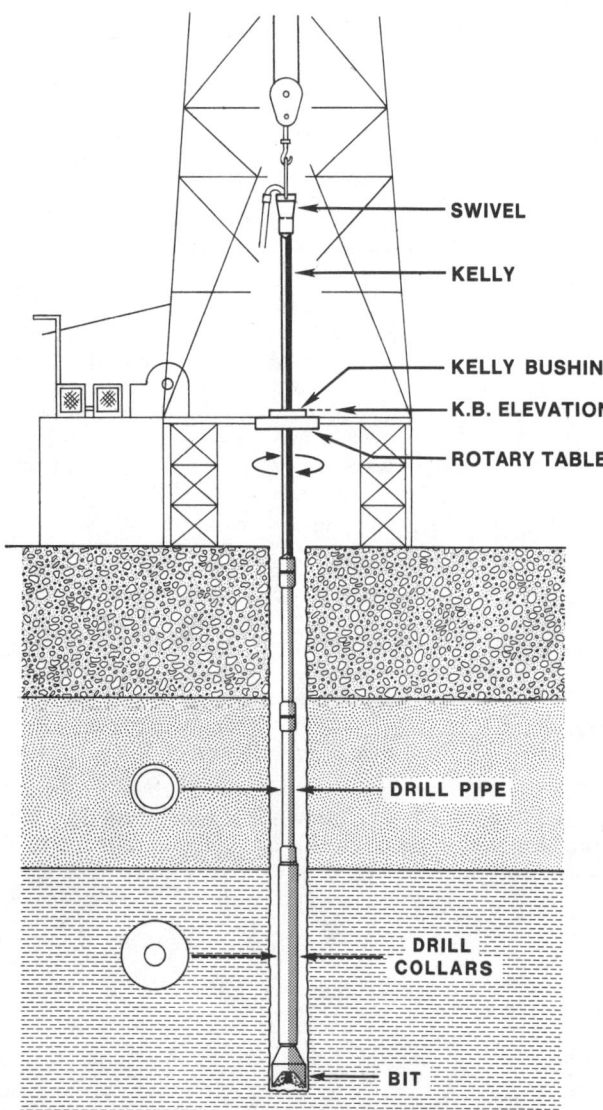

Figure 5.5. DRILL STRING AND BIT. The figures at bottom left indicate the comparative sizes of the drill pipe and drill collars.

sided piece of pipe called the **kelly**. The shape of the kelly allows it to transfer the rotating motion of the **rotary table** to the drill string. The kelly is mounted in a similarly-shaped opening in a device called the **kelly bushing.** The kelly bushing is held in the rotary table on the rig floor and supplies the necessary torque from the rig's power system to turn the drill string. The kelly is also free to move up and down through the rotary table, while it is being turned. In this manner, the drill string is allowed to steadily move down the hole as it rotates and drills deeper.

Attached to the kelly is the **drill string,** consisting of ordinary **drill pipe** and the heavier **drill collars.** Drill pipe comes in approximately 30-foot sections, called **joints,** and is threaded on each end to allow the sections to be connected together. Below the drill pipe, and immediately above the bit, are the drill collars. Drill collars differ from ordinary drill pipe in that they are thick-walled and much heavier, similar to the difference between the heavy, thick barrel of a high-powered rifle and the thin, light barrel of a shotgun. Drill collars are used to add weight on top of the drill bit, which

Figure 5.6. MILLED-TOOTH BIT.
(Photo courtesy of Hughes Tool Co.)

improves the cutting action. They are also used in certain instances to help keep the hole straight.

The **bit,** attached at the end of the drill string, is generally designed with three cone-shaped wheels, tipped with hardened teeth for cutting the rock (Figures 5.6 and 5.7). For drilling very hard formations, a special bit with a diamond-studded face often replaces the tricone rolling bit (Figure 5.8). All bits have passages that allow drilling fluid to pass through and sweep away the rock cuttings as the bit drills deeper.

C. The Fluid Circulating System

It is necessary to pump a fluid down the hole to aid in the drilling process. Generally called "drilling mud" because of its physical appearance, this fluid usually consists of water, various special chemicals, and frequently a weighting element called barite. This fluid is continuously circulated down the inside of the drill pipe, through the bottom of the bit, and back up the annular space between the drill pipe and the hole (Figure 5.9). The drilling mud serves a number of important purposes:

1. The mud carries broken rock fragments to the surface. Without the circulation of mud, the hole would quickly become clogged with rock cuttings.

2. The mud helps counterbalance any high pressure oil, gas, or water zones encountered in the formations being drilled. This is the reason for adding the weighting material, barite, to the mud.

Figure 5.7. TUNGSTEN CARBIDE INSERT BIT.

Figure 5.8. DIAMOND BIT.

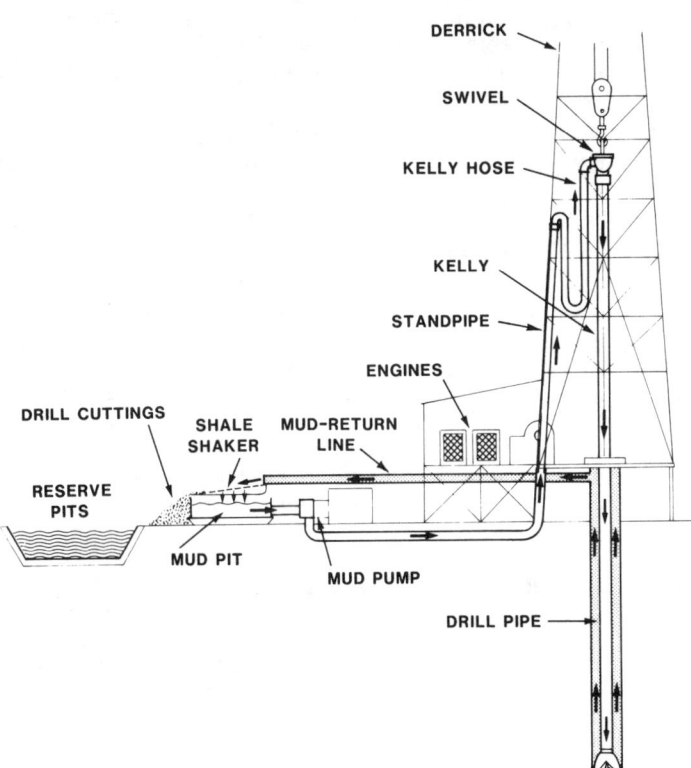

Figure 5.9. FLUID CIRCULATING SYSTEM.

Without a weighted drilling mud, high pressure zones would "blowout" or flow uncontrolled to the surface, possibly causing injury to workers and great damage to drilling equipment and the formation.

3. The column of mud in the hole contributes to wellbore stability and helps prevent the walls of the upper portion of the hole from caving in.

4. The mud also serves to lubricate and cool the bit. This facilitates the cutting action and helps prolong the life of the bit.

The surface portion of the fluid circulating system consists of a **mud pump,** which forces the mud from the **mud pits** up through the **standpipe,** into a hose called the **kelly** or **rotary hose,** and down the drill pipe to the bit (Figure 5.9). The mud passes through the bottom of the bit and returns up through the **annulus** (the space between the drill pipe and the wall of the borehole) to the surface. At the surface, it exits the hole through the **mud-return line** and flows over a vibrating screen called the **shale shaker,** which filters the rock cuttings out of the liquid mud. The rock cuttings are dumped into earthen excavations, or **reserve pits,** where such material is stored. Often the mud is additionally filtered by **desanders** and **desilters,** which remove fine-grained solids from the mud. The mud is then put into tanks, where it is recirculated back through the pump and down the hole.

Drilling mud plays an indispensable role in the process of drilling a well. During drilling operations, its properties are carefully monitored by a "mud engineer" to make sure the mud has the proper weight and chemicals to perform its various functions. Other types of fluids can be used in the circulating system, such as oil or even air, but these are not as common as water-based mud.

D. The Hoisting System

To the casual observer, perhaps the most familiar portion of a rotary drilling rig is the **mast** or **derrick**. On large rigs, the derrick is often more than 150 feet tall and can be seen from a considerable distance. The derrick or mast is part of the rig's hoisting system, another vital component of the entire system. Because drill pipe, drill collars, and the drill bit need to be lifted in and out of the hole, a hoisting system must be designed to lift loads which often weigh several hundred thousand pounds.

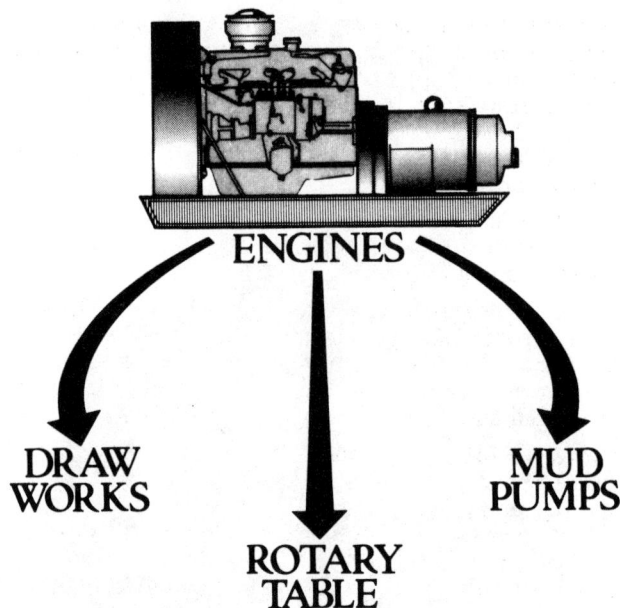

Figure 5.11. POWER SYSTEM.

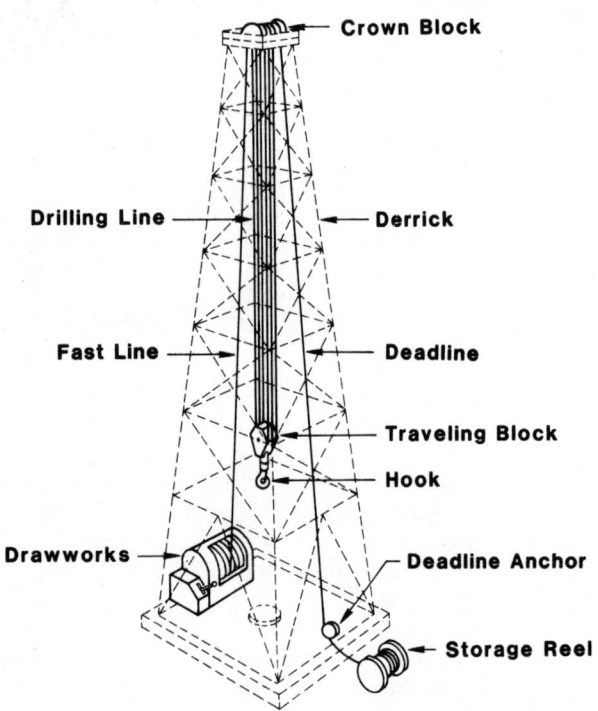

Figure 5.10. HOISTING SYSTEM.

The basic components of the hoisting system are shown in Figure 5.10. The **drawworks** is a large spool of cable driven by the rig's engines to raise or lower the pulley system that is hung in the derrick. The drawworks also contains heavy-duty brakes to restrict the speed of lowering heavy strings of pipe into the hole. The upper stationary set of pulleys, called the **crown block**, is mounted on the top of the derrick. The lower set, the **traveling block,** is moveable and is suspended in the derrick by a wire rope called the **drilling line**. The drilling line is usually threaded six or eight times (depending on the weight it needs to support) around the crown block and through the traveling block. The **fast line** is that part of the drilling line which runs from the drawworks to the crown block, and which moves as the traveling block is lowered or raised. The **deadline** runs from the crown block to the **storage reel** and is secured by the **deadline anchor.** Extra line is kept on the storage reel by the side of the rig.

The hoisting system is used to suspend the drill string in the hole, maintaining the proper weight on the bit. It is also used to pull the drill string out of the hole and to lower it back to bottom; this is called "tripping."

E. The Power System

All of the equipment described so far is operated by the power system, which is the heart of the entire rig operation. A good, dependable power source is essential to the continuous operation of any rig. A central power plant, as shown in Figure 5.11, commonly operates the rotary table, hoisting system, and fluid circulating system. The engines generally run on diesel fuel or natural gas.

A typical large rig usually has more than 2,000 horsepower available in the power system for operating all the rig equipment. Electric power for lighting, etc., is supplied by small, auxiliary generators.

F. The Blowout Prevention System

It has become ever more important that good mechanical equipment be available to prevent high pressure fluids in deep wells from escaping to the surface. A diagram of this type of equipment is shown in Figure 5.23, and is known as the **blowout preventer** or **BOP stack**. This equipment is the most important safety system on the rig, and its use is considered mandatory in most parts of the United States. The operation of blowout prevention equipment is described in Part B of the section on special drilling operations.

IV. DRILLING OPERATIONS

A. The Drilling Contract

In the United States, oil and gas companies normally do not own drilling rigs. There exist specialized companies which are in the business of drilling oil wells. These companies own drilling rigs which they move from site to site for oil companies in need of drilling services. The companies that own drilling rigs are known as **drilling contractors**; the companies that hire the drilling rigs are known as the **operators**. A contract must be drawn up between the drilling contractor and the operator before drilling begins. In the contract, the parties specify how the costs for drilling the well will be billed to the operator, and detail the special drilling obligations and responsibilities of the contractor.

There are a number of different types of drilling contracts in use. The three most common are **daywork**, **footage**, and **turnkey** contracts.

In the **daywork** contract, the contractor charges the operator a certain amount per day for the use of the rig and crews. The operator must specify all procedures and equipment to be used, while the contractor performs the work as requested.

In the **footage** contract, the contractor charges the operator a certain amount per foot of hole drilled. The contractor is obligated to drill the well to a specified depth for the agreed upon price per foot. The contractor assumes the risk of drilling, in that he must absorb the cost of any difficulties encountered during drilling.

In a **turnkey** contract, the contractor agrees to drill the well for the operator for a fixed amount, but the contractor must again absorb the extra cost from any special problems encountered during drilling. This contract is used in areas where drilling problems are minimal and drilling costs are uniform.

B. Drilling Procedures

At this point, we will examine the step-by-step procedure which is used in drilling a well. This generally follows a logical sequence in both routine land and offshore drilling operations.

The process usually begins by staking the location. On land, a survey crew goes into the field and surveys precisely the location which has been selected by the geologist or geophysicist. Offshore, a survey ship uses instruments to take bearings from either orbiting satellites or other offshore platforms and wells to locate

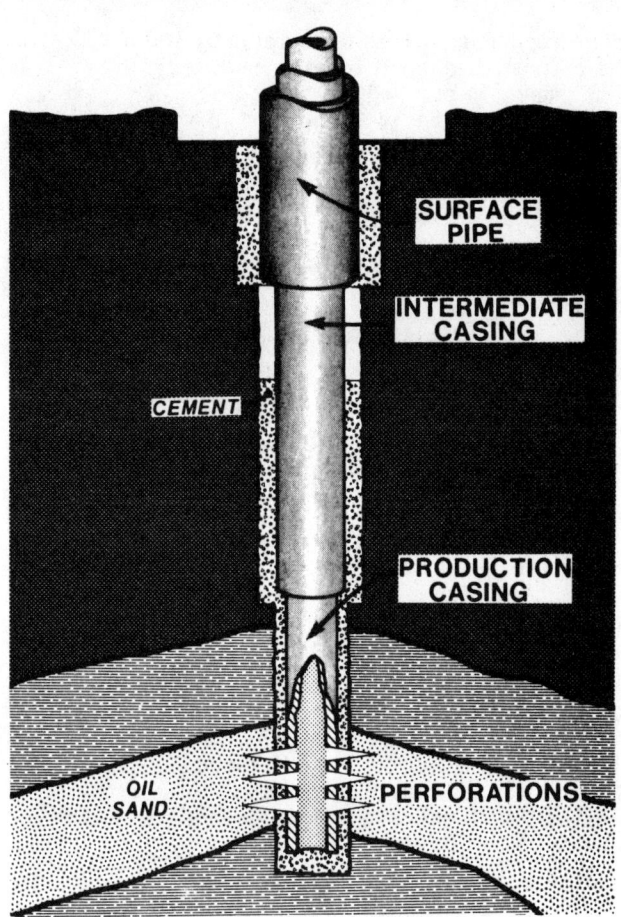

Figure 5.13. CASING STRINGS CEMENTED IN THE HOLE.

the well site. The site is marked with a buoy until the rig can be set up over the spot.

The location must then be prepared for the drilling rig. On land, this usually involves earth work: the land is leveled, earthen pits are excavated and lined with plastic to serve as reserve pits, and an access road is constructed so equipment can be brought to the site. If the location is in a swamp, it is often necessary to dredge out a canal into the area where the well is being drilled. In marshy or soft ground, board roads have to be laid and a board location made.

The drilling rig is then brought in and "rigged up." This can take anywhere from several hours to weeks, depending on how complicated the drilling equipment is. The **substructure**, which supports the mast, is assembled first; then the mast is brought in and raised on top of it (Figure 5.12). Other rigging-up operations include erecting or setting up stairways, walkways, guardrails, storage facilities, living quarters, and auxiliary equipment. Since water is necessary to a drilling operation, a water well has to be dug or a water supply line installed. Offshore, the rig is usually brought in by tugboat and anchored in place. The well is now ready to be "spudded," a term referring to the process of starting the hole.

First, **surface pipe** is set. A shallow hole is drilled into the ground, often several hundred feet deep. A string of surface pipe or **casing** is inserted into the hole and cemented in place (Figure 5.13). The pipe is usually in 30- to 40-foot lengths and is screwed together as it is run into the hole. The surface pipe is 10 to 20 inches in

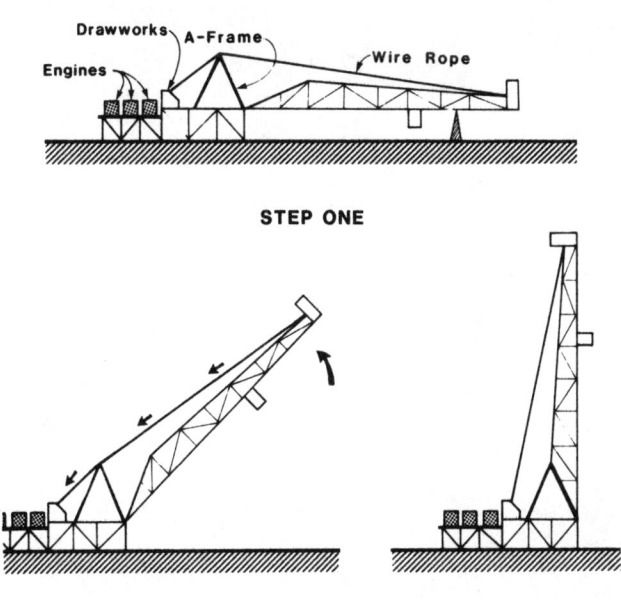

Figure 5.12. TYPICAL MAST RAISING OPERATION. The mast is raised using the rig's power system, the drawworks, and a wire rope line.

diameter, which allows the drill string and bit to pass through it for deeper drilling. The cement is pumped down the inside of the pipe, followed by a **plug** used to wipe the cement from the inside of the casing (Figure 5.14). Drilling mud is pumped in on top of the plug to displace the cement to bottom and out into the annular space between the casing and the wellbore. Once the cement has set, drilling operations are ready to resume.

On offshore locations, a **drive pipe** is set into the soft sea floor before drilling starts. This is a large pipe which is pounded into the mud and soft formations of the ocean floor. The drill bit is run through this drive pipe to begin the drilling.

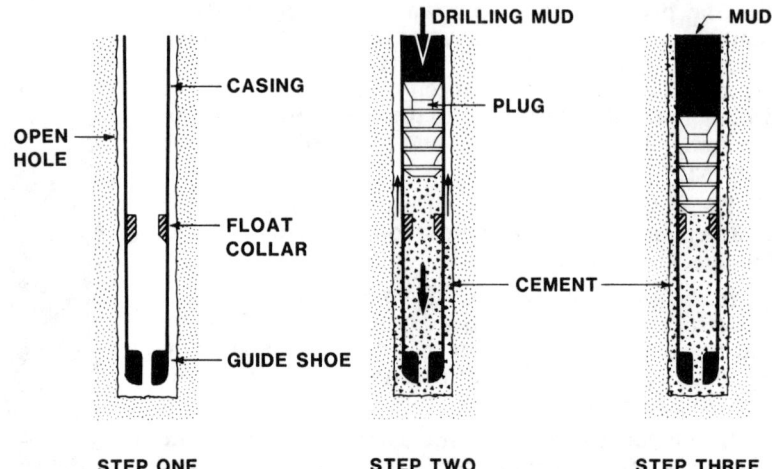

Figure 5.14. **CEMENTING CASING.** Step one: casing is lowered into the hole. Step two: cement and wiper plug are pumped down the casing. Step three: plug and cement are in place.

After the surface pipe has been set, deeper drilling begins. A smaller bit is run down the inside of the casing and drills through the plug and the **guide shoe** at the bottom of the casing. Routine drilling then continues to the desired depth.

Periodically, the drill bit wears out, and the entire drill string and bit must be pulled from the well so it can be replaced. This procedure is called "tripping" the pipe or "round tripping" to change bits. Several round trips may be made in the course of drilling a well, either to replace the bit or to run tests on the well.

As will be discussed in a later section, special problems are often encountered, such as lost circulation or the expectation of drilling into a high pressure zone. When this happens, it is sometimes necessary to set **intermediate casing** strings. The intermediate casing is lowered down through the inside of the surface pipe and cemented in place (Figure 5.13). Multiple strings of casing may be necessary before the well reaches the planned total depth. Each subsequent string of casing is placed inside the previous one and cemented in place in the same manner.

Upon reaching the desired depth, or TD, the well is evaluated to determine whether or not it has located an oil or gas formation. The hole is analyzed with electric logs and other geological evaluation techniques discussed in the chapter on Exploration. If the hole is found to be a potential producer, the final string of casing, called the **production casing,** is run into the well (Figure 5.13). This casing is cemented in place in a manner similar to that used for all previous casing strings. The production casing is the final casing, making the well a permanent vehicle for the transmission of oil or gas to the surface.

At this point, the drilling rig is usually rigged down and removed from the drill site. A smaller rig, called a **workover** or **completion rig,** is moved in to complete the well (Figure 5.15). This rig is less expensive to operate, and is considerably smaller and more portable than the standard drilling rig. These smaller workover rigs are used for lightweight work done in the casing, such as perforation and well stimulation (these will be described later). Occasionally, the large drilling rig may be used for completion operations.

C. Drilling Personnel

A number of people are required to run the machinery and oversee the work in a typical drilling

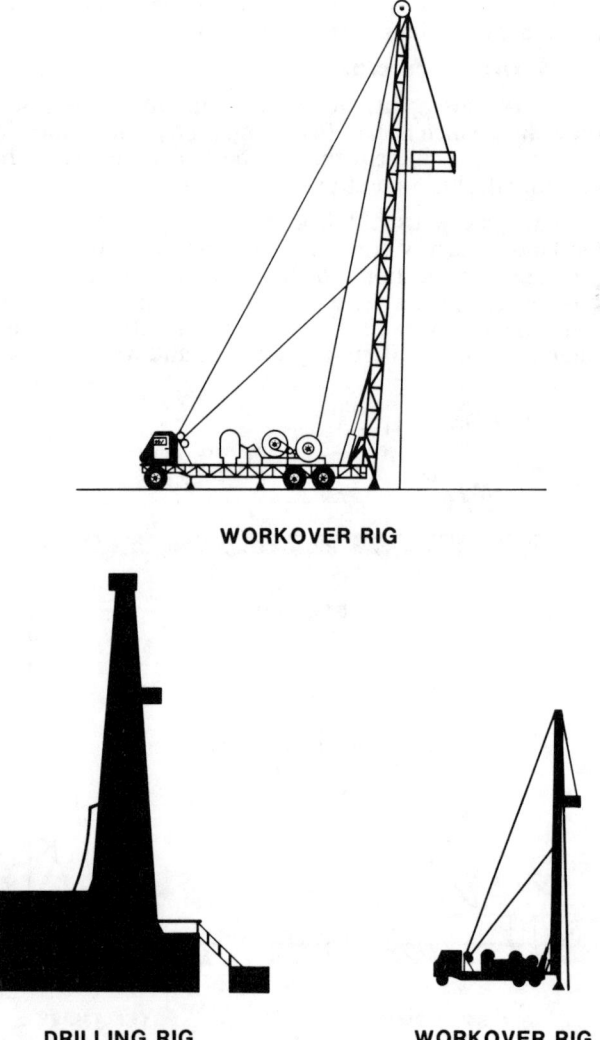

Figure 5.15. **WORKOVER RIG.** The relative sizes of a workover rig and a standard drilling rig are shown by the silhouetted figures.

operation. Many specialized jobs have developed in this part of the industry, and each individual's job title is characteristic of the work he or she is performing. Some of the more common members of a drilling crew and their job descriptions are listed below.

1. **Drilling Superintendent**—Overall manager of field drilling operations. Supervises the activities of several drilling rigs and personnel working in a given area.
2. **Toolpusher**—Second in command or often synonymous with Drilling Superintendent, depending on the size of the company. Direct field supervisor of drilling operations, generally responsible for one or more rigs to make sure that the proper materials and personnel are available to accomplish necessary jobs.
3. **Driller**—The employee in charge of the rig and crew for a given tour (shift of duty) and who primarily operates the drilling equipment.
4. **Roughneck**—Sometimes called the **rotary helper**, this worker is part of the crew who assist the driller in "floor work," i.e., jobs on the rig floor, including making connections, pulling pipe, and rig maintenance and cleanup.
5. **Derrickman**—The crew member whose work station is the pipe rack in the uppermost part of the derrick (also known as the **monkeyboard**) when drill pipe is being raised from or lowered into the hole.
6. **Mud Engineer**—The person responsible for maintaining the proper chemical composition of the drilling mud.

"Oilpatch people" have a colorful history, as described in several books listed in the bibliography after Chapter One. Early drillers, roughnecks, and others were often called "vagabonds" because of the mobile nature of the drilling business. The jobs still necessitate frequent change of location, but regional drilling operations have helped decrease some of the travel time required of drilling personnel. Historically, most field jobs have been held by men, but some women are now found on American drilling crews. In China, drilling crews are often made up entirely of women.

V. TYPES OF DRILLING RIGS

Various types of rotary drilling rigs have been developed to accommodate the different onshore and offshore areas which are currently being explored for petroleum. These different types and their specific uses are described briefly in this section.

A. Onshore Rigs

Today, the most common land rig utilizes the **cantilever mast,** shown earlier in Figure 5.12. This structure has the advantage of easy and quick assembly. The normal process of "rigging up," or assembling the rig at the drilling site, involves transporting the mast in sections to the site, assembling these sections on the ground into the complete mast, and raising the entire structure as one piece into the vertical position. This operation is achieved by using the cable and drawworks on the rig (Figure 5.12). Because this rig can be easily transported and used again, there is a significant savings in time and material, when compared to the older method of constructing each new derrick from the ground up. During the "boom town" era, rig building crews flourished, and their handiwork remains intact in some of the wooden and steel derricks still standing over old wells, especially in east Texas.

B. Offshore Rigs

The search for petroleum began moving offshore after the end of World War II, and new types of rigs were needed for this novel environment. The sea is much more costly to drill in and requires more safety measures. Several different types of rigs have come into use for offshore work. They may be generally classified as **floating rigs, bottom-supported rigs,** and **platform rigs.**

1. Floating Rigs
a. Drill Ships

This rig is used in deep water drilling and is floated to the desired location. Drill ships are very portable because they do not have any fixed base which must be assembled at the drilling site. They are used almost exclusively for drilling exploratory wells.

Drill ships are often constructed from regular ocean-going vessels, which are then fitted with drilling equipment. They are generally self-propelled and use the original propulsion system of the ship. A comparison of drill ships with other offshore vessels is shown in Figure 5.16.

Two important modifications, which differ from land-based drilling operations, are made on the drill ship. The blowout preventers are mounted on the sea floor instead of directly beneath the rig floor, as on a land rig (Figure 5.17). The preventers are attached by underwater divers or by a remote-controlled coupling system guided by underwater television cameras to a piece of pipe, which has been cemented into the ocean floor. The preventers are positioned on the sea floor to enable the drill ship to abandon the location during a storm. They can be left behind to seal the

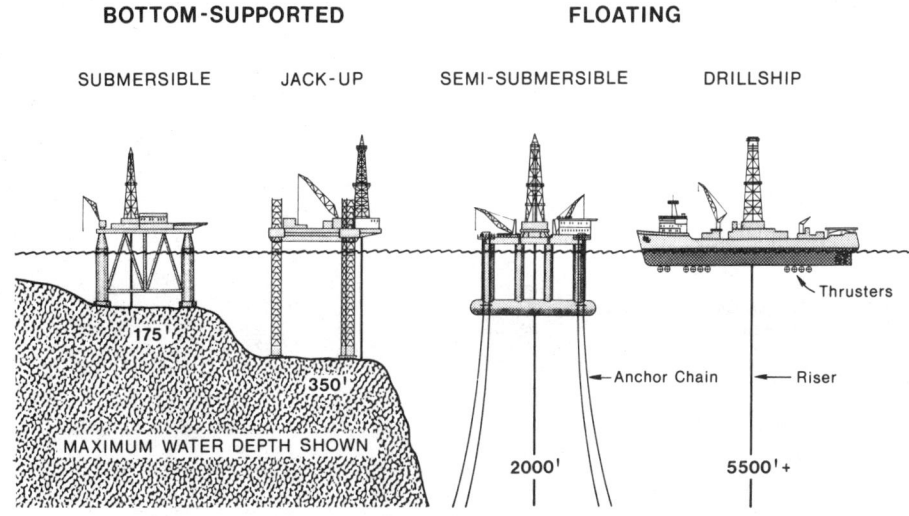

Figure 5.16. TYPES OF OFFSHORE DRILLING UNITS.

c. Semisubmersible Rigs

The semisubmersible rig is another type of floating rig (Figures 5.16 and 5.18). This rig contains a large substructure, which can be partially flooded; the lower portion will remain below the waves and surface turbulence of the sea, providing a massive and stable platform for the drilling rig. Like the drill ship, the semisubmersible rig is often used in deep water for drilling exploratory wells. It also requires anchors to steady it over the hole. In very deep waters, **propellors** or **thrusters** are sometimes mounted on the sides of the substructure and are used to take the place of the anchors. If the platform begins to drift to one side of the well, a computer-controlled location monitor is used to activate the appropriate propellor or thruster and push the platform back over the well. As with drill ships, a riser is used to connect the floating platform with the ocean floor.

The semisubmersible rig must generally be towed to and from a drill site by sea-going tugs. Semisubmersibles require more time than drill ships to move from one location to another; however, they are also more stable than drill ships and permit more efficient operations in areas where storms and rough seas are common.

2. Bottom-Supported Rigs

a. Jack-Up Rigs

The jack-up rig includes a fixed platform, which is bottom-supported by huge legs lowered to the sea floor (Figures 5.16 and 5.19). When the jack-up rig is to be moved, its legs are raised, allowing the rig to float on its hull. It can then be towed to a new location by tugboats. The jack-up rig is generally used to drill exploratory wells in shallow water.

b. Submersible Rigs

Another bottom-supported rig type, the submersible, also provides a fixed base for drilling operations (Figure 5.16). A large substructure beneath the rig allows the rig to float, so it can be towed to different locations. At the site, the substructure is

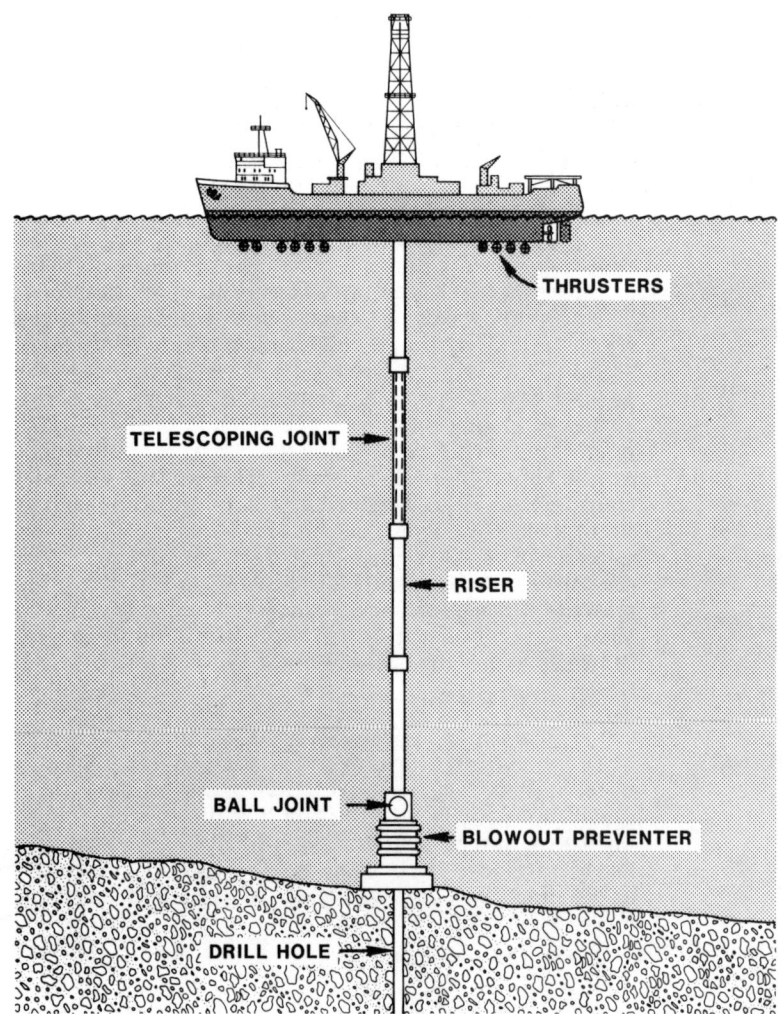

Figure 5.17. OFFSHORE DRILL SHIP.

well and prevent an uncontrolled discharge of fluid into the ocean.

The second major difference with drill ships is the **riser** (Figure 5.17). The riser is a large, heavy steel pipe used to connect the blowout preventers with the drill ship. Risers are typically two or three feet in diameter and large enough to allow the drill pipe and bit to pass through them. With the riser in place, drilling operations can be conducted as if there were a continuous wellbore from top to bottom. The drilling mud is pumped down the inside of the drill pipe and returns through the annular space between the drill pipe and the riser. Control lines are run down the outer side of the riser from the surface, and are used to actuate the blowout preventers.

The drill ship is held in place over the well by thrusters or anchors. Some motion of the ship can be tolerated, since the drill pipe is flexible. The riser often includes a **telescoping joint** to allow for vertical motion, together with a **ball joint** at the sea floor to allow for lateral motion. The strength and flexibility of the riser is the limiting factor in deep water drilling.

b. Drilling Barges

The drilling barge is similar to the drill ship, but it is not self-propelled and must be towed to the drilling site. Drilling barges are generally used in shallow coastal waters or inland swamp locations, whereas the drill ship is used in deep water areas.

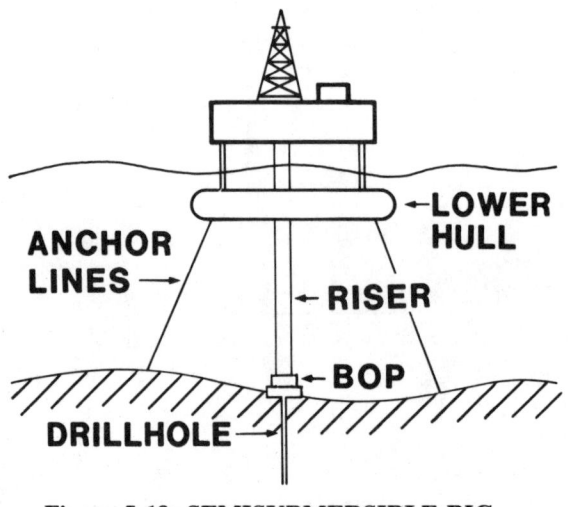

Figure 5.18. SEMISUBMERSIBLE RIG.

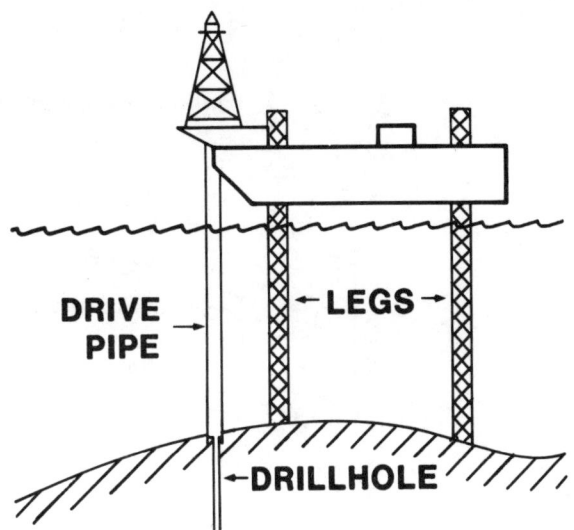

Figure 5.19. JACK-UP RIG.

flooded with water, causing it to rest on bottom and providing the fixed base. To remove the rig, the water is pumped from the substructure and the floating platform is towed away. The submersible rig is used primarily in shallow waters of depths less than 150 feet.

Today, the most common submersible rigs are barges, which are floated to the drill site and then flooded, causing them to rest on bottom. These rigs are commonly used in shallow swamp areas.

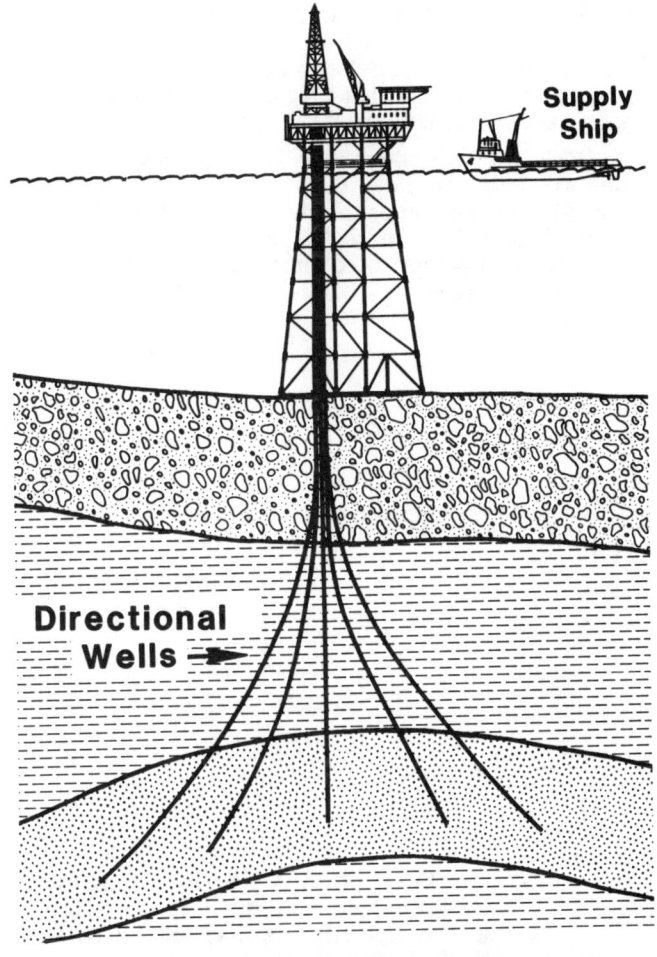

Figure 5.20. FIXED PLATFORM OFFSHORE RIG.

TABLE 5.1. OFFSHORE RIGS

Type	Common Use	Capacity	Common Water Depth Limits
Floating Rigs			
Drill ships	Exploratory Drilling	Deep waters	5,000 feet
Drilling Barges	Exploratory and development drilling	Shallow waters, swamps	
Semisubmersibles	Exploratory Drilling	Deep waters	2,000 feet
Bottom-Supported Rigs			
Jack-up Rigs	Exploratory Drilling	Shallow waters	350 feet
Submersibles	Exploratory and development drilling	Swamps	
Platform Rigs	Development drilling and production operations	Shallow or deep waters	1,000 feet

3. Platform Rigs

Drilling rigs are also commonly mounted on fixed, permanent platforms, where mobility is not required. As shown in Figure 5.20, these platform rigs are used when drilling a number of wells from the same location, usually to develop a newly discovered field. These platforms are generally constructed onshore, floated to the drill site, and then flooded and sunk in an upright position. Permanent or fixed platforms are also used for production operations as well as drilling. (This will be described in the chapter on Production.) The drilling rigs are sometimes removed from the platform after all wells have been drilled.

The function and capacities of the various types of offshore rigs are summarized in Table 5.1.

VI. SPECIAL DRILLING OPERATIONS

A number of special drilling operations are used when unusual problems occur in rotary drilling. The process of drilling a well is seldom completed without at least a few problems. These normally involve mechanical difficulties with the drilling equipment or trouble encountered in the formation being drilled.

A. Lost Circulation

Frequently, the formations being drilled cannot withstand the heavy weight of the drilling mud in the hole. As a result, mud escapes or is lost from the borehole into a porous, low pressure formation. This is particularly true of highly permeable, fractured formation rock (Figure 5.21). The result is that the amount of drilling fluid returning from the hole is less than the amount being pumped into it. The problem is remedied by "plugging up" the permeable zone, often called a **thief zone,** with "lost circulation material," which usually consists of a solid plugging element added to the drilling mud. Common plugging materials include cottonseed hulls, ground seashells, cellophane, asphalt, wood fiber, and pulverized walnut shells.

B. High Pressure Zones

The most catastrophic problem encountered in drilling is a **blowout.** When a high pressure zone is found in a well, the well is said to have taken a "kick," referring to the attempt by the high pressure fluid to flow into the wellbore. The situation is called a blowout

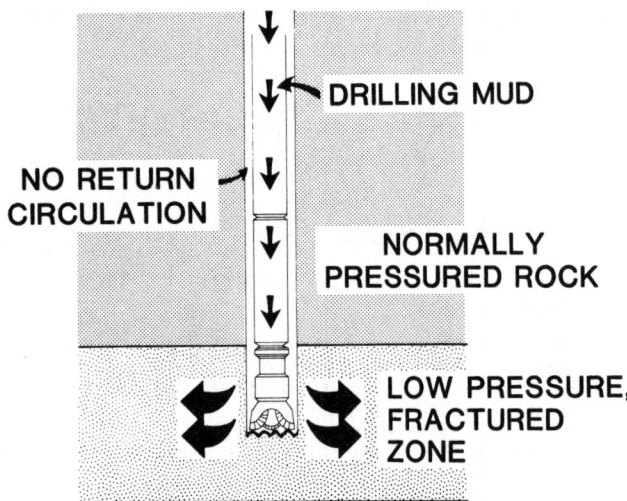

Figure 5.21. LOST CIRCULATION.

only if the "kick" goes unchecked, and the formation fluids escape uncontrolled to the surface. A blowout can be very serious, because it often involves fire and sometimes great loss of life and property.

Figure 5.22 shows how a blowout can occur under circumstances that are actually opposite those of lost circulation. A formation of abnormally high pressure is encountered while drilling the well, causing the flow of fluid from the formation into the wellbore. This

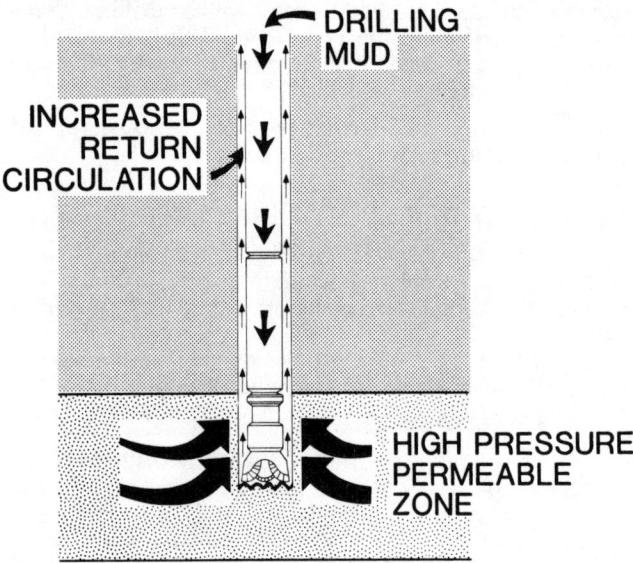

Figure 5.22. BLOWOUT.

situation is caused by insufficient pressure from the drilling mud to counterbalance the formation pressure. Well control difficulties are identified at the surface by an increased flow of mud from the well.

When this occurs, drilling operations are stopped and the well is shut in with the **blowout preventers** (Figure 5.23). These are mounted at ground level just beneath the rig floor. Closing the preventers stops the flow of fluid from the well by sealing off the annular space between the drill pipe and wellbore. Note that the blowout preventer (or BOP stack) consists of three different types of seals (Figure 5.23). The **annular**

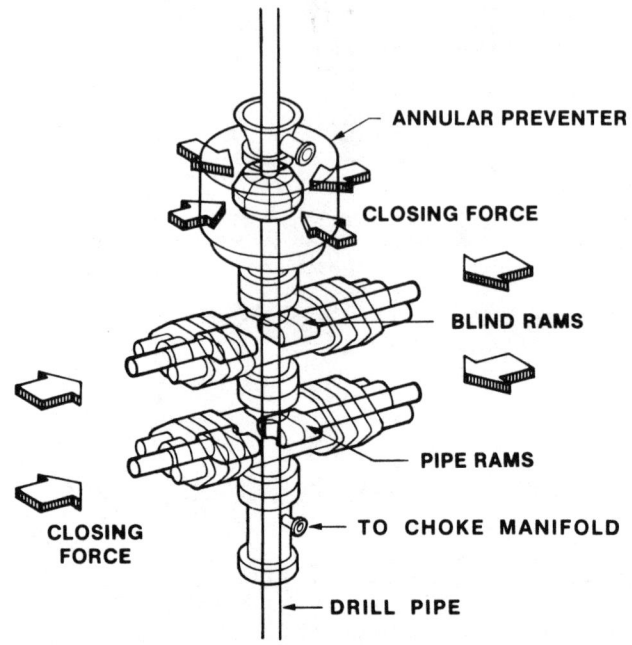

Figure 5.23. BLOWOUT PREVENTER.

preventer, located on the top of the stack, is often the first seal actuated when a high pressure flow of fluid from the well is discovered. It is an expandable, steel-reinforced rubber seal which closes off the annular space between the wellbore and drill pipe. If this seal is not sufficient, the **pipe rams** are then closed. These consist of a steel gate made to fit tightly around the drill pipe, sealing off the annular space. The **blind rams** are a similar device, except that they completely seal off the hole in the event that there is no pipe in the hole.

Once the preventers are closed, heavy mud is pumped into the hole through the drill pipe and circulated by the pumps through the **choke line** (Figure 5.24). Weighting material is added until the mud pressure is sufficient to counterbalance the formation pressure and the entry of fluid into the wellbore is stopped.

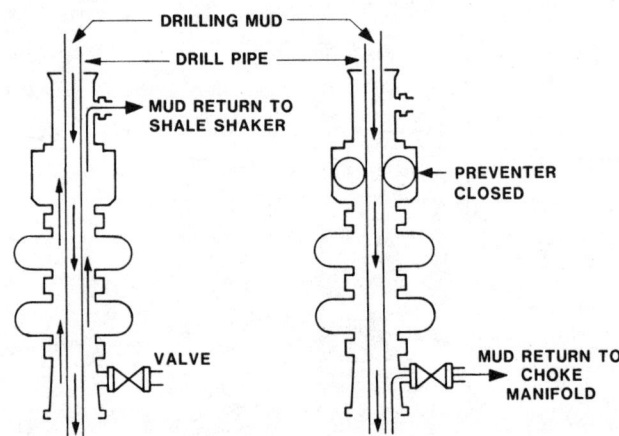

Figure 5.24. CONTROLLING A "KICK." Left: normal drilling operations, BOP open. Right: drilling operations suspended to control "kick," BOP closed.

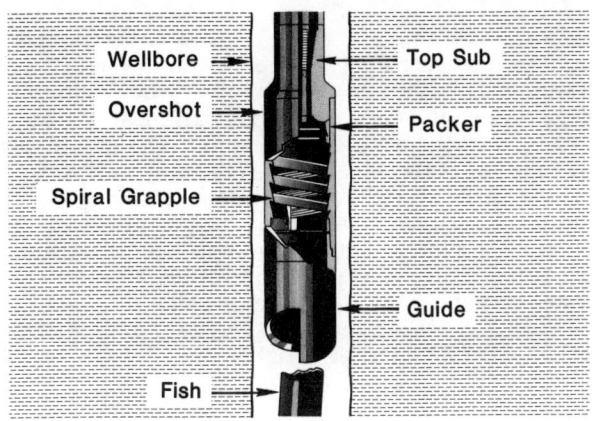

Figure 5.25. OVERSHOT FISHING TOOL.

C. Fishing

Fishing refers to the attempt to recover tools or pipe lost in the hole during drilling operations. Anything lost in the hole is known as a "fish." The drill pipe or drill collars are the most common type of fish lost in the hole, because they occasionally break during drilling. Figure 5.25 illustrates one type of fishing tool used to recover lost drill pipe. The **grapple** shown in the diagram fits over the fish, engaging it so it can be retrieved. Many different types of fishing tools have been designed to aid in the recovery of junk or fish lost in the hole.

D. Coring

Coring is a special operation used to recover samples of solid rock from formations of interest to the geologist or petroleum engineer. (This procedure is also explained and illustrated in the section on subsurface geology in Chapter Three.) A donut-shaped bit, called a **core bit** or **core head** (Figure 5.26), attached to the bottom of the drill string, is used to cut a cylindrical rock sample from the formation. This sample is held inside the drill string in a **core barrel.** When a sufficient amount of core has been cut, the core assembly is raised, causing the rock to break off and leaving the core trapped inside the core barrel. Core samples can be of great value in determining the type of rock in a formation, as well as the type of fluids the rock contains.

E. Drill-Stem Testing (DST)

In drill-stem testing, the drill string itself is used as a flow tube to test a formation for oil or gas content. This test is used if a zone with questionable potential for oil or gas production is encountered. The test is run by pulling the bit out of the hole and placing the test tools on the end of the drill string. These DST tools are then run back to the bottom of the hole to test the zone of interest. The **packer,** an expandable rubber sleeve on the outside of the drill pipe, is expanded (usually by putting weight on the packer) and forms a seal against the side of the hole. By sealing off the formation from the mud in the hole, the formation fluids flow into the wellbore and up the inside of the drill pipe. A sample of the formation fluids can be taken, and producing rates and formation pressures can be measured. (Refer to the chapter on Exploration for a more detailed discussion of DSTs.)

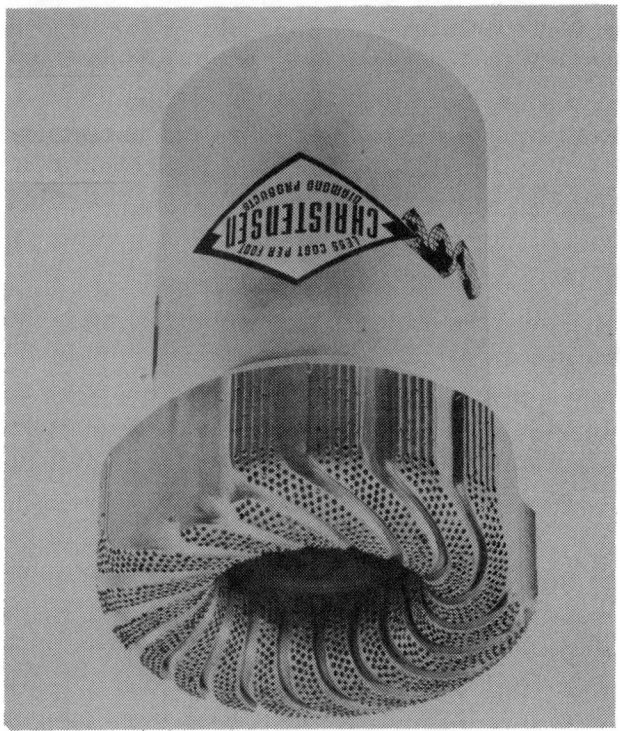

Figure 5.26. CORE BIT.

F. Directional Drilling

It is often difficult to place a drilling rig directly over the spot where a well should be drilled. This is particularly true offshore, where a number of wells must be drilled from the same location (Figure 5.20), or in an

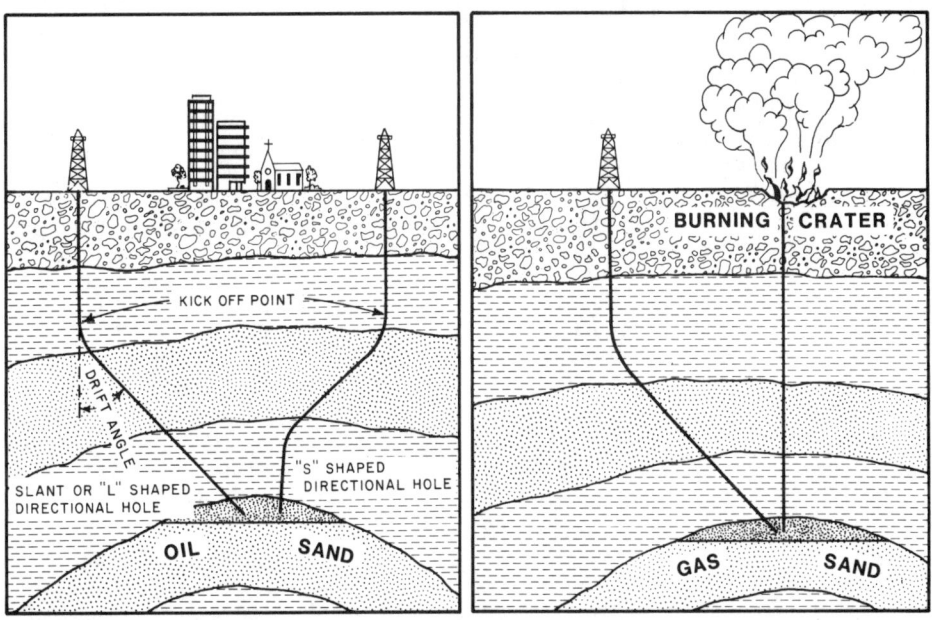

Figure 5.27. DIRECTIONAL DRILLING USES. Left: drilling under a populated area. Right: drilling a relief well to control a blowout.

area where a rig cannot be set up, such as a swamp or a heavily populated area (Figure 5.27). In these instances, wells are dug directionally from the surface location to the desired point underground. Directional drilling is occasionally used to reach the bottom of a well that has blown out. By using this procedure, the formation can be plugged with mud pumped into the second well (Figure 5.27).

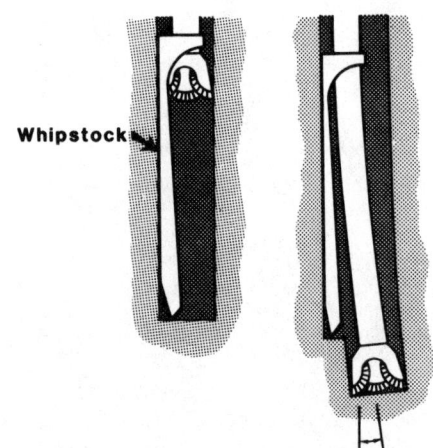

Figure 5.28. WHIPSTOCK METHOD FOR DIRECTIONAL DRILLING.

Through the experience gained over the years, the techniques for directional drilling have become increasingly more sophisticated. The earliest method used was **whipstocking,** which involves placing a wedge-shaped piece of steel at the bottom of the hole to force the bit off at a desired angle (Figure 5.28). Progressively greater angles can be built by placing additional whipstocks in the hole as it is drilled deeper.

A more recent development in directional drilling has been the use of **downhole motors** (Figure 5.29). A **bent sub,** used to start the hole off at an angle, has a motor located beneath it. The **turbine drill** is one type of downhole motor, which contains blades that are rotated by the drilling fluid as it is pumped past them, similar to the action of the wind on a windmill. The motor then turns the bit. In this manner, only the drill bit is rotated, not the drill string. Positive-displacement downhole mud-powered motors are more commonly used and operate in a manner similar to the turbine bit. Again, a bent sub is used, and the positive-displacement motor, located just above the bit, rotates the bit.

VII. COMPLETION METHODS

After the well is drilled, it must be completed and prepared for production. The term, "completion methods," refers to those processes whereby the well is prepared to produce oil or gas. A number of different types of completion methods are used today. The more common methods are described below.

A. Cased Hole Completions

The most common type of completion today is the cased hole completion. As illustrated earlier in Figures 5.13 and 5.14, the final string of casing, called the production casing or the **long string,** is inserted through the production zone and cemented. The casing serves several purposes: it keeps the formation from caving into the wellbore; it provides a permanent vehicle for the transmission of oil or gas to the surface; and it isolates the producing zone from the other zones in the well. The latter prevents foreign fluids of other zones from migrating into the producing formation.

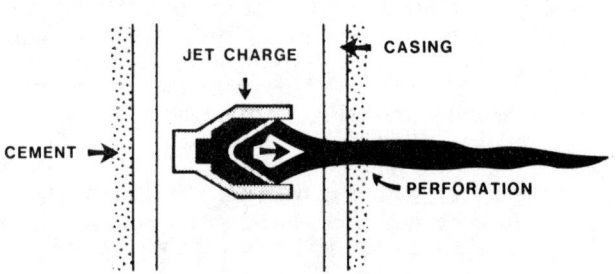

Figure 5.30. JET-PERFORATING.

When the casing is set and cemented in place, it is full of mud from the drilling and cementing processes. This mud is usually replaced with a non-damaging fluid (oil, water, or a better grade of mud), and the well is then **perforated.** To perforate a well, a device is lowered on a wireline into the well and used to shoot holes through the steel casing and cement into the formation. Figure 5.30 illustrates the technique known as **jet-perforating.** A small high-explosive charge, similar to the bazooka in World War II, is lowered into the well and used to shoot a series of small holes into the formation. These holes or perforations are shot into the casing adjacent to that part of the formation containing the oil or gas. The perforations allow the oil and gas to flow through the steel casing, into the wellbore, and up to the surface.

B. Open Hole Completions

This less common method is often used when setting a permanent casing string through the producing zone is not necessary. An open hole completion involves drilling down to the top of the producing zone, setting

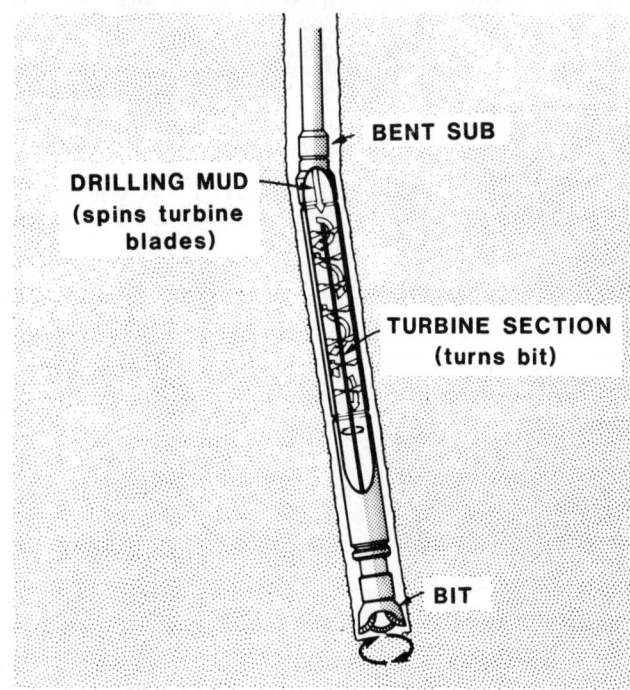

Figure 5.29. DIRECTIONAL DRILLING USING A TURBINE DRILL.

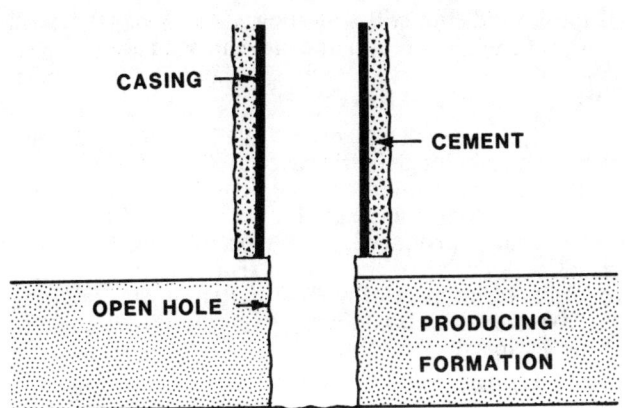

Figure 5.31. OPEN HOLE COMPLETION.

casing, and cementing it in place just above the pay zone (Figure 5.31). The well is drilled into the producing zone with a smaller bit, and an open hole or uncased section is left for the oil and gas to flow into the well. This technique can only be used in good, competent formation rock, or excessive cave-ins will result. An open hole completion is often used to prevent mud and cement damage to the producing zone. Although this method is not used to a great extent today, many older wells were completed in this manner, while some new wells still use this technique to prevent damage to the producing zone.

C. Slotted Liner Completions

This completion method is common in certain parts of the country. As with open hole completions, the well is drilled to the top of the producing zone, then casing is set and cemented in place. After the well is drilled into the pay zone, a **slotted liner** is run into the

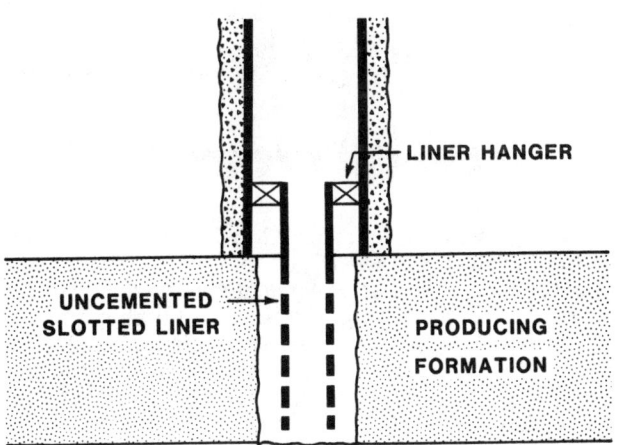

Figure 5.32. SLOTTED LINER COMPLETION.

well and connected to the production casing by a **liner hanger** (Figure 5.32). The slotted liner is used to keep the rock from caving in and closing off the bottom of the well. The technique is not very successful in soft rock, which tends to break up into fine particles that can pass through the slots. However, it is useful in certain areas, because it allows an entire section of rock to produce into the well without having to case and cement the zone.

D. Completions Requiring Sand Control

In areas such as the west coast of the United States and the Gulf Coast, formations often are very soft, causing loose formation material to be carried into the well with the oil and gas. These types of formations are often called **unconsolidated sands**. Problems can result, because the sand which flows into the well with the oil and gas plugs the borehole and can cause problems at the surface. Several techniques can help prevent or control the production of sand from such wells. **Gravel packing** and **sand consolidation** are two of the most common.

One method of **gravel packing** is shown in Figure 5.33. This technique consists in drilling through the unconsolidated formation, setting and cementing the casing, and perforating the casing, as in normal cased hole completions. Some of the soft formation behind the perforations is washed out to form a cavity behind

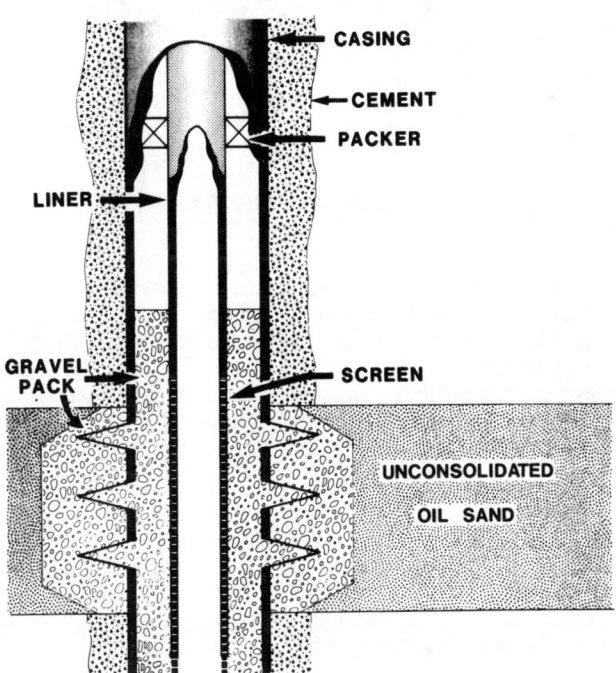

Figure 5.33. SAND CONTROL BY GRAVEL PACKING.

the casing. A gravel pack is then placed in the well. This usually consists of a coarse sand, similar to that used on construction sites, which is pumped down the casing and out into the cavity behind the perforations. It acts as a filter to help keep the unconsolidated sand in the producing formation away from the wellbore. A screen and liner is set down into the gravel pack, when the gravel is placed in the well. The openings in the screen are small enough to prevent the gravel pack from coming through, but the oil and gas can flow through the gravel pack and the screen into the well. Because the gravel pack is held in place by the screen, and the producing formation is held in place by the gravel pack, this method filters out the sand and allows the oil or gas to be produced.

Figure 5.34 illustrates another gravel packing technique which does not require setting casing all the way through the producing zone. The producing interval is penetrated with a smaller bit. After the new open hole has been cleaned or conditioned, a screened liner and packer, often called a liner hanger, are lowered to the bottom on a string of **tubing**. A gravel slightly larger than the grain size of the producing formation is pumped down the tubing, through a back-pressure valve

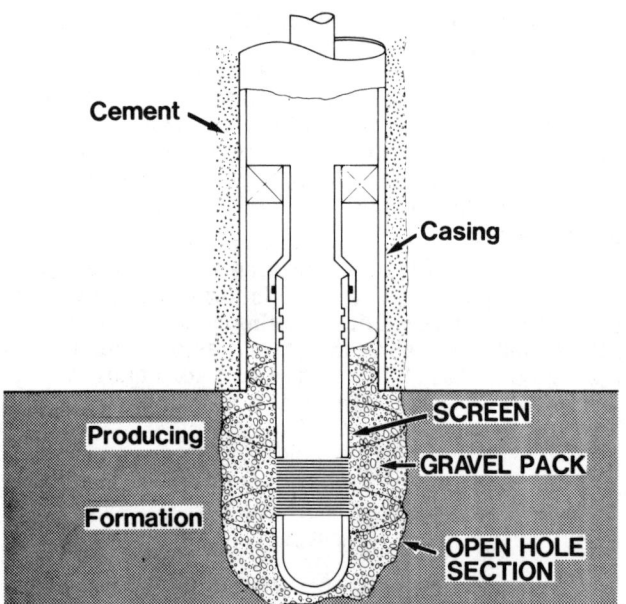

Figure 5.34. OPEN HOLE GRAVEL PACK.

located at the end of the liner assembly, and up between the open hole and the outside of the liner. When the gravel is in place, the packer is set and the tubing is disengaged from the liner assembly. The gravel pack prevents the plugging or bridging of the screened liner by loose or unconsolidated formation sands and assures maximum well productivity.

Another method for controlling loose sand formations, by means of **sand consolidation,** is shown in Figure 5.35. The loose sand in the formation is held together by an epoxy resin or chemical bonding agent, which has been pumped down the well, through the casing perforations, and into the producing sand. If properly applied, this epoxy resin will prevent the flow of loose sand, but will still allow flow passages for oil and gas through the resin and into the wellbore.

VIII. WELL STIMULATION

Once the well has been drilled, cased, and completed, the oil and gas may have sufficient force to flow to the surface without further assistance. However, many wells require some additional treatment before oil and gas can be produced. Simply perforating the well or installing a liner is often insufficient. Two of the most common techniques of well stimulation are **acidizing** and **fracture treating.** Each technique is designed to remedy a specific problem.

A. Acidizing

A formation containing oil and gas sometimes becomes contaminated with foreign material during the drilling process, or is already contaminated from indigenous material, such as calcium carbonate or scale, contained in the rock. These contaminants frequently restrict the permeability of the well and do not allow the oil and gas to flow freely into the wellbore. Often in such situations, an acid is pumped into the well to help dissolve the impediment.

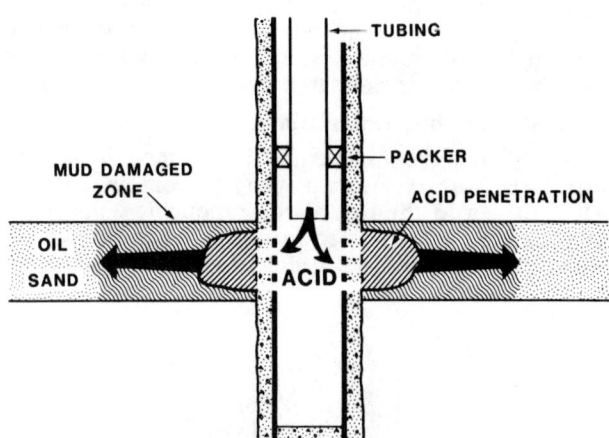

Figure 5.36. ACIDIZING.

In the situation represented in Figure 5.36, drilling mud which has been forced out into the rock has caused a blockage. This restrictive material is dissolved by inserting a smaller string of pipe, called **tubing,** into the cased well. A packer or rubber seal is placed between the tubing and casing, above the perforations. Acid is then pumped down the inside of the tubing, out through the perforations, and into the rock. The packer keeps the acid from going back up the annular space between the tubing and casing. Once the acid is pumped into the rock, it dissolves the impairment, in this case drilling mud, in the rock.

During an acidizing procedure, it is very important to use the right chemicals to prevent any damage to the well. Three common types of acid used in acidizing operations are listed below.

1. Hydrochloric Acid (HCL)

This acid is common in the petroleum industry and is effective in dissolving scale and carbonate material. It is often used for sandstones that contain carbonate material between the sand grains. When this material is dissolved, oil and gas can flow out of the well more freely. Hydrochloric acid is also used to clean old

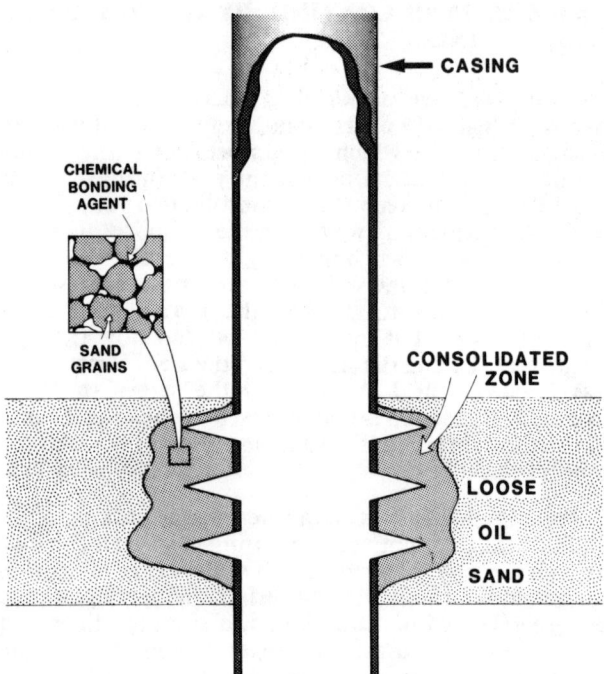

Figure 5.35. SAND CONTROL BY CONSOLIDATION.

wells, which may have formed scale from years of water and oil production.

2. Acetic Acid (H$_{AC}$)

Acetic acid is a slow-reacting acid, which also dissolves scale and carbonate materials. The slow reacting time allows this acid to be pumped farther into the formation before it expends itself dissolving material in the rock. Hydrochloric acid reacts so rapidly that it often penetrates only a short distance into the formation before it is spent. Acetic acid, although it is weaker and not as effective in dissolving scale, can reach further into the rock and provide a deeper cleaning of the rock.

3. Hydrofluoric Acid (HF)

This acid is often mixed with other acids to dissolve clays. It is helpful in dissolving certain components of drilling mud which have infiltrated the producing zone; consequently, it is often referred to as **mud acid.**

B. Fracture Treating

The permeability of the reservoir rock is often so low that it is difficult for the oil and gas to flow into the well. As a result, the rock must be broken open to create a highly permeable channel from the reservoir into the wellbore. Fracture treating is the basic method used to increase the permeability of a rock formation (Figure 5.37).

First, a string of tubing and a packer or rubber seal are inserted into the well. A high pressure fluid is then pumped at high rates into the well by pump trucks connected to the top of the well. This activity causes the rock to be split open.

The **frac fluid** that is pumped into the well is generally a thickened or gelled water containing a **propping agent.** This propping agent, which can be sand, small glass beads, nutshells, or aluminum beads, is carried into the rock as the fluid splits the rock open further from the wellbore. This is illustrated in Figure 5.37 by the small diagram showing the expanded view of a propped fracture. After the frac treatment is completed and all the frac fluid and sand have been pumped into the formation, the high pressure pump trucks are stopped, and the pressure on the well is relieved. If it were not for the propping agent, the fracture in the rock would close. The propping agent holds the fracture open, and the oil and gas can migrate into the fracture, along the fracture into the wellbore, and up to the surface.

It is common to mix nitrogen with the frac water. The nitrogen is pumped into the well with the water and facilitates the flow of water back out of the well once

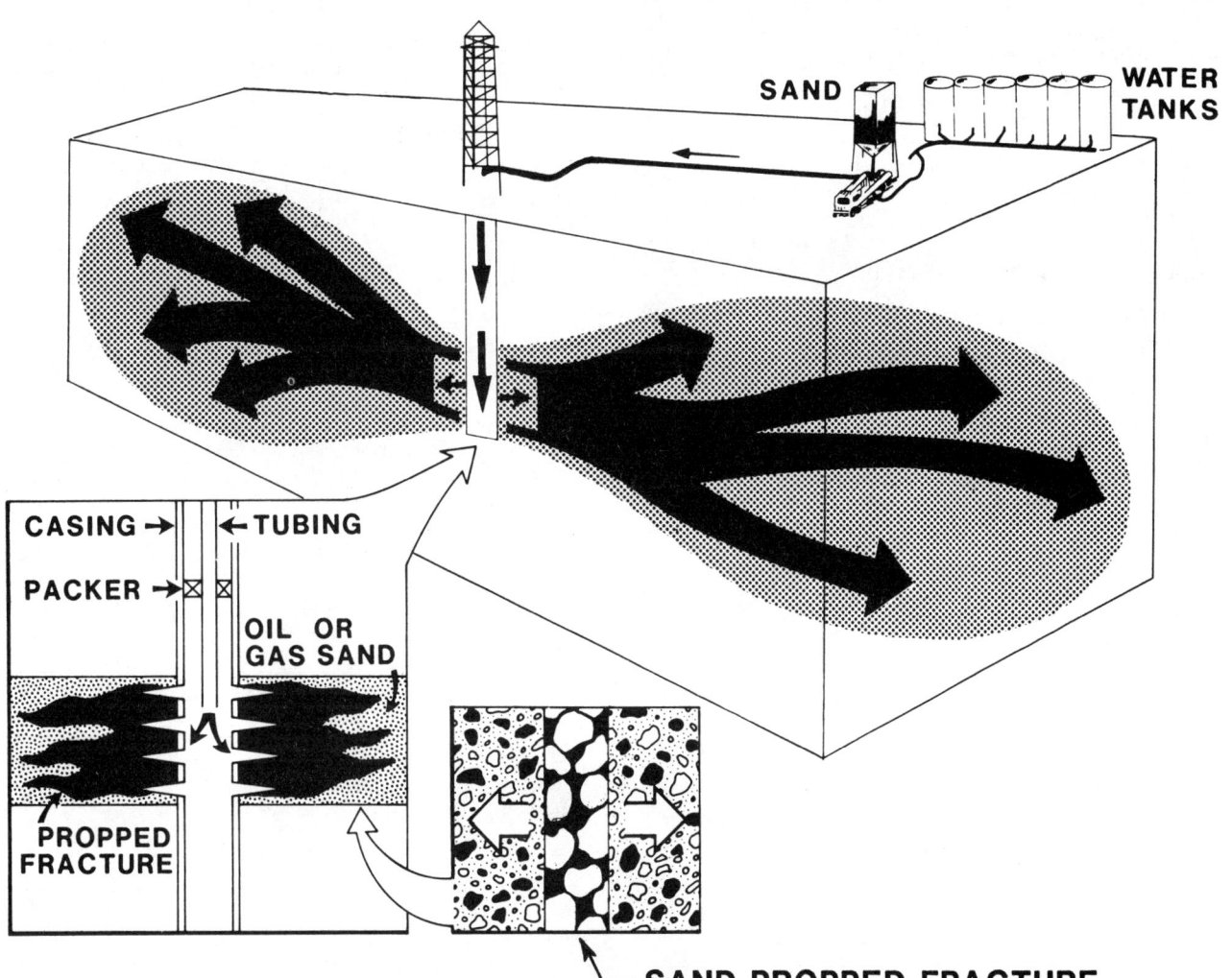

Figure 5.37. FRACTURE TREATING.

the pressure is relieved. This allows rapid removal of the water, which otherwise could be damaging to the zone. Crude oil is occasionally used instead of water for frac fluid. The crude oil is thickened and used to carry the propping agent into the well. The thickened water or oil is necessary to transport the propping agent into the fracture.

IX. DRILLING AND COMPLETION REPORTS

Almost all oil and gas companies develop drilling and completion reports to document information about the historical development of a well or completion operation. They are excellent sources of information and often are widely circulated to update everyone interested in the day-to-day progress of drilling and completion activities. Typically, these reports are developed on the job site by a company's field supervisor. Each morning, the supervisor writes a 24-hour summary report of all the activities at the well site. This report is telephoned or teletyped to the home office, where a copy of it is distributed to all interested parties. After the well has been drilled and completed, each of these daily reports is summarized on a **master drilling report**, which is retained as a permanent record of that well.

The techniques for writing these reports vary greatly from company to company. A sample drilling report has been included to acquaint the reader with the way these reports are written. They are often difficult to read, since many abbreviations are used and very few explanations are included. The reader may refer to the reference at the end of the chapter called **The Desk and Derrick Abbreviator.** This handy reference manual contains almost every known abbreviation used in these drilling and completion reports.

Several common practices used in these reports are worth pointing out. In the example which follows, notice that the current status is noted first, then the 24-hour summary is given. It is possible to misinterpret these reports by reading the current status as the beginning of the 24-hour summary. Therefore, it is often easier to read these reports by starting with the second sentence, reading through the daily activity summary, and then going back and reading the first sentence for the current status.

It is also helpful to draw a sketch or diagram of the activity being discussed in the report. This is particularly true when confusion could arise on complicated procedures. In the following example, which is highly simplified, sketches have been included to illustrate the report. It should provide a good review of the drilling procedures discussed in this chapter.

DRILLING REPORT
RED MESA FEDERAL NO. 1
SECTION 8, TOWNSHIP 4S, RANGE 98W
RIO BLANCO COUNTY, COLORADO

5-16-76   STAKING LOCATION. NEW EXPLORATION WILDCAT. 4000 FT. DOUGLAS CREEK TEST. LOCATION - 1200 FT. FNL, 3000 FEL, SECTION 8, TOWNSHIP 4S, RANGE 98W, RIO BLANCO COUNTY, COLORADO.

5-17-76   PREPARING LOCATION.

5-18-76   MI & RU. DRILLING RIG NO. 56. 6037 FT. GROUND LEVEL BEFORE GRADING. 6034 FT. GROUND LEVEL AFTER GRADING. 6044 FT. KELLY BUSHING.

5-19-76   PREPARING TO DRILL OUT. DRILLED 105' W/10-3/4" BIT. SET 102 FT. OF 8-5/8" OD CASING AND CEMENTED WITH 30 SKS. REGULAR CEMENT. HAD FULL RETURNS, PLUG DOWN AT 1:00 A.M. WOC 16 HRS.

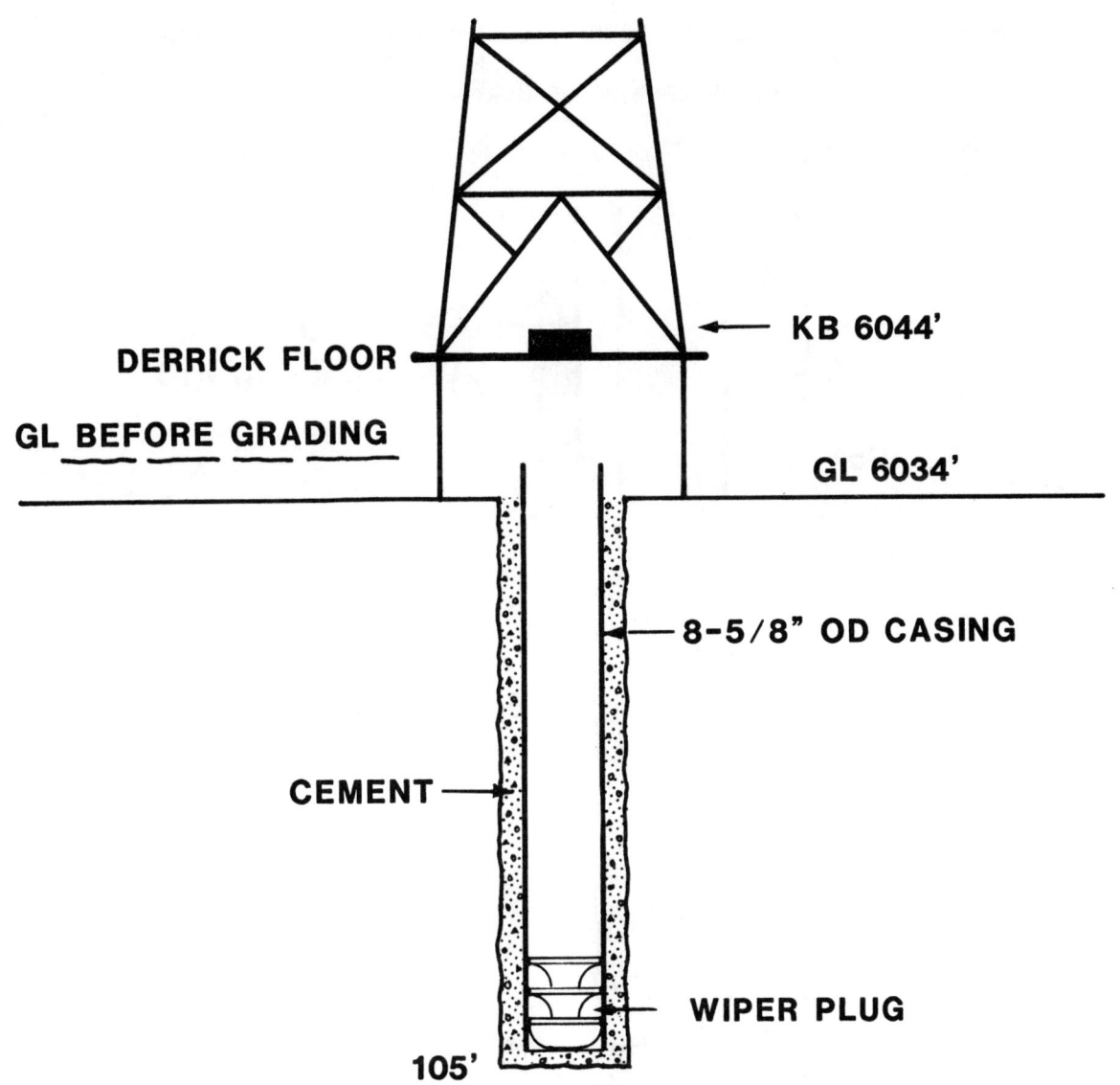

Figure 5.38.

5-20-76	DRLG. 420', SAND AND SHALE. TESTED CSG. TO 600# WITHOUT LOSS. 1/4° @ 102'. DRILLED OUT OF CASING WITH 7-7/8" BIT. MUD WT. 8.5 VISC. 29 FWG.
5-21-76	DRLG. 1385', OIL SHALE. MUD WT. 9.0. VISC. 28.
5-22-76	DRLG. 1785', OIL SHALE. MUD WT. 9.0. VISC. 29.
5-23-76	DRLG. 2190', OIL SHALE. MADE TRIP AT 2120'. 1° @ 2120'. MUD WT. 9.5. VISC. 28.
5-24-76	DRLG. 3215', SILT AND SANDSTONE. LOST 70% RETURNS @ 2500'. NOW DRILLING W/15 TO 50% RETURNS.
5-25-76	DRLG. 3725', SAND. DRLG. W/85% RETURNS. MUD WT. 9.5. VISC. 28.
5-26-76	TD 4025'. 1-1/2° @ 4020'. FINISHED 7-7/8" HOLE @ 1 P.M. CIRCULATED HOLE 3 HOURS. PULLED OUT AND LAID DOWN DRILL PIPE. RU LOGGING TRUCK. RAN CALIPER SURVEY, SONIC LOG AND INDUCTION LOG. FOUND LOGGER'S TD AT 4026'.
5-27-76	SET 5-1/2" OD, PRODUCTION CASING AT 4025' W/1000 SKS., CLASS G CEMENT. HAD FULL RETURNS, PLUG DOWN AT 5:15 P.M. 5-27-76. PREPARING TO MORT.
5-28-76	MORT.

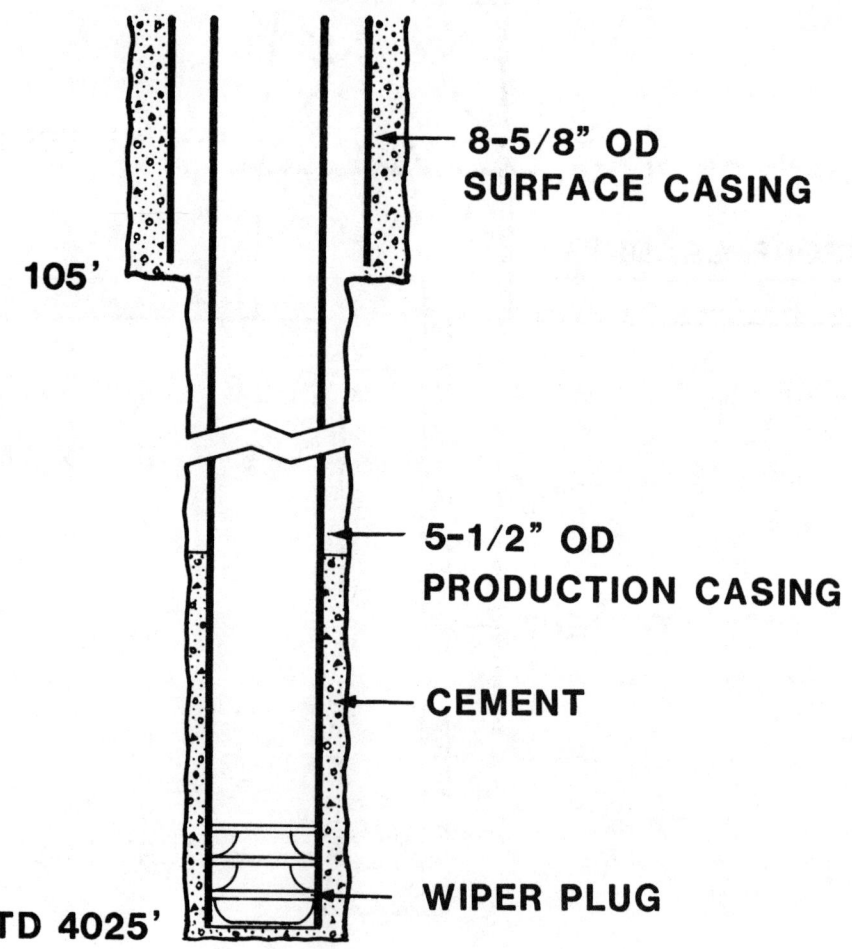

Figure 5.39.

COMPLETION REPORT
RED MESA FEDERAL NO. 1
SECTION 8, TOWNSHIP 4S, RANGE 98W
RIO BLANCO COUNTY, COLORADO

6-5-76   MI & RU. COMPLETION RIG NO. 11.

6-6-76   CIRCULATING HOLE. RUN-IN 2-7/8" TBG AND 4-3/4" BIT. DRILL CEMENT FROM 3825' TO 4010' (SHOE AT 4022') PBTD AT 4010'. CIRCULATE HOLE CLEAN.

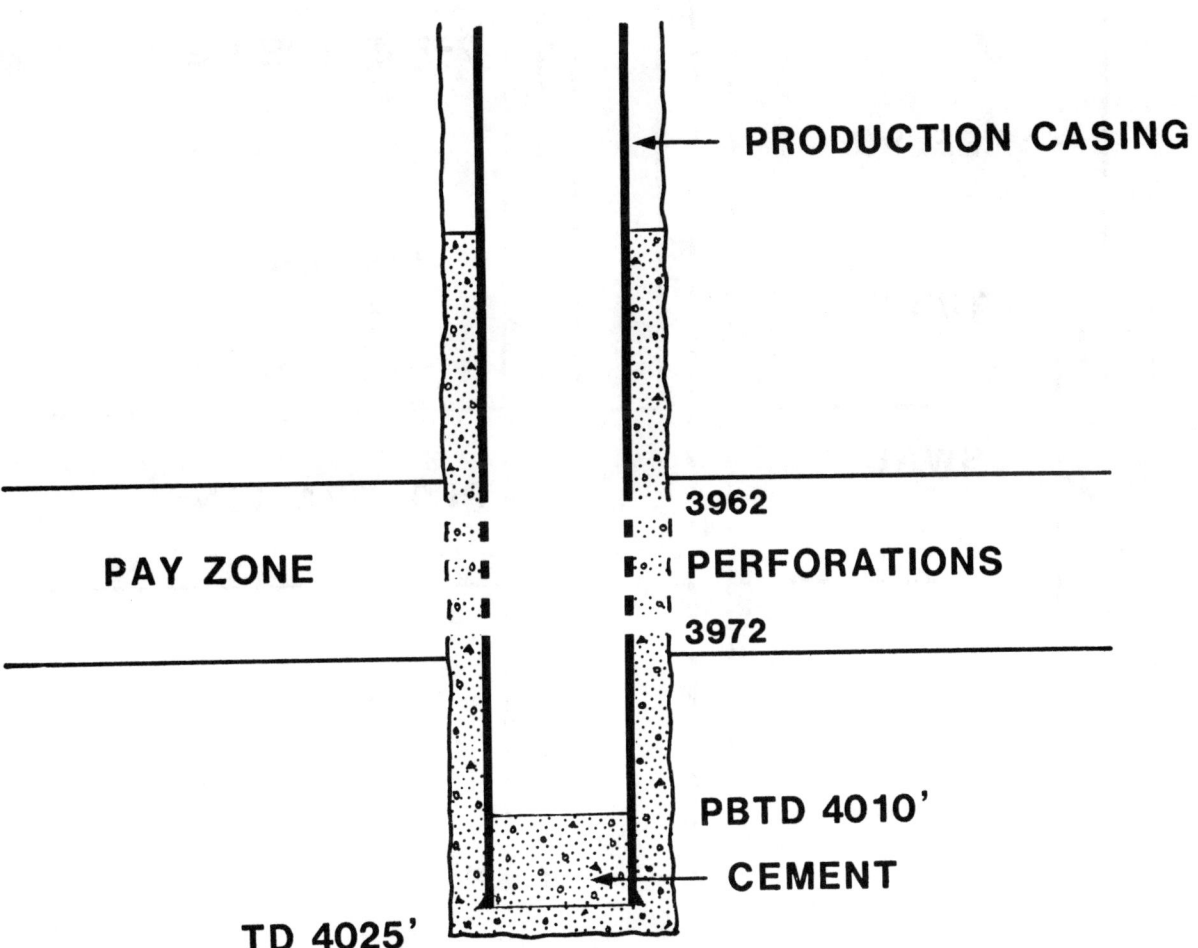

Figure 5.40.

6-7-76   PREPARING TO SWAB WELL. PULL OUT AND LAY DOWN 2-7/8"
         TUBING. RU PERFORATING TRUCK AND PERFORATE INTERVAL
         3962' TO 3972' WITH 2 JETS/FT. NO PRESSURE RESPONSE
         AFTER PERFORATING. RUN-IN WELL WITH OPEN ENDED 2-7/8"
         TUBING AND PACKER. BOTTOM OF TUBING AT 3950'. SET PACKER
         AT 3930'.

6-8-76   SWABBING. RECOVERED 50 BBLS. WATER WITH TRACE OF OIL.
         NO FLUID APPEARS TO BE ENTERING WELL BORE.

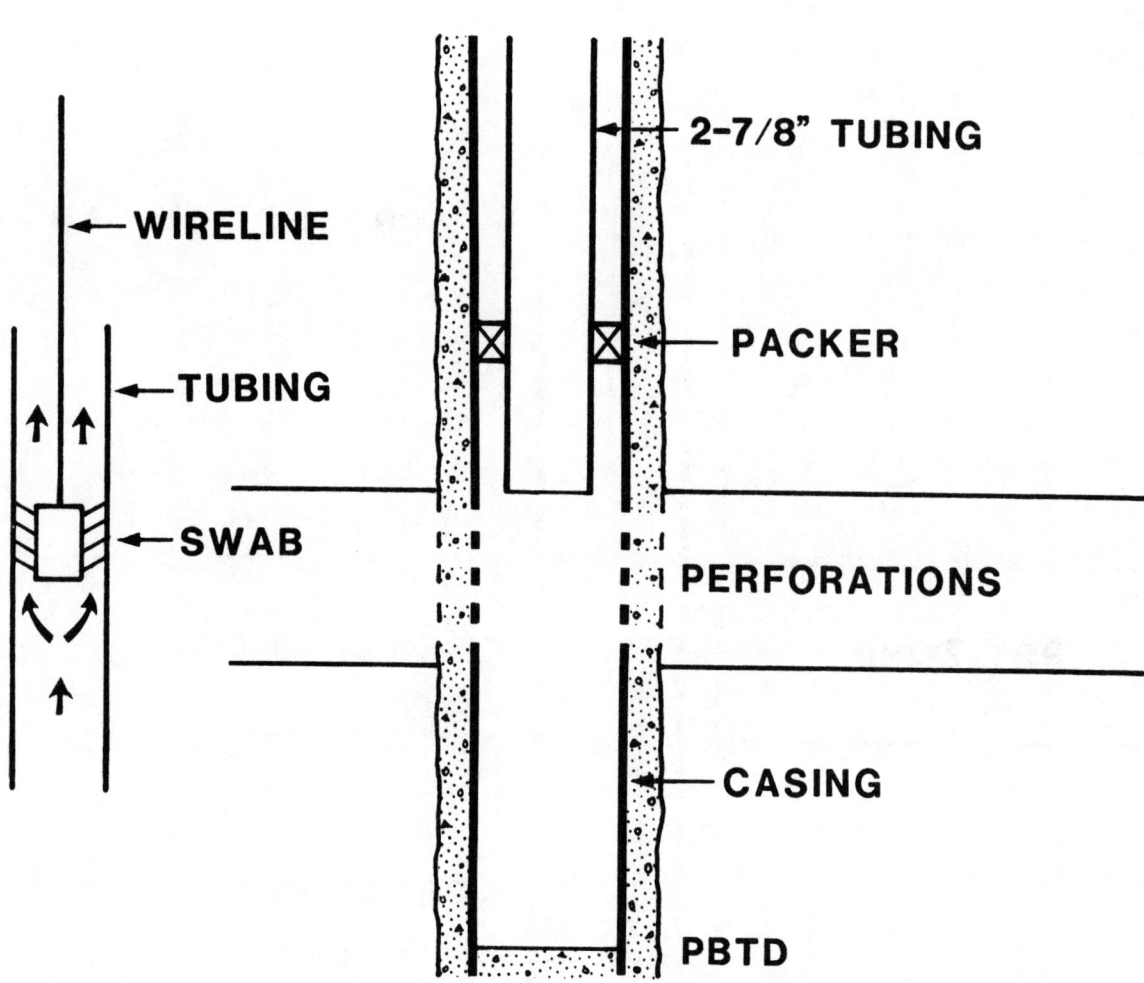

Figure 5.41.

6-9-76   SWABBING. RU ACIDIZING TRUCK. PUMPED 3000 GALS. 15% HCL INTO FORMATION FOLLOWED BY WATER. FORMATION BROKE AT 1500 PSI. SI WELL FOR HALF-HOUR AND THEN BEGAN SWABBING. SWABBED 50 BBLS. OF ACID WATER.

6-10-76 SWABBING. SWABBED BACK ALL LOAD WATER AND ACID. SWAB RATE LAST 2 HOURS 4 BARRELS PER HOUR WITH 50% OIL CUT.

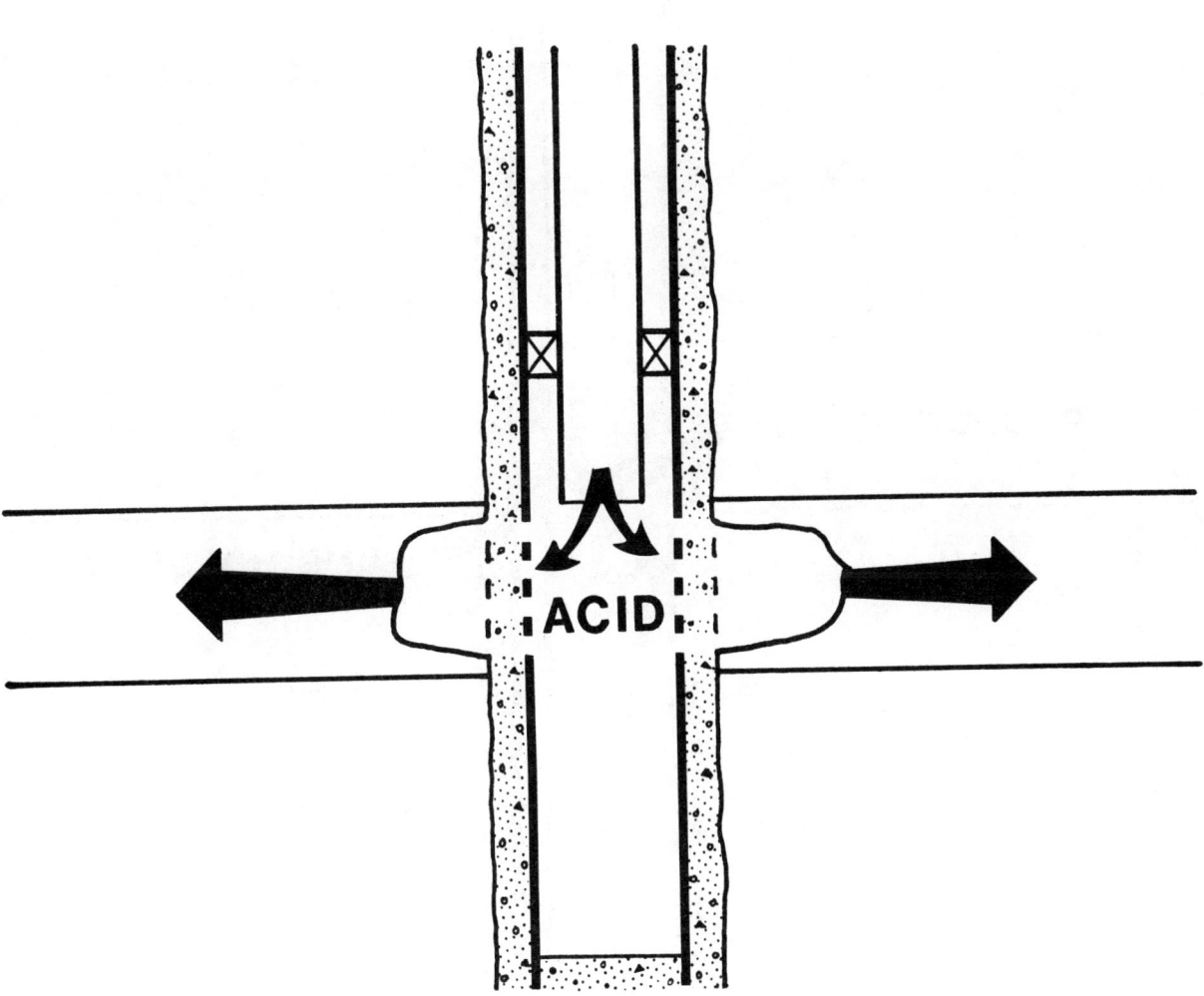

Figure 5.42.

6-11-76 WAITING ON PUMPING UNIT. RAN INSERT PUMP AND SUCKER RODS. RD COMPLETION UNIT. WELL READY FOR PRODUCTION.

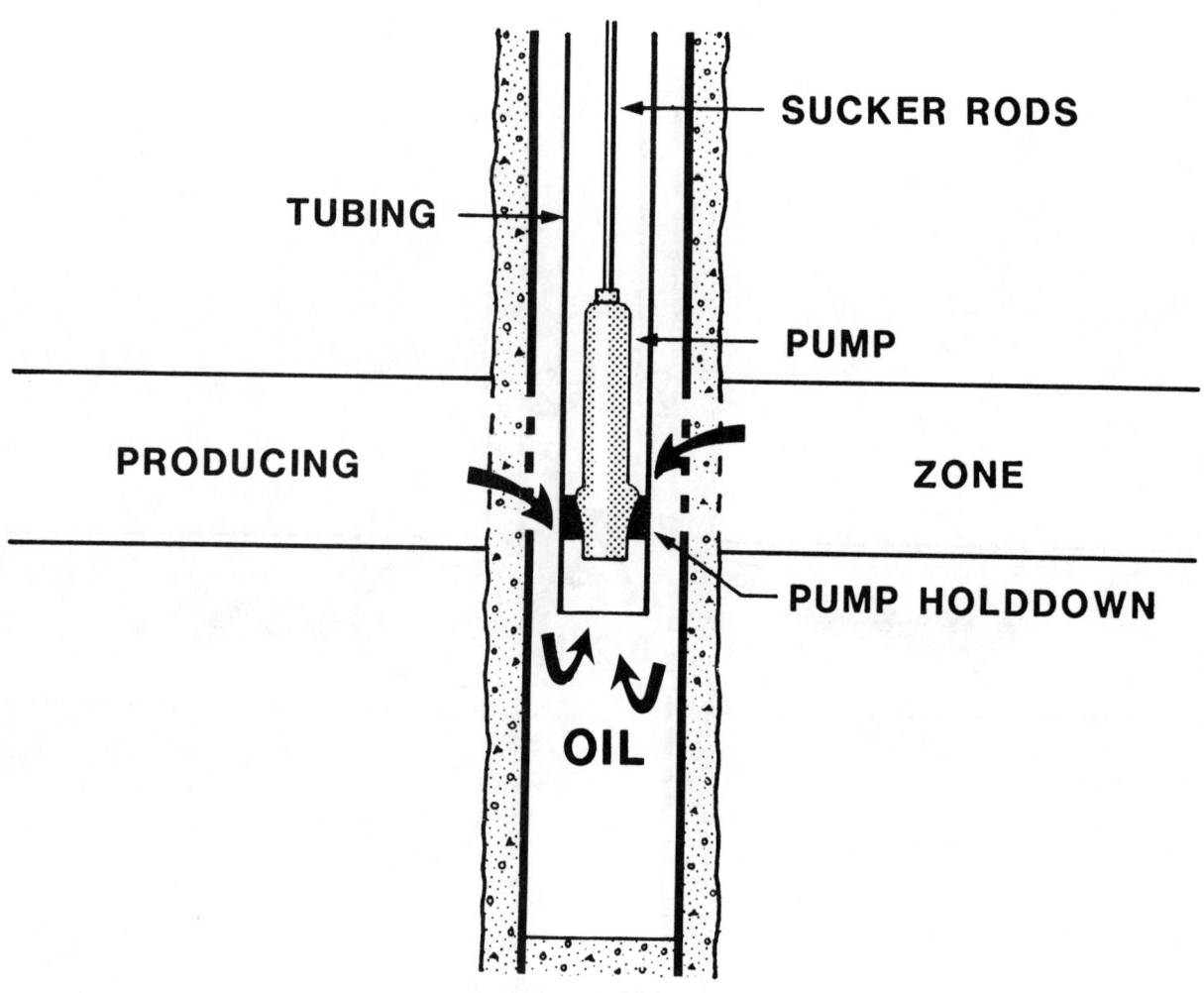

Figure 5.43.

The following glossary, taken from **A Primer of Oil-Well Drilling**, is used with the permission of Petroleum Extension Service, The University of Texas at Austin (PETEX). We wish to express our thanks to the Petroleum Extension Service for this permission.

DRILLING AND COMPLETION—GLOSSARY

—A—

abandon *v:* to cease producing oil and gas from a well when it becomes unprofitable. A wildcat well may be abandoned after it has proven nonproductive. Several steps are involved in abandoning a well: part of the casing is removed and salvaged; one or more cement plugs are placed in the borehole to prevent migration of fluids between the different formations penetrated by the borehole; and the well is abandoned. In many states, it is necessary to secure permission from official agencies before a well may be abandoned.

acid fracture *v:* to part or open fractures in productive, hard-limestone formations by using a combination of oil and acid or water and acid under high pressure. See *formation fracturing.*

acidize *v:* to treat oil-bearing limestone or other formations, using a chemical reaction with acid, to increase production. Hydrochloric or other acid is injected into the formation under pressure. The acid etches the rock, enlarging the pore spaces and passages through which the reservoir fluids flow. The acid is held under pressure for a period of time and then pumped out, and the well is swabbed and put back into production. Chemical inhibitors combined with the acid prevent corrosion of the pipe.

adjustable choke *n:* a choke in which a conical needle and seat vary the rate of flow. See *choke.*

air-actuated *adj:* powered by compressed air, as the clutch and brake system in drilling equipment.

air drilling *n:* a method of rotary drilling that uses compressed air as the circulation medium. The conventional method of removing cuttings from the wellbore is to use a flow of water or drilling mud. Compressed air removes the cuttings with equal or greater efficiency. The rate of penetration is usually increased considerably when air drilling is used. However, a principal problem in air drilling is the penetration of formations containing water, since the entry of water into the system reduces the ability of the air to remove the cuttings.

American Petroleum Institute *n:* 1. founded in 1920, this national oil trade organization is the leading standardizing organization on oilfield drilling and producing equipment. It maintains departments of transportation, refining, and marketing in Washington, D.C., and a department of production in Dallas. 2. (slang) indicative of a job being properly or thoroughly done (as, "His work is strictly API").

angle of deflection *n:* in directional drilling, the angle, expressed in degrees, at which a well is deflected from the vertical by a whipstock or other deflecting tool. See *whipstock.*

annular blowout preventer *n:* a large valve, usually installed above the ram preventers, that forms a seal in the annular space between the pipe and wellbore or, if no pipe is present, on the wellbore itself. Compare *ram blowout preventer.*

annular space *n:* 1. the space surrounding a cylindrical object within a cylinder. 2. the space around a pipe in a wellbore, the outer wall of which may be the wall of either the borehole or the casing; sometimes termed the annulus.

API *abbr:* American Petroleum Institute.

—B—

back off *v:* to unscrew one threaded piece (as a section of pipe) from another.

back up *v:* to hold one section of an object (as pipe) while another is being screwed into or out of it.

bail *n:* a cylindrical steel bar (similar to the handle or bail of a bucket, only much larger) that supports the swivel and connects it to the hook. Sometimes, the two cylindrical bars that support the elevators and attach them to the hook are called bails. *v:* to recover bottom-hole fluids, samples, or drill cuttings by lowering a cylindrical vessel called a bailer to the bottom of a well, filling it, and retrieving it. See *bailer.*

bailer *n:* a long cylindrical container, fitted with a valve at its lower end, used to remove water, sand, mud, or oil from a well.

bailing line *n:* cable attached to the bailer, passed over a sheave at the top of the derrick, and spooled on a reel. See *sheave.*

barge *n:* any one of many types of flat-decked, shallow-draft vessels, usually towed by a boat. A complete drilling rig may be assembled on a drilling barge, which usually is submersible; that is, it has a submersible hull or base that is flooded with water at the drilling site. Drilling equipment, crew quarters, and so forth are mounted on a superstructure above the water level.

barite or **baryte** *n:* barium sulfate, $BaSO_4$; a mineral used to increase the weight of drilling mud. Its specific gravity is 4.2 (i.e., it is 4.2 times heavier than water). See *barium sulfate* and *mud.*

barium sulfate *n:* 1. a chemical combination of barium, sulfur, and oxygen. Also called barite. See *barite.* 2. a tenacious scale that is very difficult to remove.

barrel *n:* a measure of volume for petroleum products. One barrel is the equivalent of 42 U.S. gallons or 0.15899 cubic metres. One cubic metre equals 6.2897 barrels.

basket sub *n:* a fishing accessory run above a bit or mill to recover small pieces of metal or junk in a well.

belt *n:* a flexible band or cord connecting and passing about each of two or more pulleys to transmit power or impart motion.

bit *n:* the cutting or boring element used in drilling oil and gas wells. The bit consists of the cutting element and the circulating element. The circulating element permits the passage of drilling fluid and utilizes the hydraulic force of the fluid stream to improve drilling rates. In rotary drilling, several drill collars are joined to the bottom end of the drill-pipe column. The bit is attached to the end of the drill collar. Most bits used in rotary drilling are roller-cone bits.

bit breaker *n:* a heavy plate that fits in the rotary table and holds the drill bit while it is being made up in or broken out of the drill stem. See *bit.*

bit record *n:* a report on each bit used in a drilling operation that lists the bit type, the amount of footage the bit has drilled, and the nature of the formation penetrated.

blind ram *n:* an integral part of a blowout preventer that serves as the closing element. Its ends do not fit around the drill pipe but seal against each other and shut off the space below completely.

block *n:* any assembly of pulleys on a common framework; in mechanics, one or more pulleys, or sheaves, mounted to rotate on a common axis. The crown block is an assembly of sheaves mounted on beams at the top of the derrick. The drilling line is reeved over the sheaves of the crown block alternately with the sheaves of the traveling block, which is hoisted and lowered in the derrick by the drilling line. When elevators are attached to a hook or the traveling block, and when drill pipe is latched in the elevators, the pipe can be raised or lowered in the derrick or mast. See *crown block, elevator, hook, reeve, sheave,* and *traveling block;* also see *drilling block.*

blooey line *n:* the discharge pipe from a well being drilled by air drilling. The blooey line is used to conduct the air or gas used for circulation away from the rig to reduce the fire hazard as well as to transport the cuttings a suitable distance from the well. See *air drilling.*

blowout *n:* an uncontrolled flow of gas, oil, or other well fluids into the atmosphere. A blowout, or gusher, occurs when formation pressure exceeds the pressure applied to it by the column of drilling fluid. A kick warns of an impending blowout. See *formation pressure, gusher,* and *kick.*

blowout preventer *n:* the equipment installed at the wellhead to prevent the escape of pressure either in the annular space between the casing and drill pipe or in an open hole (i.e., hole with no drill pipe) during drilling-completion operations. The blowout preventer is located beneath the rig at the land's surface on land rigs or at the water's surface on jack-up or platform rigs and on the seafloor for floating offshore rigs. See *annular blowout preventer* and *ram blowout preventer*.

boll-weevil *n:* (slang) an inexperienced rig or oil-field worker, sometimes shortened to "weevil."

bomb *n:* a thick-walled container, usually steel, used to hold samples of oil or gas under pressure. See *bottom-hole pressure*.

bond *n:* the state of one material adhering or being joined to another material (as cement to formation). *v:* to adhere or be joined to another material.

boot *n:* a tubular device placed in a vertical position, either inside or outside a larger vessel, through which well fluids are conducted before entering the larger vessel. A boot aids in the separation of gas from wet oil. Also called a flume or conductor pipe.

BOP *abbr:* blowout preventer.

borehole *n:* the wellbore; the hole made by drilling or boring. See *wellbore*.

bottom-hole *adj:* pertaining to the lowest or deepest part of a well.

bottom-hole choke *n:* a device with a restricted opening placed in the lower end of the tubing to control the rate of flow. See *choke*.

bottom-hole pressure *n:* the pressure in a well at a point opposite the producing formation, as recorded by a bottom-hole-pressure bomb. See *bottom-hole-pressure bomb*.

bottom-hole-pressure bomb *n:* a bomb used to record the pressure in a well at a point opposite the producing formation. See *bomb*.

Bourdon tube *n:* a flattened-metal tube bent in a curve, which tends to straighten under pressure. By the movements of an indicator over a circular scale, a Bourdon tube indicates the pressure applied to it.

box *n:* the female section of a tool joint. See *tool joint*.

brake *n:* a device for arresting the motion of a mechanism, usually by means of friction, as in the drawworks brake. Compare *electrodynamic brake* and *hydromatic brake*.

break out *v:* 1. to unscrew one section of pipe from another section, especially drill pipe while it is being withdrawn from the wellbore. During this operation, the tongs are used to start the unscrewing operation. See *tongs*. 2. to separate, as gas from liquid.

breakout cathead *n:* a device attached to the shaft of the drawworks that is used as a power source for unscrewing drill pipe; usually located opposite the driller's side of the drawworks. See *cathead*.

breakout tongs *n:* tongs that are used to start unscrewing one section of pipe from another section, especially drill pipe coming out of the hole. Also called lead tongs. See *tongs*.

bring in a well *v:* to complete a well and put it in producing status.

buck up *v:* to tighten up a threaded connection (as two joints of drill pipe).

bullet perforator *n:* a tubular device that, when lowered to a selected depth within a well, fires bullets through the casing to provide holes through which the well fluids may enter.

—C—

cable *n:* a rope of wire, hemp, or other strong fibers. See *wire rope*.

cable-tool drilling *n:* a drilling method in which the hole is drilled by dropping a sharply pointed bit on the bottom of the hole. The bit is attached to a cable, and the cable is picked up and dropped, picked up and dropped, over and over, as the hole is drilled.

cased *adj:* pertaining to a wellbore in which casing is run and cemented. See *casing*.

cased hole *n:* a wellbore in which casing has been run.

casing *n:* steel pipe placed in an oil or gas well as drilling progresses to prevent the wall of the hole from caving in during drilling and to provide a means of extracting petroleum if the well is productive.

casing centralizer *n:* a device secured around the casing at regular intervals to center it in the hole. Casing that is centralized allows a more uniform cement sheath to form around the pipe.

casing coupling *n:* a tubular section of pipe that is threaded inside and used to connect two joints of casing.

casing elevator: *n:* see *elevator*.

casinghead *n:* a heavy, steel, flanged fitting that connects to the first string of casing and provides a housing for the slips and packing assemblies by which the intermediate strings of casing are suspended and the annulus sealed off. Also called a spool. See *annular space*.

casing shoe *n:* also called a guide shoe. See *guide shoe*.

casing string *n:* the entire length of all the joints of casing run in a well. Casing is manufactured in lengths of about 30 feet, each length or joint being joined to another as casing is run in a well. See *combination string*.

catch samples *v:* to obtain cuttings for geological information as formations are penetrated by the bit. The samples are obtained from drilling fluid as it emerges from the wellbore or, in cable-tool drilling, from the bailer. Cuttings are carefully washed until they are free of foreign matter, dried, and labeled to indicate the depth at which they were obtained. See *bailer, cable-tool drilling,* and *cuttings*.

cathead *n:* a spool-shaped attachment on a winch around which rope for hoisting and pulling is wound. See *breakout cathead* and *makeup cathead*.

catline *n:* a hoisting or pulling line powered by the cathead and used to lift heavy equipment on the rig. See *cathead*.

caving *n:* collapse of the walls of the wellbore, also called sloughing.

cellar *n:* a pit in the ground to provide additional height between the rig floor and the wellhead to accommodate the installation of blowout preventers, rathole, mousehole, and so forth. It also collects drainage water and other fluids for subsequent disposal.

cement casing *v:* to fill the annulus between the casing and hole with cement to support the casing and prevent fluid migration between permeable zones.

cement channeling *n:* an undesirable phenomenon that can occur when casing is being cemented in a borehole. The cement slurry fails to rise uniformly between the casing and borehole wall, leaving spaces void of cement. Ideally, the cement should completely and uniformly surround the casing and form a strong bond to the borehole wall.

cementing *n:* the application of a liquid slurry of cement and water to various points inside or outside the casing. See *primary cementing, secondary cementing,* and *squeeze cementing*.

chain drive *n:* a drive system using a chain and chain gears to transmit power. Power transmissions use a roller chain, in which each link is made of side bars, transverse pins, and rollers on the pins. A double roller chain is made of two connected rows of links, a triple roller chain of three, and so forth.

chain tongs *n:* a tool consisting of a handle and releasable chain used for turning pipe or fittings of a diameter larger than that which a pipe wrench would fit. The chain is looped and tightened around the pipe or fitting, and the handle is used to turn the tool so that the pipe or fitting can be tightened or loosened.

check valve *n:* a valve that permits flow in one direction only.

choke *n:* an orifice installed in a line to restrict the flow and control the rate of production. Surface chokes are part of the Christmas tree and contain a choke nipple, or bean, with a small-diameter bore that serves to restrict flow. Chokes are also used to control the rate of flow of the drilling mud out of the hole when the well is closed in with the blowout preventer and a kick is being circulated out of the hole. See *adjustable choke, blowout preventer, bottom-hole choke, Christmas tree, kick, nipple,* and *positive choke*.

choke line *n:* an extension of pipe from the blowout preventer assembly used to direct well fluids from the annulus to the choke manifold.

choke manifold *n:* the arrangement of piping and special valves, called chokes, through which drilling mud is circulated when the blowout preventers are closed to control the pressures encountered during a kick. See *choke* and *blowout preventer*.

Christmas tree *n:* the control valves, pressure gauges, and chokes assembled at the top of a well to control the flow of oil and gas after the well has been drilled and completed.

circulate *v:* to pass from one point throughout a system and back to the starting point. For example, drilling fluid is circulated out of the suction pit, down the drill pipe and drill collars, out the bit, up the annulus, and back to the pits.

circulation *n*: the movement of drilling fluid out of the mud pits, down the drill stem, up the annulus, and back to the mud pits.

combination string *n*: a casing string that has joints of various collapse resistance, internal yield strength, and tensile strength designed for various depths in a specific well to best withstand the conditions of that well. In deep wells, high tensile strength is required in the top casing joints to carry the load, whereas high collapse resistance and internal yield strength are needed for the bottom joints. In the middle of the casing, average qualities are usually sufficient. The most suitable combination of types and weights of pipe helps to ensure efficient production at a minimum cost.

come out of the hole *v*: to pull the drill stem out of the wellbore. This withdrawal is necessary to change the bit, change from a core barrel to the bit, run electric logs, prepare for a drill-stem test, run casing, and so on.

company man *n*: also called company representative. See *company representative*.

company representative *n*: an employee of an operating company whose job is to represent the company's interests at the drilling location.

complete a well *v*: to finish work on a well and bring it to productive status. See *well completion*.

compound *n*: a mechanism used to transmit power from the engines to the pump, drawworks, and other machinery on a drilling rig. It is composed of clutches, chains and sprockets, belts and pulleys, and a number of shafts, both driven and driving. *v*: to connect two or more power-producing devices (as engines) to run one piece of driven equipment (as the drawworks).

conductor pipe *n*: 1. a short string of large-diameter casing used to keep the top of the wellbore open and to provide a means of conveying the up-flowing drilling fluid from the wellbore to the mud pit. 2. a boot. See *boot*.

contract depth *n*: the depth of the wellbore at which the drilling contract is fulfilled.

coupling *n*: 1. in piping, a metal collar with internal threads used to join two sections of threaded pipe. 2. in power transmission, a connection extending longitudinally between a driving shaft and a driven shaft. Most such couplings are flexible and compensate for minor misalignment of the two shafts.

crooked hole *n*: a wellbore that has deviated from the vertical. It usually occurs in areas where the subsurface formations are difficult to drill, such as a section of alternating hard and soft strata steeply inclined from the horizontal.

crown block *n*: an assembly of sheaves or pulleys mounted on beams at the top of the derrick over which the drilling line is reeved. See *block, reeve,* and *sheave*.

cuttings *n pl*: the fragments of rock dislodged by the bit and brought to the surface in the drilling mud. Washed and dried samples of the cuttings are analyzed by geologists to obtain information about the formations drilled.

—D—

daylight tour *n*: (pronounced "tower") the shift of duty on a drilling rig that starts at or about daylight; also called morning tour. Compare *evening tour* and *graveyard tour*.

deadline *n*: the drilling line from the crown-block sheave to the anchor, so called because it does not move. Compare *fast line*.

deadline tie-down anchor *n*: a device to which the deadline is attached, securely fastened to the mast or derrick substructure. Also called a deadline anchor.

degasser *n*: the equipment used to remove unwanted gas from a liquid, especially from drilling fluid.

derrick *n*: a large load-bearing structure, usually of bolted construction. In drilling, the standard derrick has four legs standing at the corners of the substructure and reaching to the crown block. The substructure is an assembly of heavy beams used to elevate the derrick and provide space to install blowout preventers, casingheads, and so forth. Because the standard derrick must be assembled piece by piece, it has largely been replaced by the mast, which can be lowered and raised without disassembly. See *crown block, mast,* and *substructure*.

derrickman *n*: the crew member who handles the upper end of the drill stem as it is being hoisted out of or lowered into the hole. He is also responsible for the conditioning of the drilling fluid and the circulating machinery.

desander *n*: a centrifugal device for removing sand from drilling fluid to prevent abrasion of the pumps. It may be operated mechanically or by a fast-moving stream of fluid inside a special cone-shaped vessel, in which case it is sometimes called a hydrocyclone. Compare *desilter*.

desilter *n*: a centrifugal device for removing very fine particles, or silt, from drilling fluid to keep the amount of solids in the fluid to the lowest possible point. Usually, the lower the solids content of mud, the faster the rate of penetration. It works on the same principle as a desander. Compare *desander*.

development well *n*: 1. a well drilled in proven territory in a field to complete a pattern of production. 2. an exploitation well. See *exploitation well*.

deviation *n*: the inclination of the wellbore from the vertical. The angle of deviation, angle of drift, or drift angle is the angle in degrees that shows the variation from the vertical as revealed by a deviation survey. See *deviation survey*.

deviation survey *n*: an operation made to determine the angle from which a bit has deviated from the vertical during drilling. There are two basic deviation-survey, or drift-survey, instruments: one reveals the angle of deviation only; the other indicates both the angle and direction of deviation.

diamond bit *n*: a drilling bit that has a steel body surfaced with industrial diamonds. Cutting is performed by the rotation of the very hard diamonds over the rock surface.

diesel-electric power *n*: the power supplied to a drilling rig by diesel engines driving electric generators, used widely offshore and gaining popularity onshore.

diesel engine *n*: a high-compression, internal-combustion engine used extensively for powering drilling rigs. In a diesel engine, air is drawn into the cylinders and compressed to very high pressures; ignition occurs as fuel is injected into the compressed and heated air. Combustion takes place within the cylinder above the piston, and expansion of the combustion products imparts power to the piston.

directional drilling *n*: intentional deviation of a wellbore from the vertical. Although wellbores are normally drilled vertically, it is sometimes necessary or advantageous to drill at an angle from the vertical. Controlled directional drilling makes it possible to reach subsurface areas laterally remote from the point where the bit enters the earth. It involves the use of turbodrills, Dyna-Drills, whipstocks, or other deflecting tools. See *Dyna-Drill, turbodrill,* and *whipstock*.

discovery well *n*: the first oil or gas well drilled in a new field; the well that reveals the presence of a petroleum-bearing reservoir. Subsequent wells are development wells. Compare *development well*.

displacement fluid *n*: in oil-well cementing, the fluid, usually drilling mud or salt water, that is pumped into the well after the cement to force the cement out of the casing and into the annulus.

doghouse *n*: 1. a small enclosure on the rig floor used as an office for the driller or as a storehouse for small objects. 2. any small building used as an office or for storage.

double *n*: a length of drill pipe, casing, or tubing consisting of two joints screwed together. Compare *thribble* and *fourble*. See *joint*.

double board *n*: the name used for working platform of the derrickman, or monkeyboard, when it is located at a height in the derrick or mast equal to two lengths of pipe joined together. Compare *fourble board* and *thribble board*. See *monkeyboard*.

drawworks *n*: the hoisting mechanism on a drilling rig. It is essentially a large winch that spools off or takes in the drilling line and thus raises or lowers the drill stem and bit.

drill bit *n*: the cutting or boring element used for drilling. See *bit*.

drill collar *n*: a heavy, thick-walled tube, usually steel, used between the drill pipe and the bit in the drill stem to weight the bit in order to improve its performance.

driller *n*: the employee directly in charge of a drilling rig and crew. His main duty is operation of the drilling and hoisting equipment, but he is also responsible for the downhole condition of the well, operation of downhole tools, and pipe measurements.

drilling block *n:* a lease or a number of leases of adjoining tracts of land that constitute a unit of acreage sufficient to justify the expense of drilling a wildcat.

drilling contractor *n:* an individual or group of individuals that own a drilling rig or rigs and contract their services for drilling wells to a certain depth.

drilling crew *n:* a driller, a derrickman, and two or more helpers who operate a drilling rig for one tour each day. See *derrickman, driller,* and *tour.*

drilling fluid *n:* circulating fluid, one function of which is to force cuttings out of the wellbore and to the surface. While a mixture of clay, water, and other chemical additives is the most common drilling fluid, wells can also be drilled using air, gas, or water as the drilling fluid. Also called circulating fluid. See *mud.*

drilling foreman *n:* the supervisor of drilling operations on a rig; also the tool pusher or rig superintendent.

drilling line *n:* a wire rope used to support the drilling tools.

drilling rate *n:* the speed with which the bit drills the formation; usually called the rate of penetration.

drilling rig *n:* See *rig.*

drill pipe *n:* the heavy seamless tubing used to rotate the bit and circulate the drilling fluid. Joints of pipe 30 feet long are coupled together by means of tool joints.

drill ship *n:* a ship constructed to permit a well to be drilled from it at an offshore location. While not as stable as other floating structures (as a semisubmersible), drill ships, or shipshapes, are capable of drilling exploratory wells in relatively deep waters. They may have a ship hull, a catamaran hull, or a trimaran hull. See *semisubmersible drilling rig.*

drill stem *n:* all members in the assembly used for drilling by the rotary method from the swivel to the bit, including the kelly, drill pipe and tool joints, drill collars, stabilizers, and various subsequent items. Compare *drill string.*

drill string *n:* the column, or string, of drill pipe with attached tool joints that transmits fluid and rotational power from the kelly to the drill collars and bit. Often, especially in the oil patch, the term is loosely applied to include both drill pipe and drill collars. Compare *drill stem.*

drum *n:* 1. a cylinder around which wire rope is wound in the drawworks. The drawworks drum is that part of the hoist upon which the drilling line is wound. 2. a steel container of general cylindrical form. Refined products are shipped in steel drums with capacities of about 50 to 55 U.S. gallons (about 200 litres).

Dyna-Drill *n:* a downhole motor driven by drilling fluid that imparts rotary motion to a drilling bit connected to the tool, thus eliminating the need to turn the entire drill stem to make hole. The Dyna-Drill, a trade name, is used in straight and directional drilling.

dynamic positioning *n:* a method by which a floating offshore drilling rig is maintained in position over an offshore well location. Generally, several motors called thrusters are located on the hull(s) of the structure and are actuated by a sensing system. A computer to which the system feeds signals then directs the thrusters to maintain the rig on location.

—E—

electrodynamic brake *n:* a device mounted on the end of the drawworks shaft of a drilling rig. The electrodynamic brake (sometimes called a magnetic brake) serves as an auxiliary to the mechanical brake when pipe is lowered into a well. The braking effect in an electrodynamic brake is achieved by means of the interaction of electric currents with magnets, with other currents, or with themselves.

elevator *n:* a set of clamps that grips a stand, or column, of casing, tubing, or drill pipe so that the stand can be raised or lowered into the hole.

evening tour *n:* (pronounced "tower") the shift of duty on a drilling rig that starts in the afternoon and runs through the evening. Compare *daylight tour* and *graveyard tour.*

exploitation well *n:* a well drilled to permit more effective extraction of oil from a reservoir. It is sometimes called a development well. See *development well.*

—F—

fast line *n:* the end of the drilling line that is affixed to the drum or reel of the drawworks, so called because it travels with greater velocity than any other portion of the line. Compare *deadline.*

fill the hole *v:* to pump drilling fluid into the wellbore while the pipe is being withdrawn in order to ensure that the wellbore remains full of fluid even though the pipe is withdrawn. Filling the hole lessens the danger of blowout or of caving of the wall of the wellbore.

filter cake *n:* 1. compacted solid or semisolid material remaining on a filter after pressure filtration of mud with the standard filter press. Thickness of the cake is reported in thirty-seconds of an inch or in millimetres. 2. the layer of concentrated solids from the drilling mud that forms on the walls of the borehole opposite permeable formations; also called wall cake or mud cake.

fingerboard *n:* a rack that supports the tops of the stands of pipe being stacked in the derrick or mast. It has several steel fingerlike projections that form a series of slots into which the derrickman can set a stand of drill pipe as it is pulled out of the hole.

fish *n:* an object left in the wellbore during drilling operations that must be recovered before work can proceed. It can be anything from a piece of scrap metal to a part of the drill stem. *v:* 1. to recover from a well any equipment left there during drilling operations, such as a lost bit or drill collar or part of the drill string. 2. to remove from an older well certain pieces of equipment, such as packers, liners, or screen pipe, to allow reconditioning of the well.

fishing tool *n:* a tool designed to recover equipment lost in the well.

float collar *n:* a special coupling device, inserted one or two joints above the bottom of the casing string, that contains a check valve to permit fluid to pass downward but not upward through the casing. The float collar prevents drilling mud from entering the casing while it is being lowered, allowing the casing to float during its descent, which decreases the load on the derrick. The float collar also prevents a backflow of cement during the cementing operation.

floorman *n:* a drilling-crew member whose work station is on the derrick floor. On rotary drilling rigs, there are at least two and usually three or more floormen on each crew. Also called rotary helper and roughneck.

formation fracturing *n:* a method of stimulating production by increasing the permeability of the producing formation. Under extremely high hydraulic pressure, a fluid (as water, oil, alcohol, dilute hydrochloric acid, liquefied petroleum gas, or foam) is pumped downward through tubing or drill pipe and forced into the perforations in the casing. The fluid enters the formation and parts or fractures it. Sand grains, aluminum pellets, glass beads, or similar materials are carried in suspension by the fluid into the fractures. These are called propping agents or proppants. When the pressure is released at the surface, the fracturing fluid returns to the well, and the fractures partially close on the proppants, leaving channels for oil to flow through them to the well. This process is often called a frac job. See *propping agent.*

formation pressure *n:* the force exerted by fluids in a formation, recorded in the hole at the level of the formation with the well shut in. See *shut-in bottom-hole pressure.*

formation testing *n:* the gathering of data on a formation to determine its potential productivity before installing casing in a well. The conventional method is the drill-stem test. Incorporated in the drill-stem-testing tool are a packer, valves or ports that may be opened and closed from the surface, and a pressure-recording device. The tool is lowered to bottom on a string of drill pipe and the packer set, isolating the formation to be tested from the formations above and supporting the fluid column above the packer. A port on the tool is opened to allow the trapped pressure below the packer to bleed off into the drill pipe, gradually exposing the formation to atmospheric pressure and allowing the well to produce to the surface, where the well fluids may be sampled and inspected. From a record of the pressure readings, a number of facts about the formation may be inferred.

fourble *n:* a section of drill pipe, casing, or tubing consisting of four joints screwed together. Compare *double* and *thribble.* See *joint.*

fourble board *n:* the name used for the working platform of the derrickman, or monkeyboard, when it is located at a height in the derrick equal to approximately four lengths of pipe joined together. Compare *double board* and *thribble board*. See *monkeyboard*.

fracturing *n:* shortened form of formation fracturing. See *formation fracturing*.

—G—

gas-cut mud *n:* a drilling mud that has entrained formation gas giving the mud a characteristically fluffy texture. When entrained gas is not released before the fluid returns to the well, the weight or density of the fluid column is reduced. Because a large amount of gas in mud lowers its density, gas-cut mud must be treated to lessen the chance of a blowout.

gas show *n:* the gas that appears in drilling-fluid returns, indicating the presence of a gas zone.

graveyard tour *n:* (pronounced "tower") the shift of duty on a drilling rig that starts at or about midnight. Compare *daylight tour* and *evening tour*.

guide shoe *n:* a short, heavy, cylindrical section of steel filled with concrete and rounded at the bottom, which is placed at the end of the casing string. It prevents the casing from snagging on irregularities in the borehole as it is lowered. A passage through the center of the shoe allows drilling fluid to pass up into the casing while it is being lowered and cement to pass out during cementing operations. Also called casing shoe.

gun-perforate *v:* to create holes in casing and cement set through a productive formation. A common method of completing a well is to set casing through the oil-bearing formation and cement it. A perforating gun is then lowered into the hole and fired to detonate high-powered jets or shoot steel projectiles (bullets) through the casing and cement and into the pay zone. The formation fluids flow out of the reservoir through the perforations and into the wellbore. See *jet-perforate* and *perforating gun*.

gusher *n:* an oil well that has come in with such great pressure that the oil jets out of the well like a geyser. In reality, a gusher is a blowout and is extremely wasteful of reservoir fluids and drive energy. In the early days of the oil industry, gushers were common and many times were the only indication that a large reservoir of oil and gas had been struck. See *blowout*.

—H—

hoist *n:* an arrangement of pulleys and wire rope or chain used for lifting heavy objects; a winch or similar device; the drawworks. See *drawworks*.

hoisting drum *n:* the large, flanged spool in the drawworks on which the hoisting cable is wound. See *drawworks*.

hook *n:* a large, hook-shaped device from which the swivel is suspended. It is designed to carry maximum loads ranging from 100 to 650 tons and turns on bearings in its supporting housing. A strong spring within the assembly cushions the weight of a stand (90 feet) of drill pipe, thus permitting the pipe to be made up and broken out with less damage to the tool-joint threads. Smaller hooks without the spring are used for handling tubing and sucker rods. See *stand* and *swivel*.

hopper *n:* a large funnel- or cone-shaped device into which dry components (as powdered clay or cement) can be poured in order to uniformly mix the components with water (or other liquids). The liquid is injected through a nozzle at the bottom of the hopper. The resulting mixture of dry material and liquid may be drilling mud to be used as the circulating fluid in a rotary drilling operation or may be cement slurry used to bond casing to the borehole.

hydraulic fracturing *n:* an operation in which a specially blended liquid is pumped down a well and into a formation under pressure high enough to cause the formation to crack open. The resulting cracks or fractures serve as passages through which oil can flow into the wellbore. See *formation fracturing*.

hydromatic brake *n:* a device mounted on the end of the drawworks shaft of a drilling rig. The hydromatic brake (often simply called the hydromatic) serves as an auxiliary to the mechanical brake when pipe is lowered into the well. The braking effect in a hydromatic brake is achieved by means of a runner or impeller turning in a housing filled with water.

—I—

inland barge rig *n:* a drilling structure consisting of a barge upon which the drilling equipment is constructed. When moved from one location to another, the barge floats, but when stationed on the drill site, the barge is submerged to rest on the bottom. Typically, inland barge rigs are used to drill wells in marshes, shallow inland bays, and in areas where the water covering the drill site is not too deep.

instrumentation *n:* a device or assembly of devices designed for one or more of the following functions: to measure operating variables (as pressure, temperature, rate of flow, speed of rotation, etc.); to indicate these phenomena with visible or audible signals; to record them; to control them within a predetermined range; and to stop operations if the control fails. Simple instrumentation might consist of an indicating pressure gauge only. In a completely automatic system, the desired range of pressure, temperature, and so on is predetermined and preset.

intermediate casing string *n:* the string of casing set in a well after the surface casing to keep the hole from caving and to seal off troublesome formations. The string is sometimes called protection casing.

—J—

jack-up drilling rig *n:* an offshore drilling structure with tubular or derrick legs that support the deck and hull. When positioned over the drilling site, the bottoms of the legs rest on the seafloor. A jack-up rig is towed or propelled to a location with its legs up. Once the legs are firmly positioned on the bottom, the deck and hull height are adjusted and leveled.

jet bit *n:* a drilling bit having replaceable nozzles through which the drilling fluid is directed in a high-velocity stream to the bottom of the hole to improve the efficiency of the bit. See *bit*.

jet gun *n:* an assembly, including a carrier and shaped charges, that is used in jet perforating.

jet-perforate *v:* to create a hole through the casing with a shaped charge of high explosives instead of a gun that fires projectiles. The loaded charges are lowered into the hole to the desired depth. Once detonated, the charges emit short, penetrating jets of high-velocity gases that cut holes in the casing and cement and some distance into the formation. Formation fluids then flow into the wellbore through these perforations. See *bullet perforator* and *gun-perforate*.

joint *n:* a single length (30 feet) of drill pipe or of drill collar, casing, or tubing, that has threaded connections at both ends. Several joints screwed together constitute a stand of pipe. See *stand, single, double, thribble,* and *fourble*.

junk *n:* metal debris lost in a hole. Junk may be a lost bit, pieces of a bit, milled pieces of pipe, wrenches, or any relatively small object that impedes drilling and must be fished out of the hole. *v:* to abandon (as a nonproductive well).

—K—

kelly *n:* the heavy steel member, four- or six-sided, suspended from the swivel through the rotary table and connected to the topmost joint of drill pipe to turn the drill stem as the rotary table turns. It has a bored passageway that permits fluid to be circulated into the drill stem and up the annulus, or vice versa. See *drill stem, rotary table,* and *swivel*.

kelly bushing *n:* a special device that, when fitted into the master bushing, transmits torque to the kelly and simultaneously permits vertical movement of the kelly to make hole. It may be shaped to fit the rotary opening or have pins for transmitting torque. Also called the drive bushing. See *kelly* and *master bushing*.

kelly spinner *n:* a pneumatically operated device mounted on top of the kelly that, when actuated, causes the kelly to turn or spin. It is useful when the kelly or a joint of pipe attached to it must be spun up; that is, rotated rapidly in order to make it up.

kick *n:* an entry of water, gas, oil, or other formation fluid into the wellbore. It occurs because the pressure exerted by the column of drilling fluid is not great enough to overcome the pressure exerted by the fluids in the formation drilled. If prompt action is not taken to control the kick or kill the well, a blowout will occur. See *blowout.*

—L—

latch on *v:* to attach elevators to a section of pipe to pull it out of or run it into the hole.

lead tongs *n:* (pronounced "leed")) the pipe tongs suspended in the derrick or mast and operated by a wireline connected to the breakout cathead. Also called breakout tongs.

location *n:* the place where a well is drilled.

—M—

magnetic brake *n:* also called an electrodynamic brake. See *electrodynamic brake.*

make a connection *v:* to attach a joint of drill pipe onto the drill stem suspended in the wellbore to permit deepening of the wellbore.

make a trip *v:* to hoist the drill stem out of the wellbore to perform one of a number of operations such as changing bits, taking a core, and so forth and then to return the drill stem to the wellbore.

make hole *v:* to deepen the hole made by the bit; to drill ahead.

make up *v:* 1. to assemble and join parts to form a complete unit (as to make up a string of casing). 2. to screw together two threaded pieces. 3. to mix or prepare (as to make up a tank of mud). 4. to compensate for (as to make up for lost time).

make up a joint *v:* to screw a length of pipe into another length of pipe.

makeup cathead *n:* a device attached to the shaft of the drawworks that is used as a power source for screwing together joints of pipe; usually located on the driller's side of the drawworks. See *cathead.*

mast *n:* a portable derrick capable of being erected as a unit, as distinguished from a standard derrick that cannot be raised to a working position as a unit. For transporting by land, the mast can be divided into two or more sections to avoid excessive length extending from truck beds on the highway. Compare *derrick.*

master bushing *n:* a device that fits into the rotary table. It accommodates the slips and drives the kelly bushing so that the rotating motion of the rotary table can be transmitted to the kelly. Also called rotary bushing. See *slips* and *kelly bushing.*

mechanical rig *n:* a drilling rig in which the source of power is one or more internal-combustion engines and in which the power is distributed to rig components through mechanical devices (as chains, sprockets, clutches, and shafts). It is also called a power rig.

mill *n:* a downhole tool with rough, sharp, extremely hard cutting surfaces for removing metal by grinding or cutting. Mills are run on drill pipe or tubing to grind up debris in the hole, remove stuck portions of drill stem or sections of casing for sidetracking, and ream out tight spots in the casing. They are also called junk mills, reaming mills, and so forth, depending on what use they have. *v:* to use a mill to cut or grind metal objects that must be removed from a well.

mix mud *v:* to prepare drilling fluids from a mixture of water or other fluids and one or more of the various dry mud-making materials (as clay, weighting materials, chemicals, etc.).

monkeyboard *n:* the derrickman's working platform. As pipe or tubing is run into or out of the hole, the derrickman must handle the top end of the pipe, which may be as high as 90 feet in the derrick or mast. The monkeyboard provides a small platform to raise him to the proper height to be able to handle the top of the pipe. See *double board, fourble board,* and *thribble board.*

morning tour *n:* (pronounced "tower") also called daylight tour. See *daylight tour.*

motorman *n:* the crew member on a rotary drilling rig responsible for the care and operation of drilling engines.

mousehole *n:* an opening through the rig floor, usually lined with pipe, into which a length of drill pipe is placed temporarily for later connection to the drill string.

mousehole connection *n:* the procedure of adding a length of drill pipe or tubing to the active string in which the length to be added is placed in the mousehole, made up to the kelly, then pulled out of the mousehole, and subsequently made up into the string.

mud *n:* the liquid circulated through the wellbore during rotary drilling operations. In addition to its function of bringing cuttings to the surface, drilling mud cools and lubricates the bit and drill stem, protects against blowouts by holding back subsurface pressures, and deposits a mud cake on the wall of the borehole to prevent loss of fluids to the formation. Although it originally was a suspension of earth solids (especially clays) in water, the mud used in modern drilling operations is a more complex, three-phase mixture of liquids, reactive solids, and inert solids. The liquid phase may be fresh water, diesel oil, or crude oil and may contain one or more conditioners. See *drilling fluid.*

mud analysis *n:* examination and testing of the drilling mud to determine its physical and chemical properties.

mud cake *n:* the sheath of mud solids that forms on the wall of the hole when the liquid from the mud filters into the formation; also called wall cake or filter cake.

mud circulation *n:* the act of pumping mud downward to the bit and back up to the surface by normal circulation or reverse circulation. See *normal circulation* and *reverse circulation.*

mud conditioning *n:* the treatment and control of drilling mud to ensure that it has the correct properties. Conditioning may include the use of additives, the removal of sand or other solids, the removal of gas, the addition of water, and other measures to prepare the mud for conditions encountered in a specific well.

mud engineer *n:* a person whose duty is to test and maintain the properties of the drilling mud that are specified by the operator.

mud gun *n:* a pipe that shoots a jet of drilling mud under high pressure into the mud pit to mix additives with the mud.

mud logging *n:* the recording of information derived from examination and analysis of formation cuttings made by the bit and mud circulated out of the hole. A portion of the mud is diverted through a gas-detecting device. Cuttings brought up by the mud are examined under ultraviolet light to detect the presence of oil or gas. Mud logging is often carried out in a portable laboratory set up at the well.

mud man *n:* also called a mud engineer. See *mud engineer.*

mud pit *n:* a series of open tanks, usually made of steel plates, through which the drilling mud is cycled to allow sand and sediments to settle out. Additives are mixed with the mud in the pit, and the fluid is temporarily stored there before being pumped back into the well. Modern rotary drilling rigs are generally provided with three or more pits, usually fabricated steel tanks fitted with built-in piping, valves, and mud agitators. Mud pits are also called shaker pits, settling pits, and suction pits, depending on their main purpose. See *shaker pit, settling pit,* and *suction pit.*

mud pump *n:* a large, reciprocating pump used to circulate the mud on a drilling rig. A typical mud pump is a single- or double-acting, two- or three-cylinder piston pump whose pistons travel in replaceable liners and are driven by a crankshaft actuated by an engine or motor. Also called a slush pump.

mud-return line *n:* a trough or pipe placed between the surface connections at the wellbore and the shale shaker, through which drilling mud flows upon its return to the surface from the hole.

mud screen *n:* also called a shale shaker. See *shale shaker.*

—N—

needle valve *n:* a globe valve that incorporates a needle-point disk to produce extremely fine regulation of flow.

nipple *n:* a tubular pipe fitting threaded on both ends and less than 12 inches long.

nipple up *v:* in drilling, to assemble the blowout-preventer stack on the wellhead at the surface.

normal circulation *n:* the smooth, uninterrupted circulation of drilling fluid down the drill stem, out the bit, up the annular space between the pipe and the hole, and back to the surface. See *mud circulation* and *reverse circulation.*

—O—

offshore drilling *n:* drilling for oil in an ocean, gulf, or sea, usually on the continental shelf. A drilling unit for offshore operations may be a mobile floating vessel with a ship or barge hull, a semisubmersible or submersible base, a self-propelled or towed structure with jacking legs (jack-up drilling rig), or a permanent structure used as a production platform when drilling is completed. In general, wildcat wells are drilled from mobile floating vessels (as semisubmersible rigs and drill ships) or from jack-ups, while development wells are drilled from platforms. See *drill ship, jack-up drilling rig, platform,* and *semisubmersible drilling rig.*

open *adj:* 1. of a wellbore, having no casing. 2. of a hole, having no drill pipe or tubing suspended in it.

open hole *n:* 1. any wellbore in which casing has not been set. 2. open or cased hole in which no drill pipe or tubing is suspended.

operator *n:* the person or company, either proprietor or lessee, actually operating an oil well or lease. Compare *unit operator.*

orifice *n:* a device with an opening in it whose diameter is smaller than that of the pipe or fitting into which it is placed to partially restrict the flow through the pipe. The difference in pressure on the two sides of an orifice plate, as determined by an orifice meter, can be used to measure the volume of flow through the pipe.

overshot *n:* a fishing tool that is attached to tubing or drill pipe and lowered over the outside wall of pipe lost or stuck in the wellbore. A friction device in the overshot, usually either a basket or a spiral grapple, firmly grips the pipe, allowing the lost fish to be pulled from the hole.

—P—

P&A *abbr:* plug and abandon.

pay sand *n:* the producing formation, often one that is not even sandstone. It is also called pay, pay zone, and producing zone.

perforate *v:* to pierce the casing wall and cement to provide holes through which formation fluids may enter or to provide holes in the casing so that materials may be introduced into the annulus between the casing and the wall of the borehole. Perforating is accomplished by lowering into the well a perforating gun, or perforator, that fires electrically detonated bullets or shaped charges from the surface. See *perforating gun.*

perforating gun *n:* a device fitted with shaped charges or bullets that is lowered to the desired depth in a well and fired to create penetrating holes in casing, cementing, and formation. See *gun-perforate.*

pin *n:* the male section of the tool joint. See *tool joint.*

pipe ram *n:* a sealing component for a blowout preventer that closes the annular space between the pipe and the blowout preventer or wellhead. See *annular space* and *blowout preventer.*

platform *n:* an immobile, offshore structure constructed on pilings from which wells are drilled, produced, or both.

plug and abandon *v:* to place a cement plug into a dry hole and abandon it.

positive choke *n:* a choke in which the orifice size must be changed to change the rate of flow through the choke. See *choke* and *orifice.*

pressure *n:* the force that a fluid (liquid or gas) exerts when it is in some way confined within a vessel, pipe, hole in the ground, and so forth, such as that exerted against the inner wall of a tank or that exerted on the bottom of the wellbore by drilling mud. Pressure is often expressed in terms of force per unit of area, as pounds per square inch (psi).

pressure gauge *n:* an instrument for measuring fluid pressure that usually registers the difference between atmospheric pressure and the pressure of the fluid by indicating the effect of such pressures on a measuring element (as a column of liquid, a Bourdon tube, a weighted piston, a diaphragm, or other pressure-sensitive device). See *Bourdon tube.*

pressure gradient *n:* a scale of pressure differences in which there is a uniform variation of pressure from point to point.

pressure-relief valve *n:* a valve that opens at a preset pressure to relieve excessive pressures within a vessel or line; also called a relief valve, safety valve, or pop valve.

preventer *n:* shortened form of blowout preventer. See *blowout preventer.*

primary cementing *n:* the cementing operation that takes place immediately after the casing has been run into the hole; used to provide a protective sheath around the casing, to segregate the producing formation, and to prevent the migration of undesirable fluids. See *secondary cementing* and *squeeze cementing.*

prime mover *n:* an internal-combustion engine that is the source of power for a drilling rig in oil-well drilling.

proppant *n:* also called propping agent. See *propping agent.*

propping agent *n:* a granular substance (as sand grains, aluminum pellets, or other material) carried in suspension by the fracturing fluid that serves to keep the cracks open when the fracturing fluid is withdrawn after a fracture treatment.

pump *n:* a device that increases the pressure on a fluid or raises it to a higher level. Various types of pumps include the reciprocating pump, centrifugal pump, rotary pump, jet pump, sucker-rod pump, hydraulic pump, mud pump, submersible pump, and bottom-hole pump.

pump pressure *n:* fluid pressure arising from the action of the pump.

—R—

ram *n:* the closing and sealing component on a blowout preventer. One of three types—blind, pipe, or shear—may be installed in several preventers mounted in a stack on top of the wellbore. Blind rams, when closed, form a seal on a hole that has no drill pipe in it; pipe rams, when closed, seal around the pipe; shear rams cut through drill pipe and then form a seal. See *blind ram, pipe ram,* and *sheer ram.*

ram blowout preventer *n:* a blowout preventer that uses rams to seal off pressure on a hole that is with or without pipe. It is also called a ram preventer. See *blowout preventer* and *ram.*

rathole *n:* 1. a hole in the rig floor 30 to 35 feet deep, lined with casing that projects above the floor, into which the kelly and swivel are placed when hoisting operations are in progress. 2. a hole of a diameter smaller than the main hole that is drilled in the bottom of the main hole. *v:* to reduce the size of the wellbore and drill ahead.

reeve *v:* to pass (as the end of a rope) through a hole or opening in a block or similar device.

reeve the line *v:* to string a wire-rope drilling line through the sheaves of the traveling and crown blocks to the hoisting drum.

reserve pit *n:* 1. (obsolete) a mud pit in which a supply of drilling fluid was stored. 2. a waste pit, usually an excavated, earthen-walled pit. It may be lined with plastic to prevent contamination of the soil.

reverse circulation *n:* the return of drilling fluid through the drill stem. The normal course of drilling-fluid circulation is downward through the drill stem and upward through the annular space surrounding the drill stem. For special problems, normal circulation is sometimes reversed, and the fluid returns to the surface through the drill stem, or tubing, after being pumped down the annulus.

rig *n:* the derrick or mast, drawworks, and attendant surface equipment of a drilling unit.

rig down *v:* to dismantle the drilling rig and auxiliary equipment following the completion of drilling operations; also called tear down.

rig up *v:* to prepare the drilling rig for making hole; to install tools and machinery before drilling is started.

roller-cone bit *n:* a drilling bit made of two, three, or four cones, or cutters, that are mounted on extremely rugged bearings. Also called rock bits. The surface of each cone is made up of rows of steel teeth or rows of tungsten carbide inserts. See *bit.*

rotary bushing *n:* also called master bushing. See *master bushing.*

rotary drilling *n:* a drilling method in which a hole is drilled by a rotating bit to which downward force is applied. The bit is fastened to and rotated by the drill stem, which also provides a passageway through which the drilling fluid is circulated. Additional joints of drill pipe are added as drilling progresses.

rotary helper *n:* a worker on a drilling rig, subordinate to the driller;

sometimes called a roughneck, floorman, or rig crewman.

rotary hose *n:* a reinforced, flexible tube on a rotary drilling rig that conducts the drilling fluid from the mud pump and standpipe to the swivel and kelly; also called the mud hose or the kelly hose. See *kelly, mud pump, standpipe,* and *swivel.*

rotary table *n:* the principal component of a rotary, or rotary machine, used to turn the drill stem and support the drilling assembly. It has a beveled-gear arrangement to create the rotational motion and an opening into which bushings are fitted to drive and support the drilling assembly.

roughneck *n:* also called a rotary helper. See *rotary helper.*

round trip *n:* the action of pulling out and subsequently running back into the hole a string of drill pipe or tubing. It is also called tripping.

roustabout *n:* 1. a worker on an offshore rig who handles the equipment and supplies that are sent to the rig from the shore base. The head roustabout is very often the crane operator. 2. a worker who assists the foreman in the general work around a producing oil well, usually on the property of the oil company. 3. a helper on a well-servicing unit.

run in *v:* to go into the hole with tubing, drill pipe, and so forth.

—S—

scratcher *n:* a device fastened to the outside of casing that removes the mud cake from the wall of the hole to condition the hole for cementing. By rotating or moving the casing string up and down as it is being run into the hole, the scratcher, formed of stiff wire, removes the cake so that the cement can bond solidly to the formation.

secondary cementing *n:* any cementing operation after the primary-cementing operation. Secondary cementing includes a plug-back job, in which a plug of cement is positioned at a specific point in the well and allowed to set. Wells are plugged to shut off bottom water or to reduce the depth of the well for other reasons. See *primary cementing* and *squeeze cementing.*

semisubmersible drilling rig *n:* a floating, offshore drilling structure that has hulls submerged in the water but not resting on the seafloor. Living quarters, storage space, and so forth are assembled on the deck. Semisubmersible rigs are either self-propelled or towed to a drilling site and either anchored or dynamically positioned over the site or both. Semisubmersibles are more stable than drill ships and are used extensively to drill wildcat wells in rough waters such as the North Sea. See *dynamic positioning.*

set casing *v:* to run and cement casing at a certain depth in the wellbore. Sometimes, the term "set pipe" is used when referring to setting casing.

settling pit *n:* the mud pit into which mud flows and in which heavy solids are allowed to settle out. Often auxiliary equipment (as desanders) must be installed to speed this process.

shaker *n:* shortened form of shale shaker. See *shale shaker.*

shaker pit *n:* the mud pit adjacent to the shale shaker, usually the first pit into which the mud flows after returning from the hole.

shale shaker *n:* a series of trays with shelves that vibrate to remove cuttings from the circulating fluid in rotary drilling operations. The size of the openings in the sieve is carefully selected to match the size of the solids in the drilling fluid and the anticipated size of cuttings. Also called a shaker.

shaped charge *n:* a relatively small container of high explosive that is loaded into a perforating gun. Upon detonation, the charge releases a small, high-velocity stream of particles (a jet) that penetrates the casing, cement, and formation. See *gun-perforate.*

shear ram *n:* the components in a blowout preventer that cut, or shear, through drill pipe and form a seal against well pressure. Shear rams are used in mobile offshore drilling operations to provide a quick method of moving the rig away from the hole when there is no time to trip the drill stem out of the hole.

sheave *n:* (pronounced "shiv") a grooved pully.

show *n:* the appearance of oil or gas in cuttings, samples, cores, and so forth of drilling mud.

shut down *v:* to stop work temporarily or to stop a machine or operation.

shut-in bottom-hole pressure *n:* the pressure at the bottom of a well when the surface valves on the well are completely closed. The pressure is caused by fluids that exist in the formation at the bottom of the well.

sidetrack *v:* to drill around broken drill pipe or casing that has become lodged permanently in the hole, using a whipstock, turbodrill, or other mud motor. See *directional drilling, turbodrill,* and *whipstock.*

single *n:* a joint of drill pipe. Compare *double, thribble,* and *fourble.*

slips *n pl:* wedge-shaped pieces of metal with teeth or other gripping elements that are used to prevent pipe from slipping down into the hole or to hold pipe in place. Rotary slips fit around the drill pipe and wedge against the master bushing to support the pipe. Power slips are pneumatically or hydraulically actuated devices that allow the crew to dispense with the manual handling of slips when making a connection. Packers and other downhole equipment are secured in position by slips that engage the pipe by action directed at the surface.

slurry *n:* a plastic mixture of cement and water that is pumped into a well to harden; there it supports the casing and provides a seal in the wellbore to prevent migration of underground fluids.

spear *n:* a fishing tool used to retrieve pipe lost in a well. The spear is lowered down the hole and into the lost pipe, and, when weight, torque, or both are applied to the string to which the spear is attached, the slips in the spear expand and tightly grip the inside of the wall of the lost pipe. Then the string, spear, and lost pipe are pulled to the surface.

spinning cathead *n:* a spooling attachment on the makeup cathead to permit use of a spinning chain to spin up or make up drill pipe. See *spinning chain.*

spinning chain *n:* a Y-shaped chain used to spin up (tighten) one joint of drill pipe into another. In use, one end of the chain is attached to the tongs, another end to the spinning cathead, and the third end is free. The free end is wrapped around the tool joint, and the cathead pulls the chain off the joint, causing the joint to spin (turn) rapidly and tighten up. After the chain is pulled off the joint, the tongs are secured in the same spot, and continued pull on the chain (and thus on the tongs) by the cathead makes up the joint to final tightness.

spud *v:* to move the drill stem up and down in the hole over a short distance without rotation. Careless execution of this operation creates pressure surges that can cause a formation to break down, which results in lost circulation. See *spud in.*

spud in *v:* to begin drilling; to start the hole.

squeeze cementing *n:* the forcing of cement slurry by pressure to specified points in a well to cause seals at the points of squeeze. It is a secondary-cementing method that is used to isolate a producing formation, seal off water, repair casing leaks, and so forth. See *cementing.*

stab *v:* to guide the end of a pipe into a coupling or tool joint when making up a connection. See *coupling* and *tool joint.*

stabbing board *n:* a temporary platform erected in the derrick or mast some 20 to 40 feet above the derrick floor. The derrickman or another crew member works on the board while casing is being run in a well. The board may be wooden or fabricated of steel girders floored with antiskid material and powered electrically to raise or lower it to the desired level. A stabbing board serves the same purpose as a monkeyboard but is temporary instead of permanent.

stake a well *v:* to locate precisely on the surface of the ground the point at which a well is to be drilled. After exploration techniques have revealed the possibility of the existence of a subsurface, hydrocarbon-bearing formation, a certified and registered land surveyor drives a stake into the ground to mark the spot where the well is to be drilled.

stand *n:* the connected joints of pipe racked in the derrick or mast when making a trip. On a rig, the usual stand is 90 feet long (three lengths of pipe screwed together) or a thribble. Compare *double* and *fourble.*

standpipe *n:* a vertical pipe rising along the side of the derrick or mast, which joins the discharge line leading from the mud pump to the rotary hose and through which mud is pumped going into the hole. See *mud pump* and *rotary hose.*

- 110 -

stimulation *n:* any process undertaken to enlarge old channels or create new ones in the producing formation of a well (e.g., acidizing or formation fracturing). See *acidize*.

string *n:* the entire length of casing, tubing, or drill pipe run into a hole; the casing string. Compare *drill string* and *drill stem*.

string up *v:* to thread the drilling line through the sheaves of the crown block and traveling block. One end of the line is secured to the hoisting drum and the other to the derrick substructure. See *sheave*.

stuck pipe *n:* drill pipe, drill collars, casing, or tubing that has inadvertently become immobile in the hole. It may occur when drilling is in progress, when casing is being run in the hole, or when the drill pipe is being hoisted.

sub *n:* a short, threaded piece of pipe used to adapt parts of the drilling string that cannot otherwise be screwed together because of differences in thread size or design. A sub may also perform a special function. Lifting subs are used with drill collars to provide a shoulder to fit the drill-pipe elevators. A kelly saver sub is placed between the drill pipe and kelly to prevent excessive thread wear of the kelly and drill-pipe threads. A bent sub is used when drilling a directional hole. Sub is a short expression for substitute.

submersible drilling rig *n:* an offshore drilling structure with several compartments that are flooded to cause the structure to submerge and rest on the seafloor. Most submersible rigs are used only in shallow waters.

substructure *n:* the foundation on which the derrick or mast and usually the drawworks sit; contains space for storage and well-control equipment.

suction pit *n:* the mud pit from which mud is picked up by the suction of the mud pumps; also called a sump pit and mud-suction pit.

surface casing *n:* also called surface pipe. See *surface pipe*.

surface pipe *n:* the first string of casing (after the conductor pipe) that is set in a well, varying in length from a few hundred to several thousand feet. Some states require a minimum length to protect freshwater sands. Compare *conductor pipe*.

swivel *n:* a rotary tool that is hung from the rotary hook and traveling block to suspend and permit free rotation of the drill stem. It also provides a connection for the rotary hose and a passageway for the flow of drilling fluid into the drill stem.

—T—

TD *abbr:* total depth.

thread protector *n:* a device that is screwed onto or into pipe threads to protect the threads from damage when the pipe is not in use. Protectors may be metal or plastic.

thribble *n:* a stand of pipe made up of three joints and handled as a unit. See *stand*. Compare *single*, *double*, and *fourble*.

thribble board *n:* the name used for the working platform of the derrickman, or monkeyboard, when it is located at a height in the derrick equal to three lengths of pipe joined together. Compare *double board* and *fourble board*. See *monkeyboard*.

throw the chain *n:* to flip the spinning chain up from a tool-joint box so that the chain wraps around the tool-joint pin after it is stabbed into the box. The stand or joint of drill pipe is turned or spun by a pull on the spinning chain from the cathead on the drawworks.

tight hole *n:* a well about which information is restricted and passed only to those authorized for security or competitive reasons.

tongs *n pl:* the large wrenches used for turning when making up or breaking out drill pipe, casing, tubing, or other pipe; variously called casing tongs, rotary tongs, and so forth according to the specific use. Power tongs are pneumatically or hydraulically operated tools that serve to spin the pipe up tight and, in some instances, to apply the final makeup torque. See also *chain tongs*.

tool joint *n:* a heavy coupling element for drill pipe made of special-alloy steel. Tool joints have coarse, tapered threads and seating shoulders designed to sustain the weight of the drill stem, withstand the strain of frequent coupling and uncoupling, and provide a leak-proof seal. The male section of the joint, or the pin, is attached to one end of a length of drill pipe, and the female section, or box, is attached to the other end. The tool joint may be welded to the end of the pipe or screwed on or both. A hard-metal facing is often applied in a band around the outside of the tool joint to enable it to resist abrasion from the walls of the borehole.

tool pusher *n:* an employee of a drilling contractor who is in charge of the entire drilling crew and the drilling rig. Also called a drilling rig foreman, manager, supervisor, or rig superintendent. See *drilling foreman*.

torque *n:* the turning force that is applied to a shaft or other rotary mechanism to cause it to rotate or tend to do so. Torque is measured in foot-pounds, joules, meter-kilograms, and so forth.

torque converter *n:* a connecting device between a prime mover and the machine actuated by it. The elements that pump the fluid in the torque converter automatically increase the output torque of the engine to which the torque is applied, with an increase of load on the output shaft. Torque converters are used extensively on mechanical rigs that have a compound. See *mechanical rig*.

total depth *n:* the maximum depth reached in a well.

tour *n:* (pronounced "tower") an eight-hour shift worked by a drilling crew or other oil-field workers. Sometimes twelve-hour tours are used, especially on offshore rigs. The most common divisions of tours are daylight, evening, and graveyard, if eight-hour tours are employed.

transmission *n:* the gear or chain arrangement by which power is transmitted from the prime mover to the drawworks, mud pump, or rotary table of a drilling rig. See *prime mover*.

traveling block *n:* an arrangement of pulleys, or sheaves, through which drilling cable is reeved and that moves up and down in the derrick or mast. See *block*, *crown block*, and *sheave*.

tricone bit *n:* a type of bit in which three cone-shaped cutting devices are mounted in such a way that they intermesh and rotate together as the bit drills. The bit body may be fitted with nozzles, or jets, through which the drilling fluid is discharged. A one-eyed bit is used in soft formations to drill a deviated hole. See *directional drilling* and *bit*.

trip *n:* the operation of hoisting the drill stem from and returning it to the wellbore. See *make a trip*.

turbodrill *n:* a drilling tool that rotates a bit attached to it by the action of the drilling mud on the turbine blades built into the tool. When a turbodrill is used, rotary motion is imparted only at the bit; therefore it is unnecessary to rotate the drill stem. Although straight holes can be drilled with the tool, it is used most often in directional drilling.

—U—

unit operator *n:* the oil company in charge of development and producing in an oil field in which several companies have joined together to produce the field.

—V—

valve *n:* a device used to control the rate of flow in a line, to open or shut off a line completely, or to serve as an automatic or semiautomatic safety device. Those with extensive usage include the gate valve, plug valve, globe valve, needle valve, check valve, and pressure-relief valve. See *check valve*, *needle valve*, and *pressure-relief valve*.

V-belt *n:* a belt with a trapezoidal cross section that is made to run in sheaves, or pulleys, with grooves of corresponding shape. See *belt*.

—W—

waiting on cement *adj:* pertaining to or during the time when drilling or completion operations are suspended so the cement in a well can harden sufficiently.

wall cake *n:* also called filter cake and mud cake. See *filter cake* and *mud cake*.

weevil *n:* shortened form of boll weevil. See *boll weevil*.

weight indicator *n:* an instrument near the driller's position on a drilling rig. It shows both the weight of the drill stem that is hanging from the hook (hook load) and the weight that is placed on the bit by the drill collars (weight on bit).

weighting material *n:* a material that has high specific gravity and is used to increase the density of drilling fluids or cement slurries.

wellbore *n:* a borehole; the hole drilled by the bit. A wellbore may have casing in it or may be open (i.e., uncased), or a portion of it may be cased and a portion of it may be open. Also called borehole or hole. See *cased* and *open*.

well completion *n:* the activities and methods necessary to prepare a well for the production of oil and gas; the method by which a flow line for hydrocarbons is established between the reservoir and the surface. The method of well completion used by the operator depends on the individual characteristics of the producing formation or formations. These techniques include open-hole completions, sand-exclusion completions, tubingless completions, multiple completions, and miniaturized completions.

wellhead *n:* the equipment installed at the surface of the wellbore. A wellhead includes such equipment as the casinghead and tubing head. *adj:* pertaining to the wellhead (as wellhead pressure).

well stimulation *n:* any of several operations used to increase the production of a well. See *acidize* and *formation fracturing*.

whipstock *n:* a long, steel casing that uses an inclined plane to cause the bit to deflect from the original borehole at a slight angle. Whipstocks are sometimes used in controlled directional drilling, to straighten crooked boreholes, and to sidetrack to avoid unretrieved fish. See *directional drilling, fish,* and *sidetrack*.

wireline *n:* a slender, rodlike or threadlike piece of metal, usually small in diameter, that is used for lowering special tools (such as logging sondes, perforating guns, and so forth) into the well. Compare *wire rope*.

wire rope *n:* a cable composed of steel wires twisted around a central core of hemp or other fiber to create a rope of great strength and considerable flexibility. Wire rope is used as drilling line (in rotary and cable-tool rigs), coring line, servicing line, winch line, and so on. It is often called cable or wireline; however, wireline is a single, slender metal rod, usually very flexible. Compare *wireline*.

WOC *abbr:* waiting on cement.

worm *n:* a new and inexperienced worker on a drilling rig.

DRILLING AND COMPLETION—BIBLIOGRAPHY

Abell, J.M. (ed): **Practical Aspects of Well Drilling and Production,** The Institute for Energy Development, Inc. (Oklahoma City) 1980.

American Petroleum Institute: **History of Petroleum Engineering,** API (Dallas) 1961.

Craft, Holden, and Graves: **Well Design: Drilling and Production,** Prentice-Hall, Inc. (Englewood Cliffs, N.J.) 1962.

Desk and Derrick Clubs of North America: **D & D Standard Oil Abbreviator,** 2nd Ed., The Petroleum Publishing Company (Tulsa) 1973.

ETA Offshore Seminars, Inc.: **The Technology of Offshore Drilling, Completion, and Production,** The Petroleum Publishing Company (Tulsa) 1976.

Gatlin, C.: **Petroleum Engineering: Drilling and Well Completions,** Prentice-Hall, Inc. (Englewood Cliffs, N.J.) 1960.

Harris, L.M.: **Deepwater Drilling Operations,** The Petroleum Publishing Company (Tulsa) 1974.

McGray, Arthur W. and Frank W. Cole: **Oil Well Drilling Technology,** University of Oklahoma Press (Norman) 1967.

Moore, Preston L.: **Drilling Practices Manual,** The Petroleum Publishing Company (Tulsa) 1974.

Petroleum Extension Service: **A Primer of Offshore Operations,** The University of Texas at Austin (Austin) 1976.

Petroleum Extension Service: **A Primer of Oil-Well Drilling,** 4th Ed., The University of Texas at Austin (Austin) 1979.

CHAPTER 6:

PRODUCTION

I. INTRODUCTION

Without oil and gas production, oil companies simply would not exist. When a company begins to produce oil or gas from underground formations, the cash flow is reversed for the first time. All the activities and processes discussed prior to this, from geology and exploration through drilling, have been concerned with spending money to find petroleum. Not until we reach the subject of producing oil do we begin to look at a return on the original investment.

This chapter is designed to complement the previous chapter on drilling and completion methods. In that chapter, the various types of equipment and drilling techniques used in today's industry were presented, followed by a description of the methods used to complete and stimulate oil and gas wells. This chapter continues from that point and discusses how oil and gas are brought to the surface through these wells and prepared for sale.

The main topics to be covered in this chapter are flowing wells, artificial lift, oil treating, storage and sale of oil, and saltwater disposal. Both onshore and offshore applications will be examined. A bibliography is provided to assist those who wish to investigate these subjects further. A glossary is also included at the end of the chapter to review the production terminology used in the chapter and to provide definitions of additional terms.

II. FLOWING WELLS

A flowing well may be defined as any well which has sufficient pressure in the reservoir rock to cause the oil or gas to flow to the surface through the wellbore. Flowing wells have always been considered ideal, because they require relatively little equipment or expense to bring the oil to the surface. Such wells are common in both offshore and onshore operations, and are typical of large oil fields, where sufficient pressure exists in the rock to force the oil to the surface. We will first examine some of the simple mechanical principles involved in a flowing well. Following that, we will look at dual zone completions, which are used with flowing wells having more than one pay zone.

The equipment commonly used in a flowing well consists of **tubing** and a **wellhead** (Figure 6.1). The casing has been inserted through the pay zone and set in place, as discussed in the previous chapter. It has been perforated to provide a flow path for the oil or gas to follow from the reservoir rock into the wellbore. A smaller string of pipe, called **tubing**, is inserted inside the casing and suspended from the surface by the wellhead. This tubing is not cemented in place, as is the casing, but serves as a conduit inside the casing for most oil and gas production from a well to the surface. It is smaller than the casing, usually being no more than two or three inches in diameter, and consists of steel joints or sections of pipe, 30 feet in length, screwed together to

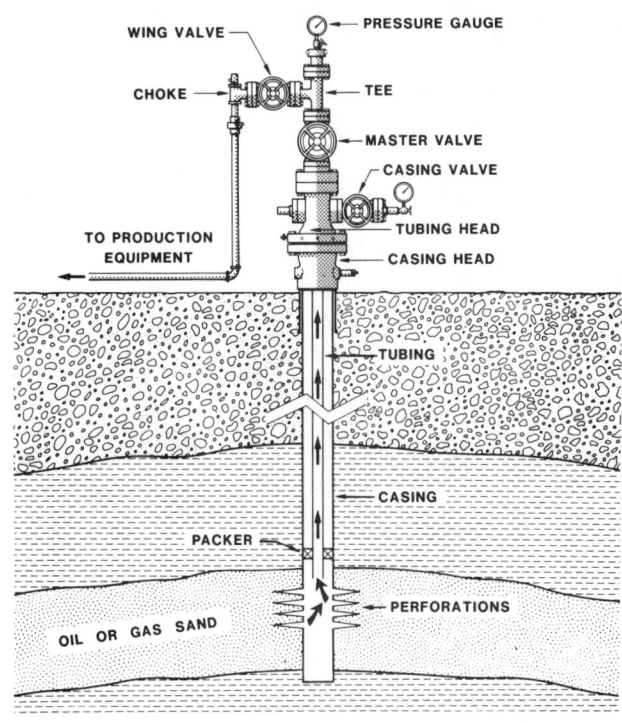

Figure 6.1. FLOWING GAS OR OIL WELL.

form a continuous flow tube from the perforations to the surface.

This tubing is usually used as a flow path instead of the casing, because the tubing is considered expendable and can be retrieved, if necessary, from the well. The tubing can be easily replaced if it becomes damaged by corrosion or the force of the fluids flowing through it. On the other hand, the casing, which has been cemented in the ground, is a permanent fixture and cannot tolerate excessive abuse or wear.

At the top of the tubing, on the surface of the ground, is a **wellhead.** This is often called the "Christmas tree," because a large wellhead with many valves resembles a tree with many ornaments hanging from it. The wellhead has several purposes: the tubing is usually suspended in the well from the bottom of the wellhead, and through its system of gauges and valves, the wellhead is able to monitor and control the high pressure often encountered within the well.

The wellhead consists of several heavy-duty valves, which have been designed to withstand pressure that the formation might exert at the surface, either during flowing operations or when the well is shut in. The large center valve in the middle of the tree is called the **master valve** (Figure 6.1). When the well is to be completely shut in, that is, when the flow is to be stopped, this valve is closed. The smaller valve on the side of the tree, called the **wing valve,** is also used for shutting in the well. Wellheads may also contain a number of other devices, such as pressure gauges, meters, chokes, and thermometers.

To produce a flowing well, the master valve, the wing valve, and the choke are opened, allowing the forces in the formation simply to drive the oil and gas to the surface. As long as sufficient reservoir pressure is available to push the oil and gas to the surface, the tubing and wellhead are all that are needed to produce petroleum.

Before proceeding further with our discussion of flowing wells, let us look at an important piece of equipment commonly used in the production of oil or gas—the **packer.** As shown in the left-hand diagram in Figure 6.2, the packer is a large, expandable, rubber sealing device, which is attached to the string of tubing and inserted into the well. The packer is small enough to be lowered into the casing to the desired point. It is then set, as shown in the center diagram, by rotating the tubing, which allows a crushing force to be exerted on the packer. This action forces the expandable rubber seal on the packer against the side of the casing, sealing off the annular space between the casing and tubing. The packer shown in Figure 6.2 is called a **retrievable packer.** As demonstrated in the right-hand diagram, the packer can be withdrawn from the well simply by rotating the tubing in the reverse direction, which permits the rubber seal to contract. The entire assembly can then be retrieved from the well.

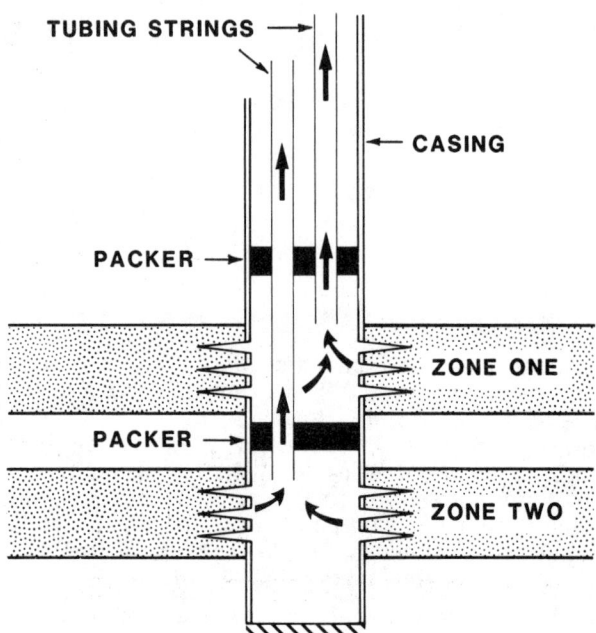

Figure 6.3. DUAL ZONE COMPLETION.

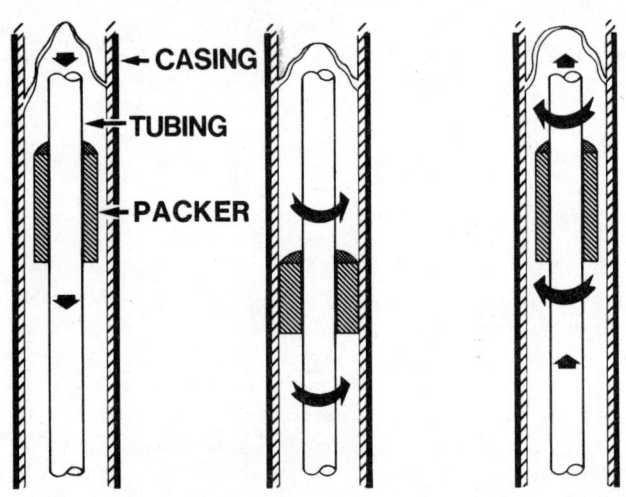

Figure 6.2. RETRIEVABLE PACKER. Left: packer and tubing are run into the well. Center: packer is set by rotating tubing and compressing packer. Right: rotation is reversed, allowing packer and tubing to be removed from the well.

Another important use of the packer is for **dual zone** or **multiple zone completions,** so called because there are two or more pay zones containing oil that the company wants to produce through one wellbore. In a dual zone completion, two strings of tubing are inserted into the well, each with a packer attached to it (Figure 6.3). A packer attached to one of the tubing strings is used to seal off the lower pay zone (pay zone two in the diagram). With the annulus sealed off by the packer, the lower pay zone can produce into a separate part of the wellbore through its own tubing string. The upper pay zone (pay zone one) produces into a second tubing string, which is also sealed off from the tubing-casing annulus by a second packer. Because of the packers, both flows of oil or gas can be brought to the surface and measured separately. Note that in the illustration, the upper packer is made so both strings of tubing fit through it, one string of tubing for zone one and the other for zone two.

There is a limit to the number of packers that can be run into a well. After placing more than two or three packers in a well, the necessary mechanical equipment becomes so complicated that it is generally not worth the trouble or risk. The packers can get stuck in the well, and become expensive to retrieve.

III. ARTIFICIAL LIFT

A. Introduction

Artificial lift implies that some type of mechanical or artificial means is necessary to bring the oil from the reservoir to the surface. Several different types of artificial lift will be discussed in this chapter.

Artificial lift is a technique which has been in use for many years throughout the world. Sooner or later, almost all oil wells require some type of artificial lift. Even though there may have been good flowing wells initially, the pressure in a reservoir gradually begins to exhaust itself. Consequently, some other method is necessary to lift the oil to the surface.

A number of different and interesting techniques have evolved over the years in the industry's use of artificial lift. Shown in Figure 6.4 is an old standard rig with a pumping unit. In the 1800s, these were common sights in the oil fields of America, and many variations on this type of equipment existed at that time. Since then, a number of refinements and advances have been made in equipment and technology. As a result, there are currently four common methods of artificial lift: **beam pumping, submersible pumping, gas lift,** and **hydraulic pumping.**

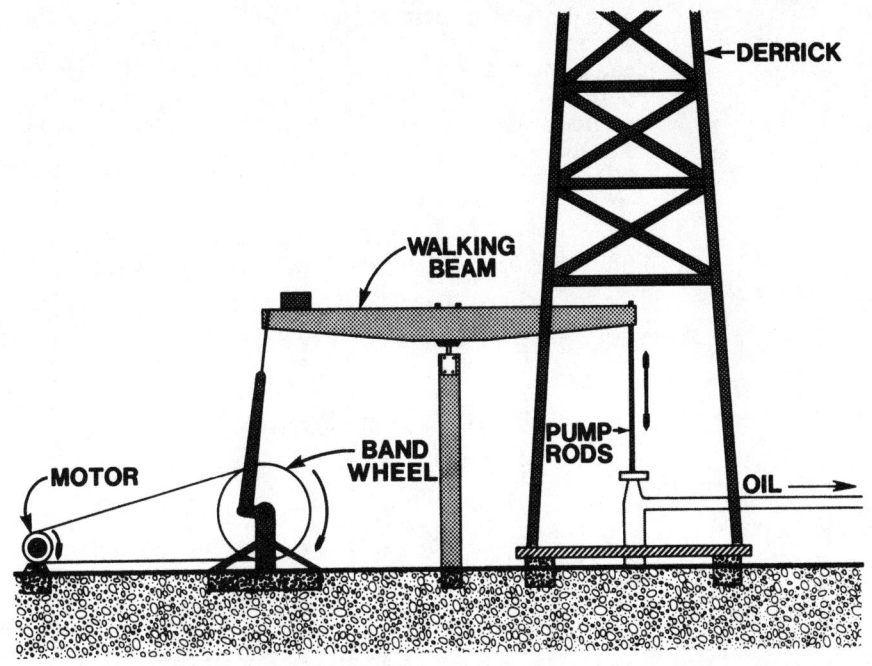

Figure 6.4. OLD STANDARD RIG WITH PUMP.

beam to rock back and forth, and in the process, move the sucker rods up and down. The entire system is run by an engine, gear box, and crank mechanism on the gear box. This type of pumping unit is common throughout the United States, and a majority of wells which require artificial lift are equipped with such devices.

C. Submersible Pumping

Submersible pumping is a more modern method of lifting fluids to the surface. A submersible pump consists of an electric motor attached to a pump which is lowered into the well on the end of the tubing string (Figure 6.6). The electric motor turns a centrifugal pump which forces oil from the bottom of the well, up through the inside of the tubing, and out at the surface. The electricity is supplied through an electric cable attached to the side of the tubing and connected to the electric motor. Submersible

B. Beam Pumping

In beam pumping, a pump is installed inside the tubing string that is inserted into the well, and a beam pumping unit is used to lift the oil to the surface (Figure 6.5). The pump is placed inside the tubing at a depth where it is submerged into the oil. This pump is similar in construction to the old water well pumps that were used on homesteads to supply water. The pump is connected to the surface by a string of **sucker rods**. These are long steel rods, usually a half inch to one inch in diameter, screwed together in 25-foot joints, which connect the pump at the bottom of the well to a pumping unit at the surface. The steel sucker rods fit inside the tubing and are stroked up and down in the tubing by the surface pumping unit, activating the pump at the bottom of the well. Each time the rods and pump are stroked, a volume of oil is lifted through the inside of the tubing and discharged at the surface. The rods connected to the pumping unit then drop, allowing more oil to flow into the tubing. This process is repeated over and over, producing an almost continuous stream of oil at the surface. The oil is pumped through the tubing between the sucker rods and the wall of the tubing.

At the surface, a large mechanical device called the **beam pumping unit** is attached (Figure 6.5). This unit consists of a large **walking beam** fixed to a pivotal post, called a **sampson post**, which enables the

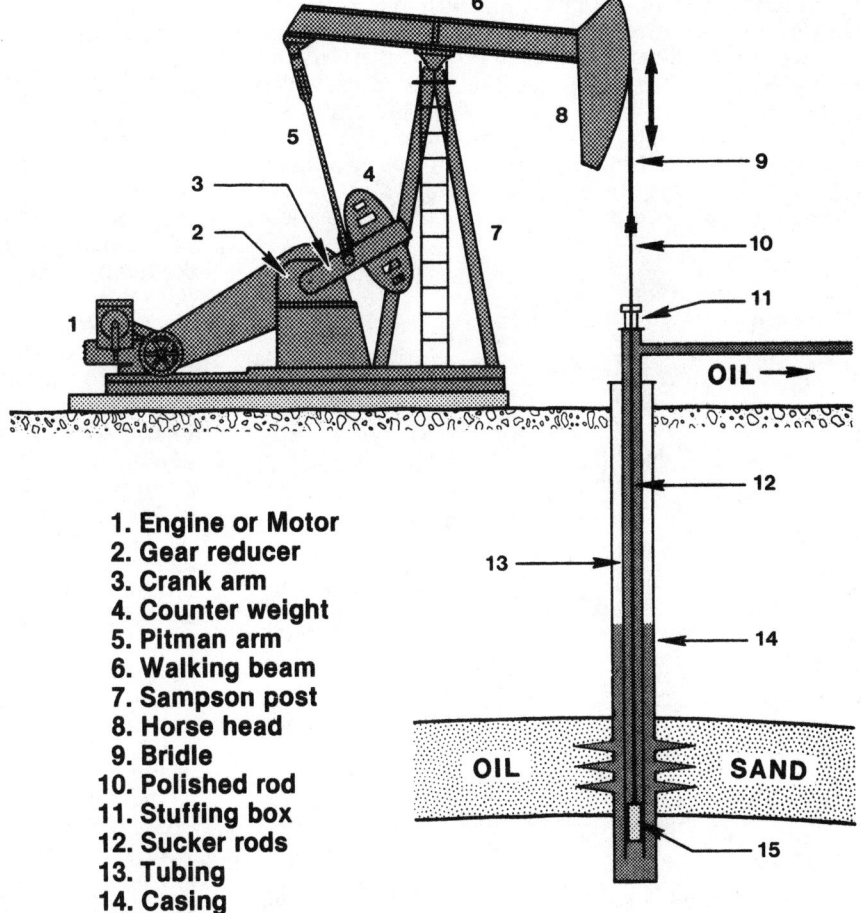

1. Engine or Motor
2. Gear reducer
3. Crank arm
4. Counter weight
5. Pitman arm
6. Walking beam
7. Sampson post
8. Horse head
9. Bridle
10. Polished rod
11. Stuffing box
12. Sucker rods
13. Tubing
14. Casing
15. Pump

Figure 6.5. BEAM PUMPING UNIT.

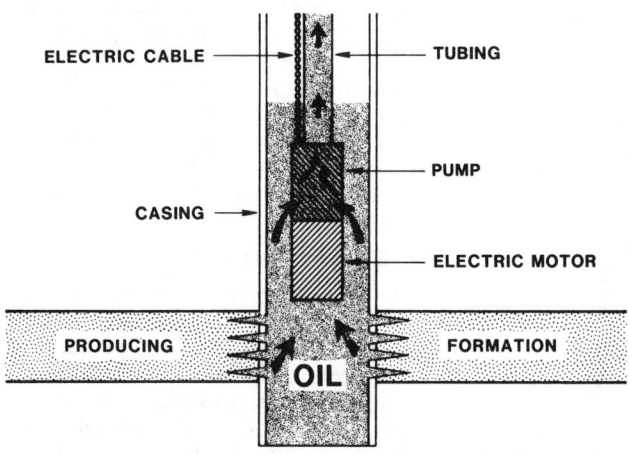

Figure 6.6. SUBMERSIBLE PUMP.

pumps are used where high volumes of fluid must be lifted to the surface. They are often used for both high volume water wells and oil wells. Submersible pumps cannot operate at extreme depths, but are most commonly used in wells no more than several thousand feet deep.

D. Gas Lift

Gas lift is a third method of bringing oil to the surface. Gas lift simply utilizes tubing which is run into the well to the point where oil flows into the wellbore (Figure 6.7). At this point, the tubing is often sealed with a packer, and a series of devices called **gas lift valves** are inserted into the sides of the tubing. The gas is injected into the well through the tubing-casing annulus and enters the tubing through the gas lift valves, where it mixes with the oil and returns to the surface. The fluid

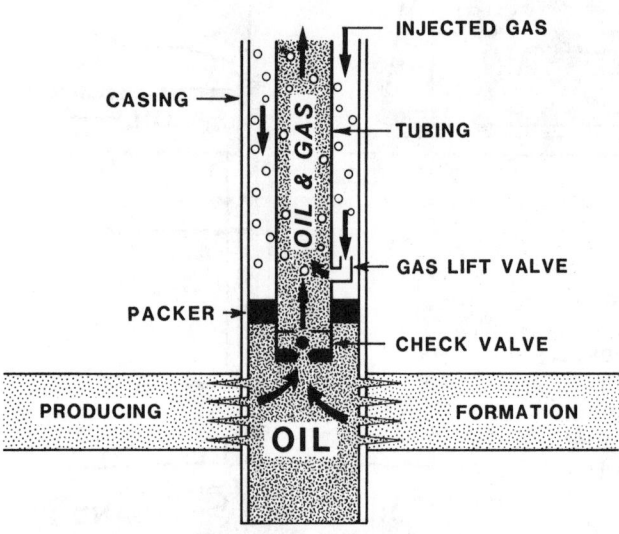

Figure 6.7. GAS LIFT.

in the tubing is made lighter by the gas, and as a result, the mixture is raised to the surface.

This method is commonly used in the Gulf Coast of the United States and in other areas of the world, where there is an abundant supply of gas to power the artificial lift mechanism. Often the gas from the oil well itself is collected as it is produced at the surface and reinjected into the well to help lift additional oil and gas to the surface. This type of lift mechanism also requires a compressor at the surface. The compressor is used to pump high pressure gas into the well, so the gas has sufficient force to enter the tubing and lift the oil to the surface.

E. Hydraulic Pumping

In many ways, hydraulic pumping is similar to gas lift. In hydraulic pumping, however, oil is used to lift

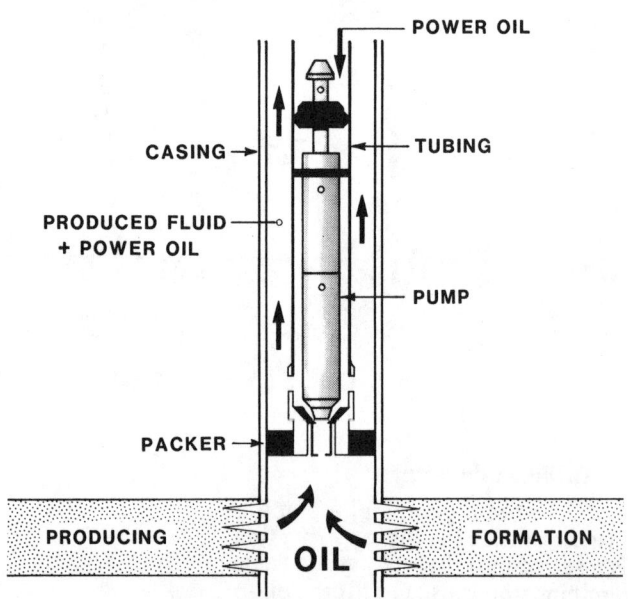

Figure 6.8. HYDRAULIC PUMPING.

the production from the pay zone instead of gas. **Power oil,** or high pressure oil, is pumped into the well through the tubing string (Figure 6.8). At the bottom of the well, the power oil enters a mechanical device, causing the mechanism to reciprocate. This mechanical device activates a pump, which lifts the oil from the producing formation, together with expended power oil, to the surface. This technique is often used to produce deep formations. In the mechanical setup for hydraulic pumping, a packer is again used to seal off the producing zone at the bottom of the well from the power oil. The pumping mechanism is included in the tubing immediately above the packer. The pump pictured in Figure 6.8 is only one of a number of different types of hydraulic pumps. Many variations exist to handle different problems.

IV. OIL TREATING

A. Introduction

Rarely is clean oil, ready for sale into a pipeline, produced from an oil well. Generally, what comes out of the well is a mixture of oil, water, gas, and even sand or solid material. Foreign material, such as water and sand, must be separated from the oil and gas before they can be sold. This process is known as **oil treating** or **oil dehydration,** in which water is removed from the oil. The amount of this foreign material is referred to as the **BS&W,** or **basic sediment and water,** content of the oil. Normally, the BS&W content must be less than one percent before the oil will be acceptable for sale into a pipeline. Pipeline companies do not want to buy water or solid materials any more than a consumer would

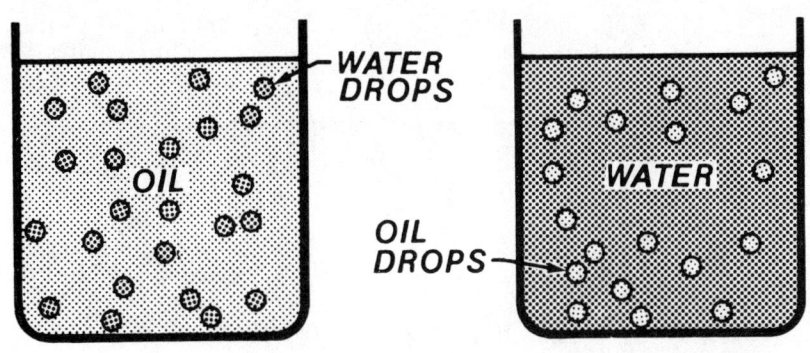
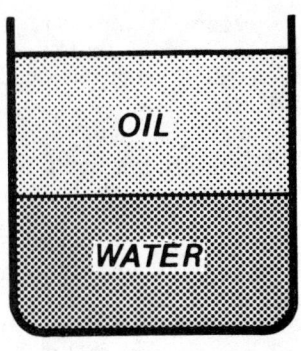

WATER-IN-OIL EMULSION **OIL-IN-WATER EMULSION** **SEPARATED OIL AND WATER**

Figure 6.9. EMULSIONS.

want to purchase hamburger with excessive amounts of fat. Some type of treating involving special equipment is usually necessary to remove these materials from the oil or gas.

Oil treating requires a knowledge of **emulsions**. Oil-water emulsions are common in the oil field and need specialized treating before the oil can be cleaned. A water-in-oil emulsion consists of water drops suspended in an oil solution (Figure 6.9). Conversely, an oil-in water emulsion consists of oil drops suspended in water. These tiny droplets often will not separate because the finely dispersed droplets are not large enough to coalesce and form into separate oil and water components. A good example of an emulsion is homogenized milk. In homogenized milk, cream has been emulsified, or finely dispersed, into the milk so it will not separate out. The same type of thing often happens to oil and water as production occurs from a formation. The oil and water become mixed together and form an emulsion to the extent that the water or oil will not easily separate.

The objective is to separate the oil from the water, or to break the emulsion (Figure 6.9). Generally, the emulsion must be heated and some emulsion-breaking chemical added to accomplish this. It has been found that, if an emulsion is warmed or heated, the emulsified fluids separate. For example, a jar of mayonnaise left out on a warm day for a length of time will begin to separate into its different components.

In the following sections, we will look at some of the equipment that is used to separate or de-emulsify oil and water mixtures. Many different types of oil treating equipment are used in the oil field. A few of the basic types will be discussed in this chapter: **separators, free-water knockouts,** and **heater-treaters.**

B. Separators

A separator is a simple device, which basically utilizes the force of gravity to separate oil-gas mixtures. In a vertical separator, as shown in Figure 6.10, the oil and gas mixture enters at the center of the vessel. The oil, which is heavier than the gas, falls to the bottom of the vessel and is taken off through the **fluid line.** The lighter gas rises to the top and is removed for separate sale.

The **oil level control** in Figure 6.10 is used to maintain a certain fluid level in the separator, so gas cannot escape out the bottom of the vessel. The **float** is connected to a control device that closes the valve on the oil flow line when the oil level drops below a designated point. After the vessel fills to a certain level, the float again rises and allows the oil valve to open and the oil to flow out of the vessel.

Separators are commonly used with high pressure wells involving large volumes of fluid, or in offshore production operations, where high pressure flowing oil and gas wells usually require separation of the gas and oil for shipment in different pipelines.

C. Free-Water Knockouts

Another rather simple device used to separate oil, water, and gas is the free-water knockout (Figure 6.11). Oil, water, and gas from the well flow into the vessel through an **inlet valve** and then are allowed to slow down in the large settling chamber. Here, any free water mixed with the oil settles to the bottom of the vessel and is drawn off and disposed of. The oil is removed through a separate line and sent to additional

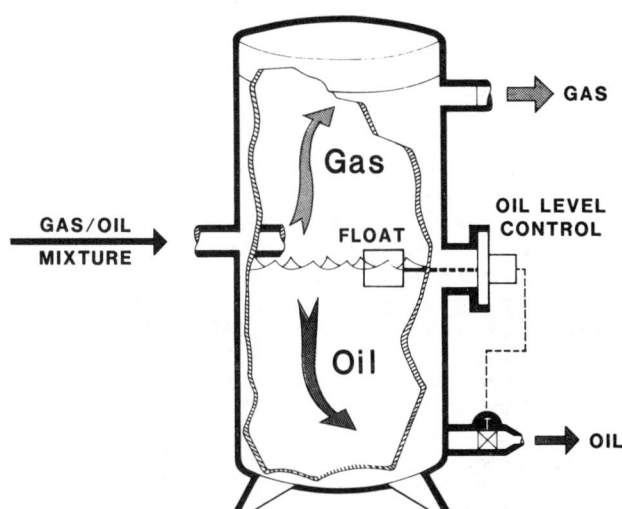

Figure 6.10. VERTICAL SEPARATOR.

oil processing equipment. The natural gas, which rises to the top of the vessel, is extracted through a third line.

The vessel is called a free-water knockout, because it is designed to eliminate only water which has not been emulsified with the oil, that is, water which is free to separate by itself. Water which has been emulsified, or dispersed into very fine drops within the oil, will not separate by itself and will not be eliminated by the free-water knockout. This water must be further processed by a device which can supply heat to the emulsion and force it to separate.

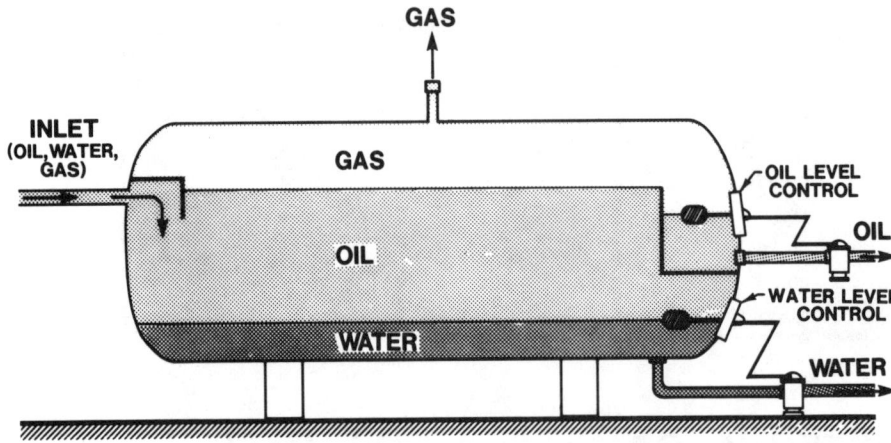

Figure 6.11. FREE-WATER KNOCKOUT.

In principle, free-water knockouts are almost identical to separators, although free-water knockouts are larger and are generally placed horizontally. As with separators, the oil and water outlets are controlled by level control valves, which prevent the vessel from draining completely and keep the gas trapped in the top of the vessel. Free-water knockouts are most commonly used for low pressure wells, but they can be used to handle high volumes of fluid.

D. Heater-Treaters

Heater-treaters, also called **emulsion treaters**, are common devices used throughout the world to separate oil and water emulsions. The heater-treater is similar to a free-water knockout, but the treater has heating capability with the inclusion of **fire tubes** (Figure 6.12). The fire tubes extend horizontally into the vessel; natural gas or oil burns inside the fire tubes and creates heat, which passes through the fire tubes and into the emulsion. The combustion of the gas within the fire tubes is used to heat the oil and water emulsion entering the vessel and passing around the outside of the tubes. As the oil and water mixture grows hotter, the emulsion breaks or separates and forms into clean oil and clean water.

The water is removed from the bottom of the heater-treater and sent to the water disposal system. The clean oil is drawn off the center of the vessel and sent to the oil storage tanks for sale. Again, level control devices are used to regulate the drainage of water and oil from the vessel. Any natural gas which occurs in the oil-water emulsion exits at the top of the heater-treater. This gas, together with any gas produced from the free-water knockout or other vessels, can subsequently be used as a fuel for the heater. If an excess quantity of gas exists, it may be sold into a separate gas pipeline.

E. Combinations

The above separating and treating devices are often used together to treat the fluids produced by a well. A typical combination of equipment is diagrammed in Figure 6.13. At the left of the diagram, oil, water, and gas are being produced as a mixture from the well. They are first sent into a free-water knockout, where any water that would readily and freely separate from the oil is extracted and sent into a saltwater disposal system. (This disposal system will be discussed in a later section.) Any free gas which separates out of the mixture is drawn off as a separate sidestream from this

Figure 6.12. HORIZONTAL HEATER-TREATER.

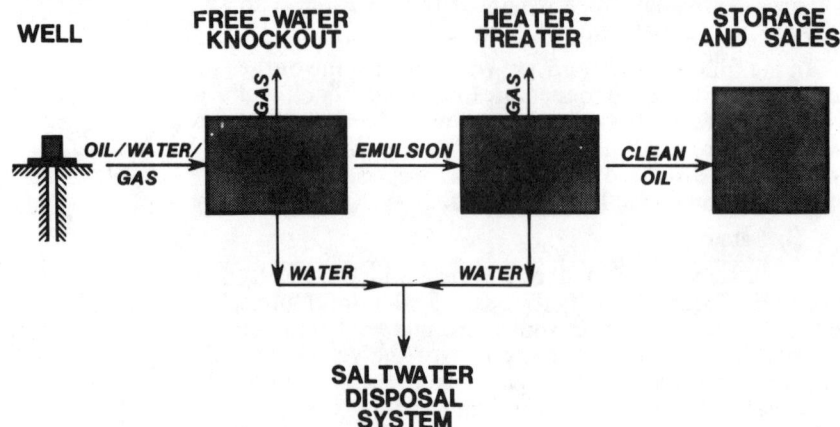

Figure 6.13. SURFACE FACILITIES FOR TYPICAL OIL WELL.

vessel. This gas will either be used as fuel or sold. The remaining oil and water emulsion leaves the free-water knockout and is transferred to the heater-treater. Here, heat is applied to break the emulsion and separate out any remaining water, which is also sent to the saltwater disposal system. Again, gas is removed at the top of the vessel for use as fuel. The clean oil stream is sent to the storage tanks, where it will be ready for shipment into a pipeline or truck sales outlet.

This is one method of dehydrating or treating the produced oil. The size and number of vessels will depend on the type of oil produced and the amount of water and gas being produced with the oil. On an offshore platform, a high pressure separator often will be used instead of the free-water knockout.

F. Other Treating Methods

In addition to the techniques described above, there are other methods of treating or dehydrating crude oil. One of the most common is the use of has been found that certain types of chemicals can be added to a water-oil emulsion or an oil-water emulsion that will cause the emulsion to break, allowing the oil and water to separate. Chemicals are often added to the oil-water emulsion at the same time that the emulsion is run through the heater-treater. This improves the overall performance of the heater-treater.

Electricity can also be used for the same purpose. An electric current applied to an emulsion as it passes through a heater-treater can facilitate the separation of the oil and water. Through the years, these techniques have been refined and modified to aid in the treatment and dehydration of crude oil. Today, chemicals and electricity are frequently used in conjunction with the more standard oil treating devices, such as heater-treaters.

V. OIL STORAGE AND SALES

A. Onshore

A typical oil storage and sales facility for an onshore operation is outlined in Figure 6.14. The clean oil passes from the heater-treater or other oil processing device to the storage tanks. In the early days of the industry, these storage tanks were either earthen pits dug out of the ground or were wooden structures. Today, however, most oil storage tanks are made of welded or bolted steel and can be a variety of sizes, ranging from 50 and 100 barrels to 10,000 barrels or larger. At a typical facility, sufficient storage is provided to accommodate oil from two or three days of production. For example, if a well produces 100 barrels a day, storage for three days would require a 300-barrel tank. Clean oil is stored in the tank until it is ready for sale. At that time, it may be withdrawn from the tank and sold by truck to a refinery or other oil purchaser, or it may be sent through a pipeline to the refinery.

The pipeline connection requires some type of metering equipment to automatically measure the oil as it passes into the pipeline from the tank. This is done by a **LACT unit,** LACT coming from **Lease Automatic Custody Transfer.** At this unit, the oil is transferred from the custody or ownership of the lease where the oil well is located to the often different ownership of the pipeline. The LACT unit consists of an automatic meter which accurately measures the amount of oil passing through it into the pipeline and continuously samples the oil for its BS&W content. Any time the BS&W content exceeds the maximum amount allowed by the pipeline company, an automatic valve on the LACT unit rejects the oil with the high BS&W content, or the bad oil, and returns it to the heater-treater for further processing. In this manner, the process of selling the oil is completely automatic and provides for better efficiency and control.

When oil is sold through a trucking company, the amount of oil in the tank must be measured before the truck withdraws it. After the oil is withdrawn, the tank is measured again. The difference between the two measurements is used to compute the exact amount of oil sold. This process is called making out a **run ticket.** It must be done accurately; even a small error in the measurement of the fluid level will result in the loss of a great deal of valuable product.

B. Offshore

The storage and sale of oil offshore is often handled differently from onshore operations due to the space limitations on a platform in the ocean. Offshore production platforms can sometimes be equipped with oil storage equipment, if the platform is located in shallow water (Figure 6.15). Platforms in shallow water can be made large enough to house processing and storage equipment. The clean oil can be sold through a pipeline extending from the platform, along the sea floor, to the land.

In deep water production operations, however, more than one platform may be necessary. It is extremely expensive to construct a large platform in deep water; therefore, the platform is only made large enough to house the drilling rig and, later, the production equipment. After the oil has been separated from the water, it is shipped through a pipeline along the sea floor to another storage platform in shallow water (Figure 6.16). The oil is then sold into a sales pipeline. Sometimes a storage platform is not necessary, and the oil is shipped directly into the sales line.

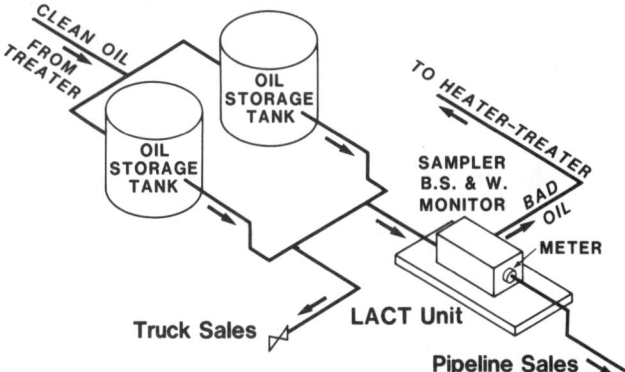

Figure 6.14. OIL STORAGE AND SALES.

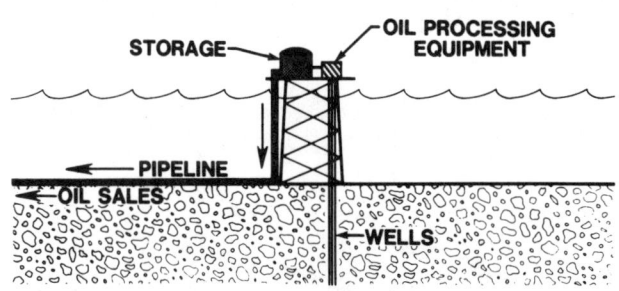

Figure 6.15. OFFSHORE PRODUCTION PLATFORM: SHALLOW WATER.

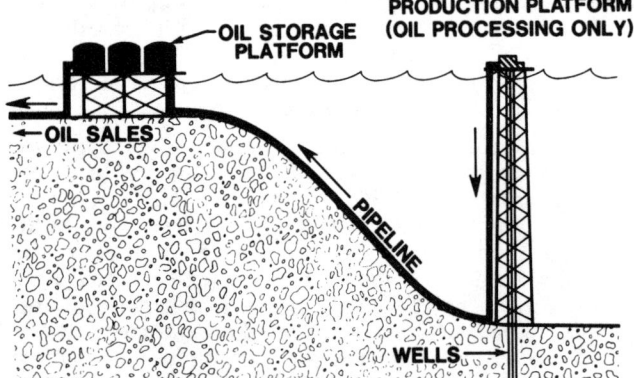

Figure 6.16. OFFSHORE PRODUCTION PLATFORM: DEEP WATER.

The installation of oil storage tanks on the ocean floor around the base of the platform is a more novel, and somewhat rare, method of storing oil offshore (Figure 6.17). A number of these platforms with underwater storage have been assembled in the North Sea. Usually, the storage tanks and the supporting towers are made of concrete, which provides weight to stabilize the platform and fix it in the sea. These tanks contain large volumes of oil, but such facilities are extremely expensive to construct.

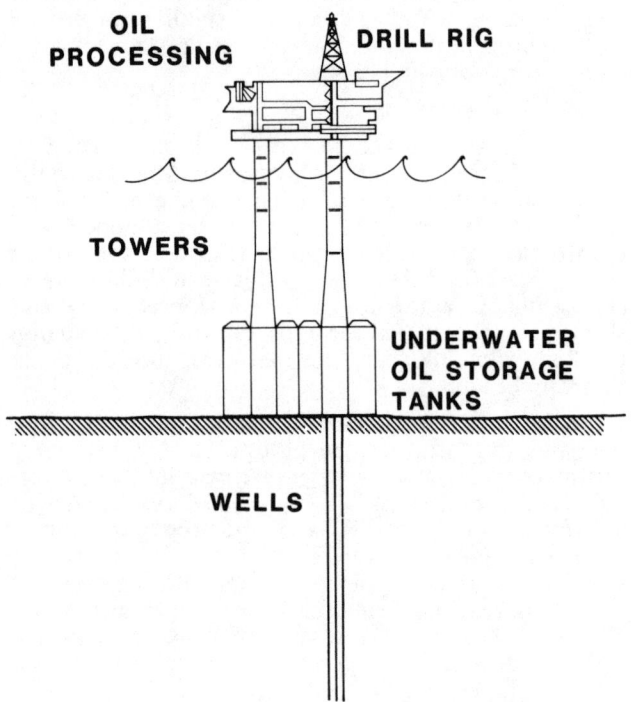

Figure 6.17. OFFSHORE PLATFORM WITH UNDERWATER OIL STORAGE.

VI. SALTWATER DISPOSAL

In addition to the necessary treating of oil, large amounts of water produced with the oil must also be processed and disposed of in a satisfactory manner. This procedure is often referred to as saltwater disposal, because the water produced from the oil well is frequently salty or brackish. This quality of the water stems from the fact that the water is taken from deep within the earth. In early geologic time, the water was probably part of an ancient sea and was trapped in the rock as it was formed. Generally, this salt water is not clean enough or fresh enough to be used for agriculture or to be disposed of at the surface. To prevent environmental pollution, the water must be reinjected into the ground, often into the same rock from which it was drawn.

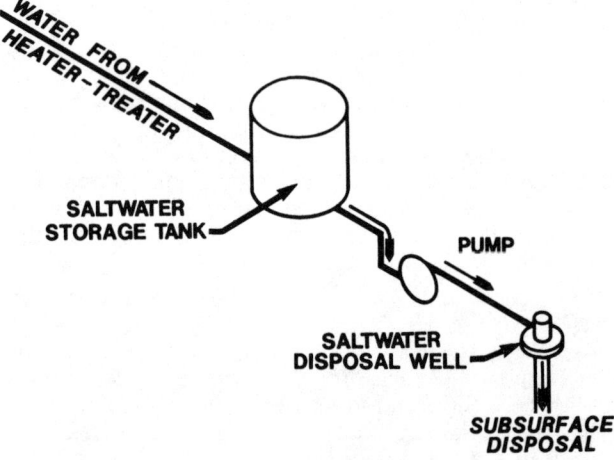

Figure 6.18. SALTWATER DISPOSAL SYSTEM.

In a saltwater disposal system, the water is taken from the oil processing equipment and usually sent to a storage tank, where it is held until a sufficient volume has been accumulated to reinject into the ground (Figure 6.18). It is then passed through a pump and into the saltwater disposal well. Similar to producing wells, this well has been drilled to the desired disposal formation and cased with steel pipe. The water is pumped down the cased hole into the formation through perforations in the casing. In this way, the salt water goes only into the formations desired and is not allowed to contaminate other formations or the surface.

Often in offshore operations, the salt water from the wells can be disposed of into the sea itself, since the two are compatible. Before this can be done, however, the salt water must first be sent through several cleaning devices which remove any traces of oil to prevent the sea from being polluted.

The following glossary is taken from **A Primer of Oil and Gas Production**, published by the American Petroleum Institute. We wish to express our thanks to the American Petroleum Institute for their permission to use this glossary.

PRODUCTION—GLOSSARY

The following list of words and phrases is included in terminology that is in common use in the oil and gas production industry. Some of the words in the list are considered slang—others common—but their usage as applied to production of oil and gas differs from more normal usage. Only the most common of the words used in the oil and gas production industry are given here. The definitions are those of widest applicability, since the word meaning may vary from area to area. The list cannot therefore be considered all-inclusive, nor as covering all local variations in word meaning.

—A—

absorption—to soak up as a sponge takes water.

adapter—a device to provide connection between two other parts.

adsorption—the accumulation of a thin layer of molecules of gas or liquid on a solid surface.

air/gas lift—lifting of liquids by injection, directly into the well, of air or gas.

allowable—the amount of oil or gas that a well is permitted by state authorities to produce during a given period.

apron ring—the first or lowest ring of plates in a tank.

associated gas—natural gas, commonly known as gas-cap gas, which overlies and is in contact with crude oil in the reservoir.

—B—

back pressure—the pressure resulting from restriction of full natural flow of oil or gas.

back-up man—the person who holds one length of pipe while another length is being screwed into or out of it.

baffles—plates which change the direction of flow of fluids.

ball and seat—the main parts of the valves in a plunger-type oil-well pump.

barrel wrench—a friction wrench used in repairing oil-well pumps.

batch—a definite amount of oil, mud, acid, or other liquid in a tank or pipe.

beam—the walking beam of a pumping unit.

beam well—a well using a pumping unit and rods to lift fluid.

bean—a choke, used to regulate flow of fluid from a well. Different sizes of beans are used for different producing rates.

bell hole—a bell-shaped hole dug beneath a pipeline to provide room for use of tools.

bird cage—to flatten and spread the strands in a wire rope.

blank flange *(also a blind flange)*—a solid disc used to dead end a companion flange.

blank liner—a liner without perforations.

blank-off—to close off, such as with a blank flange or bull plug.

bleed into—to cause a gas or liquid to mingle slowly with another gas or liquid, usually by pressure.

bleed off or bleed down—reduce pressure by letting oil or gas escape at a low rate.

bleeder—a valve or pipe through which bleeding is done.

bob tail—any short truck.

boilerhouse—to make up or fake a report without actually doing the work.

bonnet—the part of a valve that packs off and encloses the valve stem.

boomer—a link and lever mechanism which is used to tighten a chain holding a load on a truck.

bowl—a device into which fit the slips or wedges which support tubing.

bradenhead gas—commonly called casinghead gas; gas that is produced with oil or from the casing head of an oil well.

BS or BS&W—basic sediment, or basic sediment and water.

bump down—to have too long a length of rods between the pumping unit and the pump seat so that the pump hits bottom on the downstroke.

—C—

cage—the part of a pump valve which holds the ball to limit its movement.

casinghead gas—(oil well gas) is associated and dissolved gas produced along with crude oil from oil completions.

casing pressure—gas pressure built up between the casing and tubing.

cat—a crawler-type tractor.

cat walk—the narrow walkway on top of a tank battery.

chase threads—to straighten and clean threads of any kind.

cheater—a length of pipe used to increase the leverage of a wrench.

chisel tongs—pipe tongs that grip the pipe with a chisel-like insert in the jaw of the wrench.

clip—a U-bolt or similar device used to fasten parts of a wire cable together.

closed-in—a well capable of producing oil or gas, but temporarily shut in.

close nipple—a very short piece of pipe having threads over its entire length.

collar—a pipe coupling threaded on the inside.

come-along—a stretching or tightening device that crawls along a length of chain.

computer—(1) a device capable of solving problems by accepting data, performing prescribed operations on the data, and supplying the results of these operations. Various types of computers are calculators, digital computers, and analog computers. (2) In information processing, usually an automatic stored program computer.

computer control—a system whereby the end devices in the field (switches, valves, gauges, alarms, etc.) are controlled by a program placed in the computer.

computer program—a plan or routine for solving a problem on a computer.

connection—the joining of two lengths of pipe.

control panel—part of a computer system that contains manual controls—switches and devices to start, stop, measure, monitor or signal what is taking place.

coupon—small metal strip which is exposed to corrosive systems for the purpose of determining nature and severity of corrosion.

crack a valve—to barely open a valve so that it leaks just a little.

crater—to fail.

crude oil production—the volume of liquids statistically reported as crude oil, which is produced from oil reservoirs during given period of time.

cut oil—oil that contains water; also called wet oil.

—D—

dead man—a piece of wood or concrete, usually buried, to which a wire guy line is attached for bracing a mast or tower.

dead well—a well that will not flow.

debug—to detect, locate and remove mistakes from a routine or malfunctions from a computer.

depletion—a deduction allowed in computing the taxable income from oil and gas wells.

disposal wells—a well through which water (usually salt water) is returned to subsurface formations.

dissolved gas—natural gas which is in solution with crude oil in the reservoir.

dog leg—a bend in pipe, a ditch, or a well.

donkey pump—any little pump; used for many kinds of small temporary pumping operations.

dope—material used on threads of pipe or tubing to lubricate and prevent leakage.

doughnut—a ring of wedges that supports a string of pipe or a threaded, tapered ring used for the same purpose.

downcomer—a pipe through which flow is downward.

dozer—a powered machine for earthwork excavations.

dresser sleeve—a slip-type collar that is used to join plain-end pipe.

drifter—a worker who never stays long in one place.

drip—equipment designed to remove small quantities of liquids from a gas stream.

dry hole—an exploratory or development well found to be incapable of producing either oil or gas in sufficient quantities to justify completion as an oil or gas well.

dutchman—a piece of pipe that has been twisted off inside a female connection; or a short section of material, such as belting or pipe, used to lengthen existing equipment.

—E-F—

expansion loop—a bend placed in a line to absorb stretch or shrinkage.

fatigue—failure of a metal under repeated loading.

female connection—a pipe or rod coupling with the threads on the inside.

field facility—an installation designed for one or more specific and limited extraction units, scrubbers, absorbers, drip points, conventional single or multiple stage separation units, LTX low temperature separators, and other types of separation and recovery equipment.

fire wall—a wall of earth built around an oil tank to hold the oil if the tank breaks or burns.

fittings—the small pipes and valves that are used to make up a system of piping.

float—a long flat-bed semi-trailer.

flow a well hard—to let a well flow at too high a rate.

flow bean—a plug in the flow line at the well head which has a small hole drilled through it through which oil flows, and which keeps a well from flowing at too high a rate.

flow by heads—a well flowing oil at irregular intervals.

flow chart—a chart made by a recording meter which shows rate of production.

flowing well—a well which produces oil or gas without any means of artificial lift.

flow lines—the surface pipes through which oil travels from the well to storage.

flow tank—a lease storage tank to which produced oil is run.

flow treater—a single unit which acts as an oil and gas separator, an oil heater, and an oil and water treater.

fluid level—distance between well head and point to which fluid rises in the well.

flush production—the high rate of flow made by a good well right after it is drilled.

frost up—icing of equipment due to the cooling effect of expanding gas.

frozen up—said of equipment of which the components do not operate freely.

—G—

gaging nipple—a small section of pipe in the top of a tank through which a tank may be gaged.

gas plant products—natural gas liquids recovered from natural gas in gas processing plant and, in some situations, from field facilities.

gas processing plant—a facility designed (1) to achieve the recovery of natural gas liquids from the stream of natural gas which may or may not have been processed through lease separators and field facilities, and (2) to control the quality of the natural gas to be marketed.

gas well—a well capable of producing natural gas.

gathering lines—the flow lines which run from several wells to a single tank battery.

gin-pole truck—a truck equipped with a pair of poles, and hoisting equipment for use in lifting heavy machinery around a lease.

girth or girt—one of the horizontal braces between the legs of a derrick.

gone to water—describes a well in which water production is increasing.

grass gooser—a hoe or other kind of weed cutter.

guy wire—a rope or cable used to steady a mast or pole.

—H-I-J—

handy—a connection that can be unscrewed by hand.

hatch—an opening into a tank, usually through the top deck.

hay tank—a tank or enclosure filled with hay-like material used to filter oil out of water.

heat *(a connection)*—to loosen a collar or other threaded connection by striking it with a hammer.

hot oil—oil production in violation of state regulations or transported interstate in violation of federal regulations.

insulated flange—a flange which incorporates plastic pieces to separate the metal parts.

jack board—a device used to support the end of a length of pipe while another length is being screwed on.

—K-L—

kill a well—to overcome pressure in a well by use of mud or water so that surface connections may be removed.

knockout—a kind of tank or filter used to separate oil and water.

LACT—"Lease Automatic Custody Transfer," possible where measuring equipment installed at the point of transfer from lease to pipeline is so completely automated as not to require any manual activity or witnesses.

lazy board—see *jack board*.

lease condensate—a natural gas liquid recovered from gas well gas (associated and non-associated) in lease separators of field facilities.

lease separator—a facility located at the surface for the purpose of (1) separating casinghead gas from produced crude oil and water at the temperature and pressure conditions of the separator; and (2) separating gas from that portion of associated and non-associated gas which liquefies at temperature and pressure conditions of the separator.

live oil—oil that contains gas.

load binder—chain or rope used to tie down loads of equipment, or the "boomer" used to tighten the chains.

—M-N-O—

make a hand—to become a good worker.

make it up another wrinkle—to make up a connection one more turn.

male connection—a connection with the threads on the outside.

manhole—a hole in the side of a tank through which a man can enter the tank, also the cleanout plate.

marginal well—an oil or gas well the production of which is so limited in relation to production costs that profit approaches the vanishing point.

master gate—a large valve used to shut in a well.

multiple completion well—a well equipped to produce oil and/or gas separately from more than one reservoir.

natural gas liquids—those portions of reservoir gas which are liquefied at the surface in lease separators, field facilities, or gas processing plants.

non-associated gas—natural gas which is in reservoirs that do not contain significant quantities of crude oil.

off production—said of a well when it is shut in or temporarily not able to produce.

offset well—well drilled near another one.

oil country tubular goods—oil-well, casing, tubing, or drill pipe.

oil well—a well completed for the production of crude oil from at least one oil zone or reservoir.

old hand—a man who has been around the oil field for a long time.

on the line—said of a tank when it is being emptied into a pipeline.

on the pump—said of a well that is being pumped.

overproduced—said of a well that has produced more than its allowable.

—P-R—

persuader—a big tool for a small job, used to overcome some excess friction.

pig iron—what a large heavy piece of equipment is said to be made of.

pipeline oil—oil clean enough to be acceptable to transport or purchase.

plug back—to shut off lower formation in a well bore.

plunger lift—a method of lifting oil using a swab or free piston propelled by compressed gas from the lower end of the tubing string to the surface.

poor boy—homemade; something done on a shoestring basis.

potential test—a test of the maximum rate at which a well can produce oil.

power rated—rating given by a manufacturer of an engine operating at most efficient output.

power tools—equipment operated hydraulically or by compressed air for making up and breaking out drill pipe, casing, tubing, and rods.

pressure regulator—a device for maintaining pressure in a line, downstream from the valve.

production—the yield of an oil or gas well; the branch of the industry that brings the oil and gas to the surface for sale.

productive capacity—estimates of productive capacities of crude oil developed by the American Petroleum Institute Committee on Reserves and Productive Capacity represent the maximum daily rates of production which can be attained under specified conditions on March 31 of any given year.

productivity test—a test of a well's capacity to produce, usually conducted at different pumping rates or rates of flow (see *potential test*).

proration—a system of allocating production on a per well basis.

proved reserves—proved reserves of crude oil as of December 31 of any given year are the estimated quantities of all liquids statistically reported as crude oil, which geological and engineering data demonstrate with reasonable certainty to be recoverable in the future from known reservoirs under existing economic and operating conditions.

pump off—to pump so rapidly that the oil level drops below the standing valve on the pump.

put a well on—to start a well flowing or pumping.

put on pump—to install a pump or pumping unit, sucker rods, and bottom-hole pump.

rabbit—a small plug that is run through a flow line by pressure to clean the line or test for obstructions.

relief valve—a valve that will open automatically when pressure gets too high.

remote control station—a station containing equipment to control and regulate operations in the field.

riser—a pipe through which liquid travels upward.

rock a well—to bleed pressure from casing of a dead well, then from tubing, then from casing, and so on so that the well will start to flow.

—S—

saddle bearing—a bearing between the walking beam and the sampson post of a pumping unit.

sanded up—clogged by sand entering the well bore with the oil.

scraper—a device used to clean deposits of paraffin from tubing or flow lines (see *rabbit*).

settled production—a loose term used to describe oil fields that produce at nearly the same rate from day to day.

shake out—to spin a sample of oil at high speed to determine its BS&W content.

sharpshooter—a long narrow shovel used in ditch digging.

shut in—to close valves on a well so that it stops producing; said of a well on which the valves are closed.

shut-in pressure—pressure at the top of a well when it is shut in.

slack off—to lower a load or ease up on a line.

sling—a wire-rope loop for use in lifting heavy equipment.

snake out—to pull out.

snatch block—a sheave or pulley that can be opened up for putting a line over the roller or sheave.

soft rope—a small loose fiber rope.

spacing—distance between wells producing from the same pool (usually expressed in terms of acres, e.g., 10-acre spacing).

spaghetti—very small tubing or pipe.

stabilized—a well is considered "stabilized" when, in the case of a flowing well, the rate of production through a given size of choke remains constant, or, in the case of a pumping well, when the fluid column within the well remains constant in height.

strip a well—to pull rods and tubing from a well at the same time. Tubing must be "stripped" over the rods a joint at a time.

stripper—a well which produces a very small amount of oil, usually in an old field.

swab—a device that fits the inside of tubing closely that is pulled through the tubing to lift fluid from it, or to pull such a device through the tubing.

swamper—a helper on a truck.

—T—

tail out rods—to pull the bottom end of a sucker rod away from a well when laying rods down.

take a strain on—to begin to pull on a load.

tally—to measure and record length of pipe or tubing.

tank strapper—the person who measures a tank to determine the volume it holds at ¼" intervals of height.

tap—a notched tool used to cut inside threads.

telecommunications—pertaining to the transmission of signals over long distances, such as by telegraph, radio, or television.

telemetry—a system by which the variations recorded by any physical or other instrument can be shown at a distance by means of electricity.

tie-down—a device to which a guy wire or brace may be attached.

tin hat—the metal hat worn by oil-field workers to protect them from falling objects.

transducer—a device for converting energy from one form to another.

tubing job—the pulling and running of tubing.

—W—

water well—a well drilled to (1) obtain a water supply to support drilling or plant operations, or (2) obtain a water supply to be used in connection with an improved recovery program.

widow maker—anything liable to cause death or serious injury of a workman.

winch—a machine used for pulling or hoisting that does so by winding a cable around a spool.

working pressure—the pressure to which a particular piece of equipment is subjected during normal operations.

work over—to clean out or otherwise work on a well in order to increase or restore production.

PRODUCTION—BIBLIOGRAPHY

Abell, J.M. (ed.): **Practical Aspects of Well Drilling and Production,** The Institute for Energy Development, Inc. (Oklahoma City) 1980.

American Petroleum Institute: **Primer of Oil and Gas Production,** 3rd Ed., Production Department (Dallas) 1971.

Craft, Holden, and Graves: **Well Design: Drilling and Production,** Prentice-Hall, Inc. (Englewood Cliffs, N.J.) 1962.

ETA Offshore Seminars, Inc.: **The Technology of Offshore Drilling, Completion, and Production,** The Petroleum Publishing Company (Tulsa) 1976.

Frick, Thomas C. (ed.): **Petroleum Production Handbook,** 2 Vols., Society of Petroleum Engineers of AIME (Dallas) 1962.

Lapedes, Daniel N. (ed.): **Encyclopedia of Energy,** McGraw-Hill Book Company (New York) 1976.

Petroleum Extension Service: **Lessons in Well Servicing and Workover: Artificial Lift Methods, Lesson 5,** University of Texas and International Association of Drilling Contractors (Austin and Houston) 1971.

Petroleum Extension Service: **Lessons in Well Servicing and Workover: Production Rig Equipment, Lesson 6,** University of Texas and International Association of Drilling Contractors (Austin and Houston) 1971.

CHAPTER 7:
RECOVERY

I. INTRODUCTION

Before the production of a newly discovered oil or gas field can begin, the best methods for recovering the hydrocarbons must be determined. This determination is based on the type of petroleum reservoir present and the energy sources available within that reservoir.

There are three classifications used to describe the methods of recovering fluids from oil or gas reservoirs:

Primary recovery is the initial production of fluids from the reservoir using only available natural sources of energy to recover the oil or gas.

Secondary recovery includes those methods used to supplement natural reservoir energy to increase fluid recovery, generally consisting of water injection into the reservoir.

Tertiary recovery refers to those methods used to provide supplementary energy to the reservoir in addition to water injection. These include injection of steam, carbon dioxide, polymers, micellar fluids, and in situ combustion.

Primary recovery does not involve the addition of any supplementary energy to the reservoir to increase fluid recovery; it relies solely on the energy available in the reservoir itself. On the other hand, both secondary and tertiary recovery methods include the addition of energy into the reservoir to increase recovery. This additional recovery is usually accomplished by injecting some type of fluid into the reservoir through injection wells. In this chapter, recovery methods will be classified as either **primary recovery methods** or **fluid injection processes** (often called **enhanced recovery** methods). Fluid injection processes include both secondary and tertiary recovery processes.

In any discussion of recovery methods, two concepts are useful to describe how such recovery methods work. These are known as an **energy balance** and a **material balance** for the reservoir. An **energy balance** refers to the amount of energy contained in a petroleum reservoir. The energy available to drive fluids out of the reservoir is equal to the amount of natural energy present, plus the energy added by fluid injection processes, minus the energy withdrawn when fluids are taken from the reservoir. This reservoir energy balance can be defined by the following equation:

Recovery Energy Available = Natural Reservoir Energy + Injected Energy − Energy Withdrawn From the Reservoir.

The most common forms of energy in a petroleum reservoir are pressure and heat.

The amount of petroleum which may be recovered from a reservoir is related to the amount of energy present in the reservoir. Fluid injection is used primarily to increase the amount of energy available. However, certain injected fluids may also increase the mobility of reservoir fluids, thereby increasing the percentage of those fluids which can be recovered.

The second concept, **material balance**, is a mathematical expression which states that the amount of material present in a reservoir is equal to the original amount of fluid, plus natural influx, plus the amount added from injection processes, minus that withdrawn from production of the reservoir. The following equation represents that expression:

Fluid Present in Reservoir = Original Fluid Present + Natural Influx + Injected Fluid − Fluid Withdrawn.

The term "material" refers only to the fluids contained in the reservoir—specifically oil, water, and gas.

A material balance is generally expressed in terms of volumes of the fluids present. In the United States, oil and water are usually measured in units of oil field **barrels**. A barrel contains 42 U.S. gallons. Gas is measured in units of **standard cubic feet** (SCF), which refers to the amount of gas present in a cubic foot of volume at some standard atmospheric pressure and temperature, often 14.7 psia (pounds per square inch absolute) and 60°F. Since a cubic foot of gas represents a rather small unit compared to the volumes of gas often found in petroleum reservoirs, gas is commonly measured in thousands of standard cubic feet (MSCF) or millions of standard cubic feet (MMSCF). The amount of oil or gas originally present in a reservoir, or present at some time during the production of that reservoir, can be determined by using the material balance equation.

Geologists and petroleum engineers identify the best recovery methods for petroleum reservoirs. Engineers from several fields contribute to the final determination and implementation of this recovery plan. **Geological engineers** describe the reservoir from a geological point of view, that is, its area, thickness, rock quality, and the types of fluids it contains. **Reservoir engineers** determine the fluid and pressure distributions throughout the reservoir, the natural energy sources available, and the methods most useful in recovering the maximum amount of oil or gas from the reservoir. **Drilling engineers** oversee the drilling of the wells. **Logging engineers** analyze the reservoir by using different logging techniques. **Production engineers** are concerned with optimizing the use of the wells and production equipment for withdrawing fluids from the reservoir. **Facilities engineers** deal with the mechanical surface equipment necessary to handle the oil or gas once it has been withdrawn from the reservoir.

II. PRIMARY RECOVERY METHODS

A. Introduction

Most petroleum reservoirs have a certain amount of natural energy available in the reservoir which provides for the movement of fluids into the wellbore. The term **primary recovery** refers to production of oil or gas through the use of this natural energy alone. There are several different types of natural energy which aid in this primary recovery process. (Primary recovery does not mean the same thing as a flowing well, discussed in Chapter Six. Whether the fluids rise to the surface unaided due to sufficient pressure in the reservoir rock, or whether they have to be artificially lifted, is not relevant to the definition of primary recovery. Primary recovery refers specifically to the existence of natural energy within the reservoir enabling the fluids to move to the wellbore.)

The majority of petroleum reservoirs are capable of producing oil or gas by primary recovery methods alone. There are, however, many petroleum deposits which cannot be produced at all without the addition of some type of supplemental energy to stimulate production. This is particularly true of heavy oil deposits and tar sands.

This section examines the different types of oil and gas reservoirs and the natural energy sources commonly available in each.

B. Oil Reservoirs

Oil reservoirs can be divided into several major classifications: **undersaturated, solution gas, gas cap,** and **water influx** reservoirs, as well as combinations of these types. Each of these has somewhat different natural energies available for the recovery of fluids.

1. Undersaturated Reservoirs

Undersaturated reservoirs are oil reservoirs which contain little entrained gas in the oil. An analogous situation might be a soft drink that is left standing over night in an open bottle; the "fizz" (dissolved gas) contained in the liquid escapes. Much the same thing can happen in oil reservoirs. Over time, gas contained in the oil slowly escapes, leaving an undersaturated oil. Large deposits of such oil exist in various parts of this country.

Fluid expansion is one natural source of energy used to produce undersaturated reservoirs. Oil deposits are often buried beneath thousands of feet of sediments. Pressure from this overlying rock compresses the fluids contained in the rocks. When a well is drilled, pressure is released and the oil and connate water (naturally occurring water which cannot be removed from the rock) expand, causing oil to flow into the wellbore. The release of pressure in the formation also allows pore spaces in the rock to contract, which helps push expanding fluids into the wellbore (Figure 7.1).

Gravity drainage may also be significant in reservoirs which do not have sufficient natural pressure to make fluid expansion useful. As shown in Figure 7.2, the oil is drained by wells in the bottom of the reservoir,

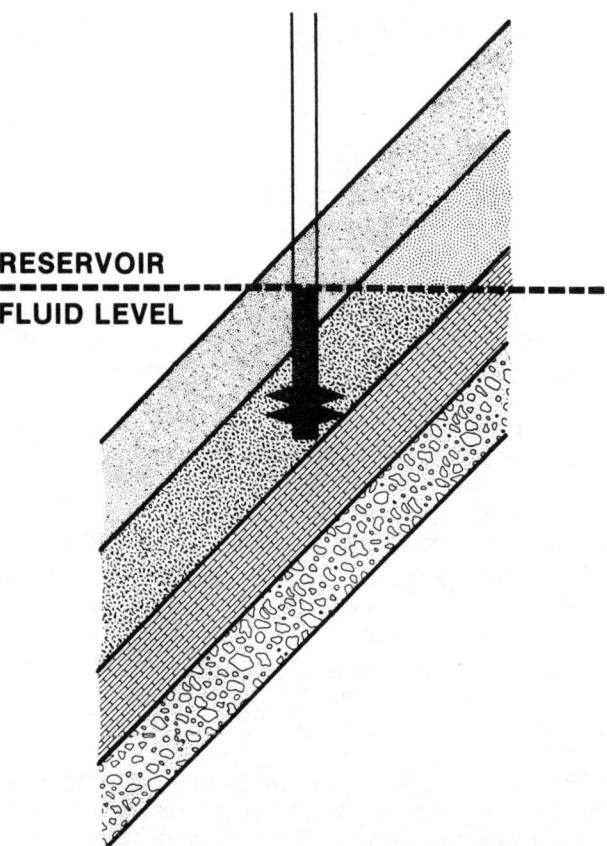

Figure 7.2. GRAVITY DRAINAGE.

in much the same way a water tank is drained. A pump lifts the fluid out of the well, since little natural pressure exists to help the well flow by its own means.

Since undersaturated reservoirs generally have little reservoir energy available, recovery efficiency from these types of reservoirs is low, and large amounts of fluid are left in the ground. Supplementary fluid injection can be used to recover additional oil. Table 7.1 summarizes some of the main characteristics of undersaturated oil reservoirs.

2. Solution Gas Reservoirs

Gas is often entrained in the oil in petroleum reservoirs. Natural pressure on the reservoir is attributed to the fact that the oil is buried deep beneath the earth's surface. As a result, gas is kept in solution with the oil and does not exist in a free, gaseous state. Since such reservoirs are generally closed traps, the gas contained in the oil (**solution gas**) has no place to escape and is therefore held in the oil.

The solution gas provides most of the energy which

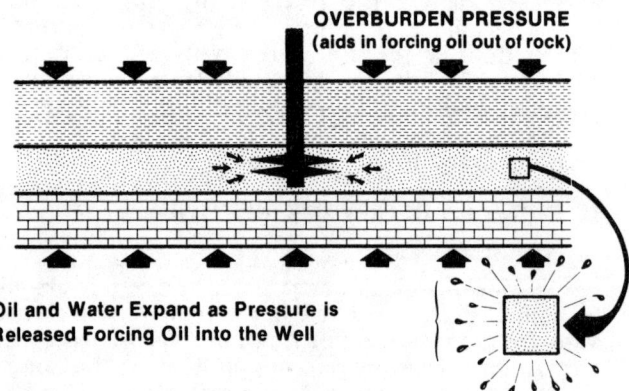

Figure 7.1. FLUID EXPANSION IN AN UNDERSATURATED RESERVOIR.

Table 7.1. UNDERSATURATED RESERVOIRS

Characteristic	Trend
Reservoir Pressure	Declines rapidly and stabilizes at a low value
Surface Gas-Oil Ratio	Small to zero (little or no gas produced with oil)
Well Behavior	Requires pumping at early stage
Expected Oil Recovery	Less than 5 percent of original oil in place on primary production

(Source: N.J. Clark, **Elements of Petroleum Reservoirs**)

drives oil into the wellbore. This process is called **solution gas drive.** When wells are drilled into solution gas reservoirs, the pressure on the formation is released, allowing the solution gas to break out of the oil and form free gas bubbles. The expansion of these gas bubbles naturally drives the oil along with the gas into the well (Figure 7.3). A similar process occurs when a warm soft drink is opened suddenly and the expanding gas in the bottle causes the liquid to overflow out of the bottle. Solution gas drive is also called **depletion drive,** since the gas breaking out of solution is rapidly produced, thus depleting the energy available to recover the oil.

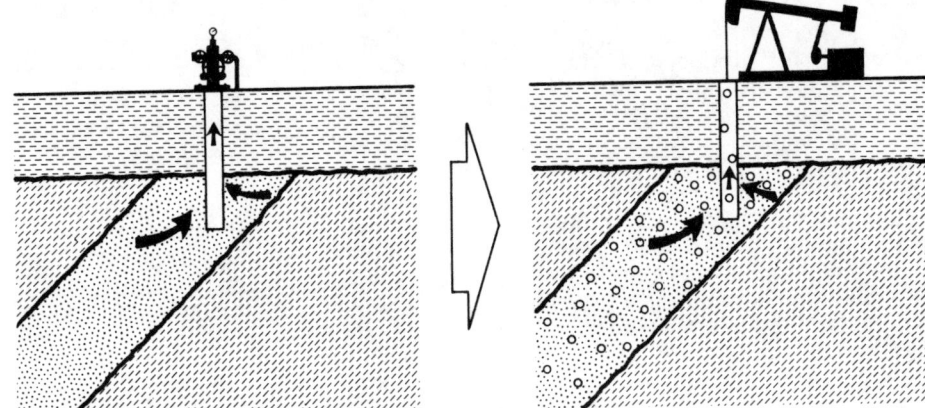

Figure 7.3. SOLUTION GAS DRIVE RESERVOIR. Left: original conditions. Right: partially depleted.

In solution gas drive reservoirs, the **bubble point** refers to the pressure at which solution gas (or dissolved gas) will break out of the oil and become free gas. The wellbore pressure must be reduced below this bubble point pressure before the free gas is available to drive the oil out of the reservoir.

The gas-oil ratio (GOR) is a measure of the amount of gas produced with the oil, and is generally expressed in terms of standard cubic feet of produced gas per barrel of produced oil (SCF/Bbl). The higher the GOR for a well, the more gas per barrel of oil is being produced.

form a **gas cap.** If the reservoir is under sufficient pressure, the energy stored in this pressurized gas cap can aid in the recovery of oil. The gas cap expands as oil is produced from the reservoir and drives remaining oil down into the producing wells. This expanding gas cap is called a **gas cap drive** (Figure 7.4).

A reservoir with gas cap drive also has solution gas in the oil, which further aids in the recovery of the oil. Because of this combined mechanism, recovery efficiency in such fields can be slightly higher than that resulting from solution gas drive alone. The major characteristics for gas cap drive reservoirs are summarized in Table 7.3.

Table 7.2. SOLUTION GAS DRIVE RESERVOIRS

Characteristic	Trend
Reservoir Pressure	Declines rapidly and continuously
Surface Gas-Oil Ratio	First low, then rises to maximum, then drops
Well Behavior	Requires pumping at early stage
Expected Oil Recovery	5 to 30 percent of original oil in place

(Source: N.J. Clark, op. cit.)

Table 7.3. GAS CAP DRIVE RESERVOIRS

Characteristic	Trend
Reservoir Pressure	Falls slowly and continuously
Surface Gas-Oil Ratio	Rises continuously in upstructure wells
Well Behavior	Long flowing life depending on size of gas cap
Expected Oil Recovery	20 to 40 percent of original oil in place

(Source: N.J. Clark, op. cit.)

Because solution gas provides only a limited amount of energy to aid in producing oil, the percentage of the oil that can be recovered by this process is low. Table 7.2 outlines some of the characteristics of a solution gas drive reservoir.

3. Gas Cap Reservoirs

Since a solution gas reservoir is formed through the concurrence of various geologic processes, gas frequently exists in amounts exceeding that which can be dissolved in the oil. This free gas collects in the top of the reservoir to

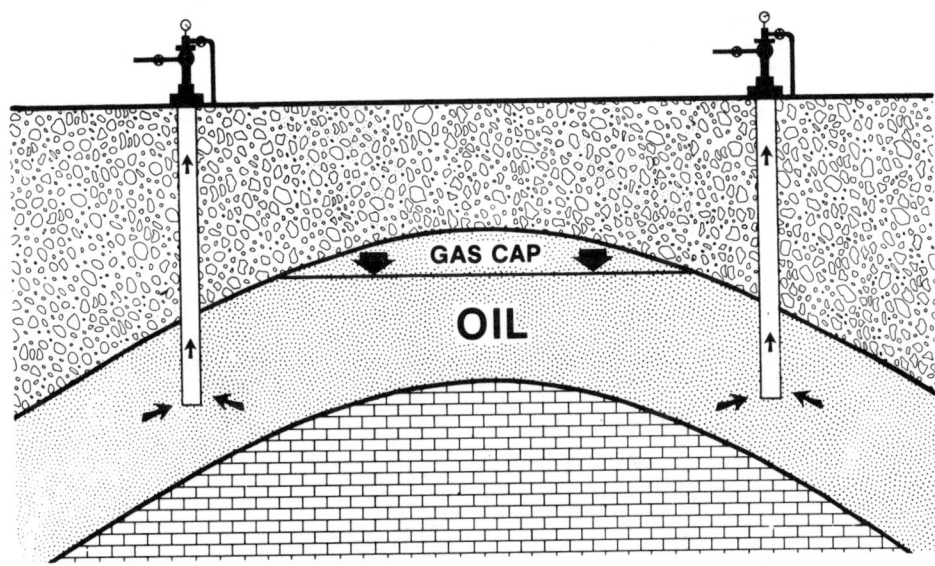

Figure 7.4. GAS CAP DRIVE RESERVOIR.

4. Water Influx Reservoirs

Perhaps the most important source of reservoir energy for producing oil during primary recovery comes from water influx into the reservoir. Water is present in most sedimentary rocks. Generally, sedimentary rocks were laid down in shallow seas or lakes and, therefore, were filled with water from the time of deposition. This water remains in the rock and is often replenished by surface waters migrating down into the rocks. The water provides pressure to the reservoir, which aids in moving the oil into the producing wells. As the oil is produced, the water rises into the reservoir to take the place of the produced oil, thereby maintaining the reservoir pressure at high levels. This movement of water into the reservoir also has the effect of displacing the oil in front of it toward the wells.

Table 7.4. WATER INFLUX RESERVOIRS

Characteristic	Trend
Reservoir Pressure	Remains high
Surface Gas-Oil Ratio	Remains low
Water Production	Starts early and increases to appreciable amounts
Well Behavior	Flows until water production becomes excessive
Expected Oil Recovery	35 to 75 percent of original oil in place

(Source: N.J. Clark, op. cit.)

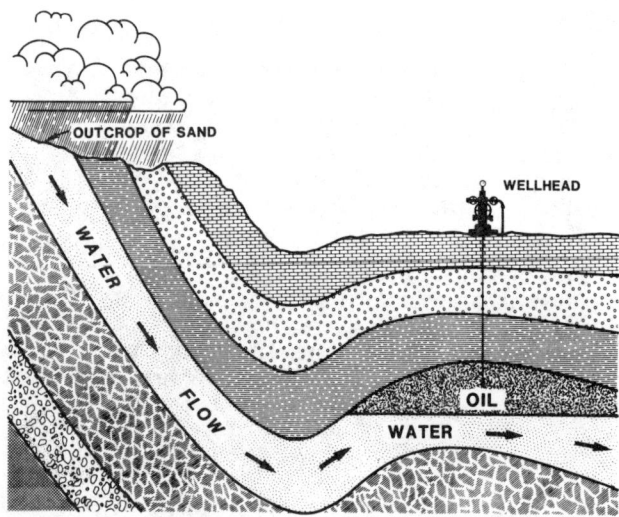

Figure 7.5. BOTTOM WATER DRIVE RESERVOIR.

There are two common types of water influx reservoirs. A **bottom water drive** reservoir is one in which water expands across the entire bottom of the reservoir (Figure 7.5). In an **edge water drive** reservoir, the water encroaches around the perimeter of the oil zone (Figure 7.6).

Because water influx reservoirs have a continuing supply of energy to help produce the oil, they often have higher recovery efficiencies than the other drive mechanisms discussed above. The important characteristics of water influx reservoirs are outlined in Table 7.4.

5. Combination Reservoirs

Because several geological processes affect the formation of petroleum reservoirs, generally more than one primary production mechanism is available. Drive mechanisms in most reservoirs are a combination of some, or all, of the previously discussed sources of reservoir energy.

One of the more frequent combinations is illustrated in Figure 7.7. The reservoir in this example is being produced both from gas cap expansion on the top and water influx on the bottom. Solution gas drive from dissolved gas in the oil also contributes to the production of oil. Under proper circumstances, the combination of various drive mechanisms can provide good recovery efficiency for these types of reservoirs.

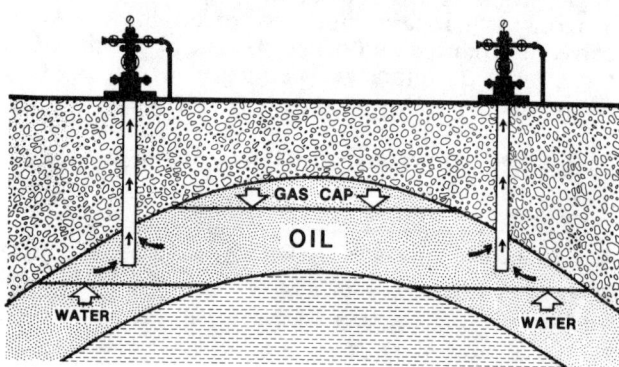

Figure 7.7. COMBINATION DRIVE RESERVOIR.

C. Gas Reservoirs

Gas reservoirs differ from oil reservoirs in that they generally contain no oil, but produce mainly gas or, sometimes, gas with varying amounts of **condensate** (lightweight hydrocarbons similar to gasoline) or water. Unlike some oil reservoirs which have poor recovery efficiencies in primary production, gas reservoirs generally produce well without the addition of supplementary energy. Primary recovery methods are usually sufficient because a sizeable amount of stored energy is contained in the compressed gas in the reservoir. In addition, gas is highly mobile, which enables it to travel easily through the rock. Gas reservoir recovery efficiencies in excess of 80 percent are not uncommon. The major types of gas reservoirs are discussed in the following sections.

1. Dry Gas Reservoirs

Dry gas reservoirs contain no liquid hydrocarbons. The gas consists of methane and ethane with only minor amounts of heavier hydrocarbon gases, such

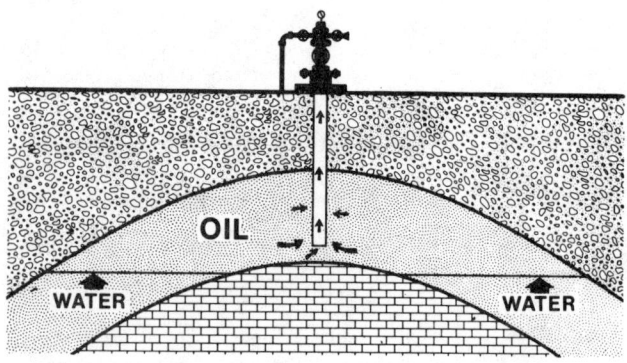

Figure 7.6. EDGE WATER DRIVE RESERVOIR.

as propane or butane. Since no liquids can be condensed from this type of gas, it is often suitable for direct use as a fuel in homes, factories, and power plants.

Dry gas reservoirs do, however, contain water. The original sedimentary rocks were generally deposited in water environments, and water was dispersed throughout the pore spaces of the rock. Gas filled the rock when the petroleum reservoir was formed, but not all of the water was displaced. Occasionally, some of the remaining water is produced at the surface with the gas. Before the gas can be used as a fuel, the water must be removed.

Both dry gas and oil reservoirs occur in the same types of traps, but the method used to produce gas is much simpler than that used to produce oil. The gas is generally contained in the reservoir at a pressure sufficient to force it into the wellbore and up to the surface. Consequently, no pumping is required.

The rate at which gas can be withdrawn from the reservoir depends on the amount of back-pressure held on the reservoir by the pipeline system into which it is produced. In the example in Figure 7.8, a dry gas reservoir has a reservoir pressure of 1,000 psi (pounds per square inch). It is producing into a well connected to a pipeline which operates at a pressure of 500 psi. The gas well is maintained at this pressure by the mechanical equipment on the wellhead. This equipment, usually containing a **gas choke,** controls the flow rate of the well. If the choke is closed to the point at which the reservoir pressure is 1,000 psi, flow from the well will cease. If it is opened as far as possible, the well will produce at the maximum rate allowed by the pipeline. Thus, the recovery of gas from a dry gas reservoir depends on the type of surface equipment used to produce it.

2. Gas Condensate Reservoirs

Gas reservoirs frequently contain heavier hydrocarbons, such as propane, butane, and pentanes. These can exist in the reservoir as either a gas or liquid, depending on the reservoir pressure and temperature. If the reservoir is very hot and at high pressure, these components will exist in a gaseous state. If the reservoir is cool and at low pressure, they will be liquid, interspersed with the lighter hydrocarbon gases comprised of methane and ethane. In either case, when these types of reservoirs are produced, the cooling of the gas which generally results from this process causes both liquids and gas to be recovered at the surface. This type of gas is called **wet gas,** meaning that it contains these hydrocarbon liquids. The liquids are called **condensate liquids,** and reservoirs producing such fluids are called **gas condensate reservoirs** or **wet gas reservoirs.**

Condensate reservoirs can also contain water, but the terms "wet" gas and "dry" gas refer to the presence or absence of hydrocarbon liquids, and **not** to water, which may or may not be produced with the gas.

The amount of liquid contained in the wet gas is measured by determining the total amount of condensate which will form if the gas is cooled to a standard temperature and maintained at a standard pressure, usually 60°F and 14.7 psi. As shown in Figure 7.9, a gas at reservoir conditions containing no condensate may form significant amounts of condensate when cooled at the surface and reduced to a lower pressure.

The number of gallons of condensate contained in 1,000 standard cubic feet of wet gas is called the **gpm** (gallons per thousand) value of that gas. The condensate content of the gas may also be expressed as the number of barrels of liquids per million standard cubic feet of gas (BPMM). The higher the gpm value of a wet gas, the more liquids it contains. These liquids are usually extracted at the surface and sold separately or made into other products, such as liquefied petroleum gas (LPG) or gasoline.

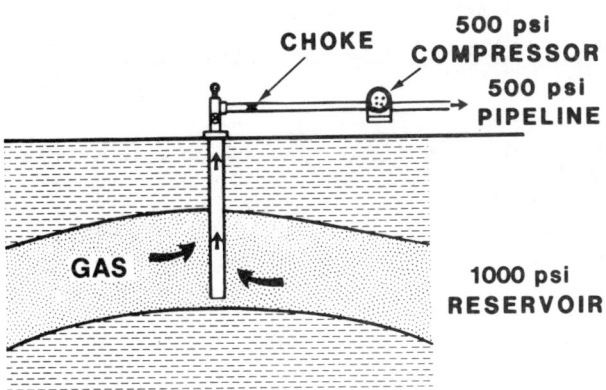

Figure 7.8. DRY GAS RESERVOIR.

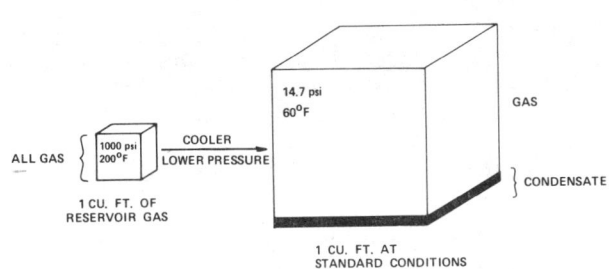

Figure 7.9. WET GAS AND GAS CONDENSATE.

In the reservoir shown in Figure 7.8, if the pipeline pressure were maintained at 500 psi as indicated, the well would continue flowing gas until the reservoir no longer had sufficient pressure to flow into the 500 psi pipeline. At this point, flow would cease and approximately 50 percent of the reservoir's gas would have been produced. If, however, the pipeline pressure was reduced, and the reservoir was allowed to flow until it reached a pressure of 100 psi, approximately 90 percent of the gas would have been recovered. Instead of reducing the pipeline pressure, compressors often are installed to draw the wellhead pressure down as far as possible and then raise the gas pressure to pipeline pressure.

Not all gas condensate reservoirs have the simple condensation relationship shown in Figure 7.9. **Retrograde condensate reservoirs** exhibit condensation characteristics opposite those normally expected. As illustrated in Figure 7.9, if a wet gas is cooled significantly and lowered in pressure, condensate forms. If only the pressure is lowered, any condensate present tends to evaporate, since cooling is generally required to form the liquids. However, in retrograde condensate reservoirs this is not true. Condensate liquids form in the reservoir as pressure is reduced, even without cooling (Figure 7.10). Retrograde condensation only occurs in deep reservoirs with pressures above approximately 2,000 psi.

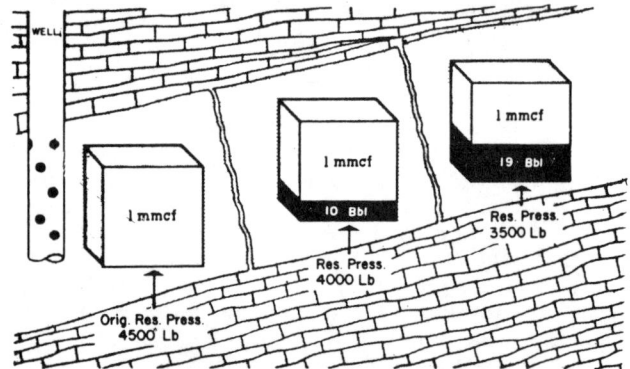

Figure 7.10. RETROGRADE CONDENSATION OF LIQUID FROM GAS IN A RESERVOIR.
(Figure courtesy of SPE of AIME)

Gas condensate reservoirs are generally produced in the same way as dry gas reservoirs. The expansive nature of the pressurized gas forces the gas and condensate into the wellbore and up to the surface. If large amounts of condensate exist with the gas, mechanical equipment may be necessary in some instances to help lift the liquids to the surface. Retrograde condensate reservoirs often require special recovery techniques. These will be discussed in the section on fluid injection processes.

3. Water Influx Reservoirs

Many gas reservoirs are subject to water influx, in much the same way oil reservoirs are. Water is often present at the bottom of gas reservoirs and exerts pressure on the gas accumulations (Figure 7.11). As gas is withdrawn from the reservoir, the water advances into the reservoir and replaces the gas. When the reservoir is finally depleted, only water and a small amount of trapped gas remain. Both dry gas and gas condensate reservoirs can be subject to this type of water influx.

Water influx gas reservoirs differ from simple gas reservoirs in that the water influx maintains approximately the original pressure in the reservoir by continually filling the vacated reservoir space. Normal gas reservoirs experience pressure depletion as the gas is withdrawn, since no fluid is available to replace the original gas and maintain the pressure of the reservoir.

III. FLUID INJECTION PROCESSES
A. Introduction

As the demand for petroleum has increased and oil prices have risen, it has become prudent and economically feasible to find ways to recover additional oil left in existing reservoirs. Often, under primary recovery methods, less than 30 percent of the original oil in place is produced. Since new oil fields in the United States have become harder to find, many companies are returning to developed reservoirs and using improved methods to recover some of the remaining oil.

These improved recovery methods, which generally involve fluid injection into the reservoir, are described in this section. Fluid injection processes for both oil and gas reservoirs will be examined.

B. Fluid Injection Processes for Oil Reservoirs
1. Water Injection

Water injection, or **waterflooding**, involves pumping water into the reservoir to stimulate production. The injected water provides pressure to force the oil out of the rock and to sweep it toward producing wells (Figure 7.12). Waterflooding, which has been used since the late nineteenth century, is the most frequently used fluid injection method and now accounts for about one half of U.S. oil production.

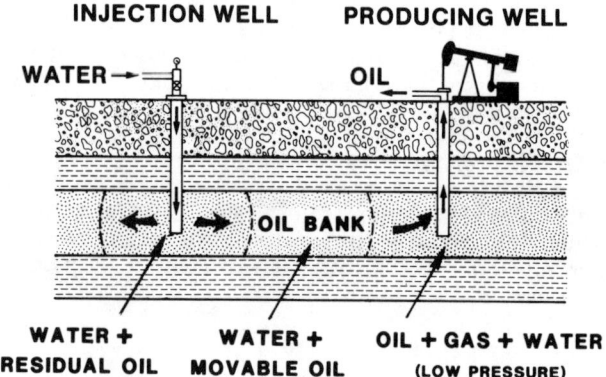

Figure 7.12. WATERFLOODING.

Waterfloods are commonly described according to the relative location of the injection and producing wells (Figure 7.13). Well locations are often chosen according to some pattern, such as a five-spot or a line-drive. Peripheral floods, in which injection wells are located around the edge of the oil accumulation, and irregular floods, in which the injection wells are placed in no definite pattern, also exist.

A number of methods have been developed to examine and predict the recoveries from waterflooding.

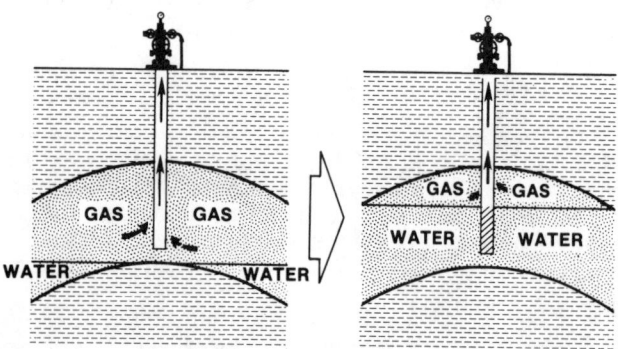

Figure 7.11. WATER INFLUX GAS RESERVOIR.
Left: original conditions.
Right: reservoir near depletion.

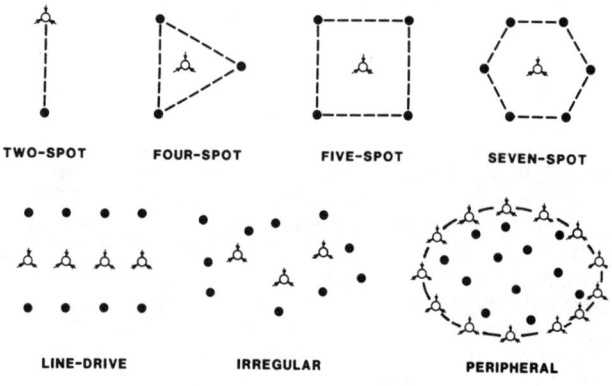

Figure 7.13. WATERFLOOD CONFIGURATIONS.

Some of the better known methods are designated by the names of the men who developed them: Buckley-Leveritt, Welge, Hurst, Stiles, Dykstra-Parsons, Craig-Geffen-Morse, and Higgins-Leighton.

Waterflooding has been attempted in almost every type of reservoir. It has been most successful in relatively homogeneous reservoirs with sufficient permeability to allow water injection at a reasonable rate.

2. Gas Injection

Gas injection has frequently been used in place of, or in conjunction with, waterflooding. Three basic types of gas injection methods have proved useful: **dispersed gas injection, attic gas injection,** and **gas flooding.** Under certain circumstances, gas injection may be considered a **miscible fluid displacement** process (see part 5).

Dispersed gas injection involves reinjecting gas at a number of locations in a field, primarily to maintain reservoir pressure. For solution gas reservoirs, this pressure maintenance can be useful in accelerating the recovery of oil. Dispersed gas injection may also be used as a method for storing natural gas.

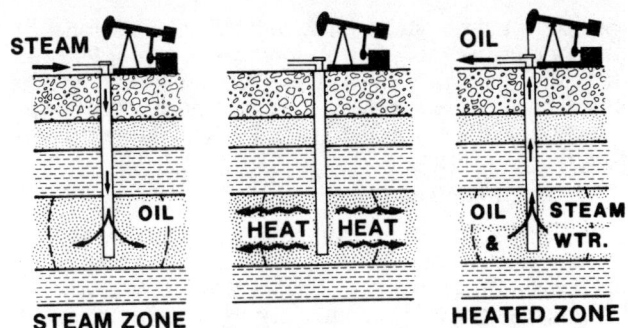

Figure 7.15. **STEAM SOAK OPERATION.**
Left: injection cycle. Center: soak cycle. Right: backflow cycle.

Steam soak operations entail injecting steam into a well (the **injection cycle**), letting it sit to allow the heat to percolate through the rock (the **soak cycle**), and finally producing the well after the soak (the **backflow cycle**). (See Figure 7.15.) During production, the heavy oil must flow through the region that has been heated during the injection cycle and the soak cycle. After flowing through the heated zone, the oil is hotter and can be produced at a higher rate. Since part of the heat input during the injection cycle is produced, and remaining heat tends to dissipate, a well is often soaked a number of times during its producing life.

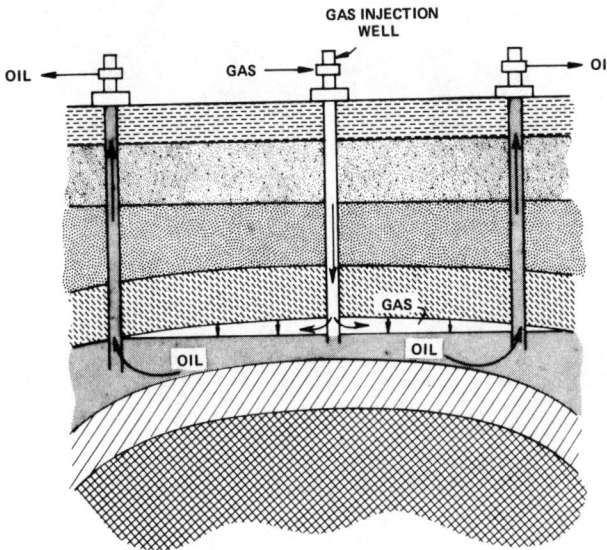

Figure 7.14. **ATTIC GAS INJECTION.**

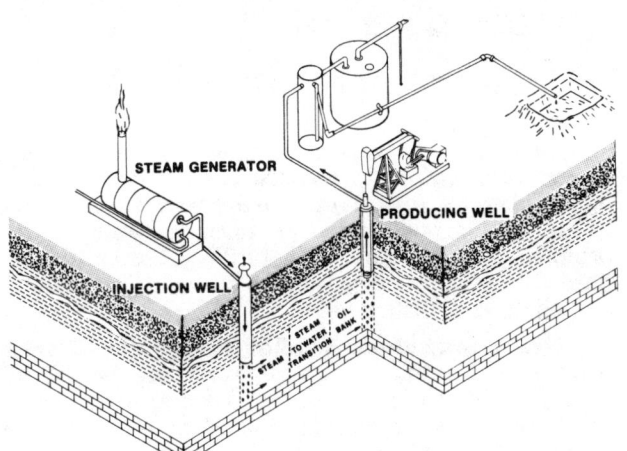

Figure 7.16. **STEAM DRIVE.**

Attic gas injection consists of injecting gas either into an existing gas cap, or "attic," or into an oil zone to create a gas cap (Figure 7.14). Attic gas injection is a method of providing or supplementing a gas cap drive to recover more oil.

Gas flooding is similar to waterflooding in that its purpose is to sweep oil from injection wells to producing wells. This procedure normally utilizes some pattern arrangement of wells, as in waterflooding.

3. Steam Injection

Steam injection methods fall into two major categories: **soaks** and **drives.** Generally, steam injection is used to provide heat in reservoir deposits of heavy oils. These heavy oils, which may be "thicker" than butter, generally become "thinner" and more mobile with the addition of heat, just as butter liquefies on a hot stove. The steam also provides pressure which causes the fluids to flow.

In the **steam drive** process, steam from injection wells pushes the oil to producing wells (Figure 7.16). In most cases, steam drives are performed from a regular pattern of injection and producing wells. As the steam drive sweeps the oil to the producing wells, the oil left behind the steam front is somewhat distilled by the steam. **Distillation** of the oil means that some of the lighter components of the oil are heated to a gaseous state and are then carried forward by the steam to the producing wells. **Steam reforming** can also occur, whereby the high temperature of the steam causes the oil to chemically break down into simpler compounds, which are also pushed to the producing wells.

Because distillation and reforming of the oil can occur with this process, steam drives are more efficient than waterfloods in recovering oil. Application of steam drives to light oil reservoirs is now being considered. At the present time, steam soak and drive operations are mainly confined to relatively shallow (400 to 2,000 feet in depth), thick (greater than 20 feet), heavy oil deposits in sandstone reservoirs. The greatest number of steam injection projects in the United States are presently found in California.

4. Combustion

Oil recovery can also be accomplished by injecting air into the reservoir and burning some portion of the oil underground. The heat generated by the burning oil reduces the viscosity of the unburned oil, thereby increasing production. The air injection also serves to keep reservoir pressure high to drive the oil to the producing wells.

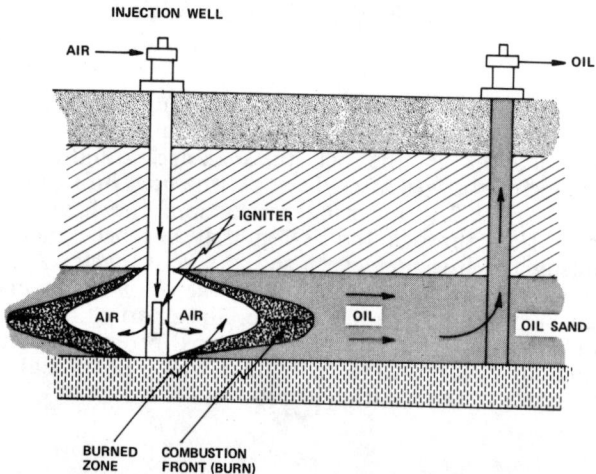

Figure 7.17. FORWARD COMBUSTION.

There are two types of combustion: **forward** and **reverse** combustion. In **forward combustion**, the zone of burning oil (called the **combustion front** or the "burn") is spread from the injection wells to the producing wells (Figure 7.17). In this process, the oil is pushed to the producing wells by means of the injected air and the combustion gases.

Reverse combustion is useful for heavier oils which offer more resistance to flow, by spreading the "burn" from the producing wells to the injection wells (Figure 7.18). In reverse combustion, since the oil must flow through a zone already heated by the "burn," the oil receives much more heat than in forward combustion.

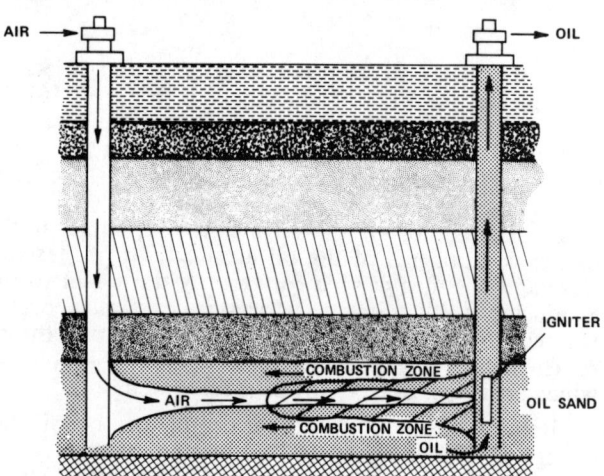

Figure 7.18. REVERSE COMBUSTION.

A special type of forward combustion, known as **quenched** or **wet combustion**, incorporates the injection of some water with the air to keep the combustion front temperature down. This process requires less air for combustion. If large amounts of water are injected during a combustion project, the process is called a **combination of forward combustion and waterflooding**, which has the acronym **COFCAW**.

Combustion methods can be applied to highly oil-saturated, intermediately thick (5 to 50 feet), shallow (400 to 2,000 feet deep) reservoirs with medium weight oils (10 to 35° API).

5. Miscible Fluid Displacement

Miscible fluid displacement methods have been developed to recover more of the original oil in place than is possible with conventional water or gas injection methods.

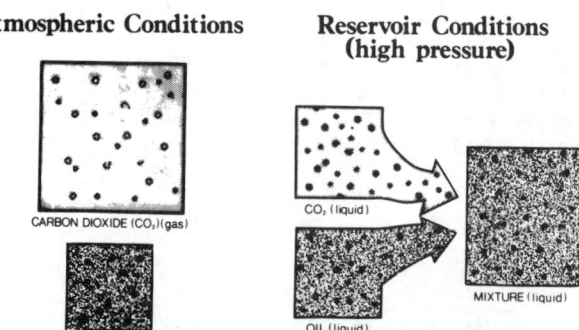

Figure 7.19. MISCIBILITY OF CARBON DIOXIDE AND OIL AT RESERVOIR CONDITIONS.

Miscibility refers to the ability of the displacing fluid and the displaced fluid to mix completely in all proportions. For example, water is miscible with alcohol since water and alcohol mix together with no separation, irrespective of the proportions of alcohol and water that are mixed. An illustration of the miscibility of oil and carbon dioxide at reservoir conditions is found in Figure 7.19. Because of miscibility between the displacing fluid (injected into the reservoir) and the oil, only a small residual oil saturation is left behind after flooding (Figure 7.20).

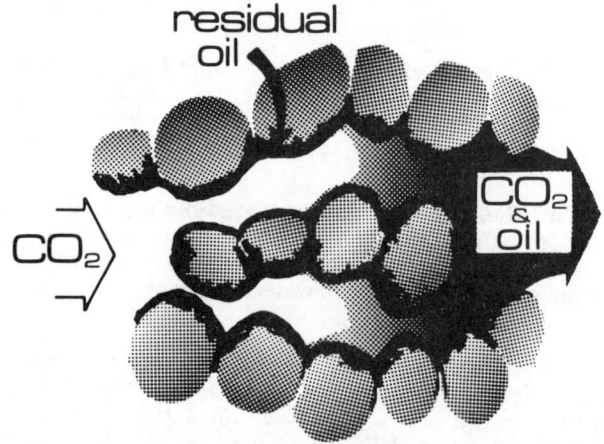

Figure 7.20. CARBON DIOXIDE RECOVERY. Carbon dioxide (CO_2) contacts residual oil, allowing the oil to flow out of the tight pore channel.

The two basic methods of miscible fluid displacement consist of **continuous injection** of a miscible fluid and injection of a small amount of miscible fluid, known as a **slug**, followed by another displacing fluid.

Continuous injection of a miscible fluid may be accomplished with high pressure dry gas, enriched gas, alcohol, carbon dioxide, flue gas, or inert gas. All of these fluids can be miscible with oil; for example, at high pressures, dry gas is miscible with most oils. However, since increasing the gas to high pressures is expensive, and since high pressures cannot always be used safely, lower pressure enriched gas may be injected to form a miscible displacement. Enriched gas is simply gas that contains more hydrocarbon liquids.

Some alcohols, such as methyl alcohol (wood alcohol) or ethyl alcohol are miscible in oils. Depending on availability and economic considerations, carbon dioxide (the "fizz" in soft drinks), flue gas (the gases generated by burning oil), or inert gas (such as air with the oxygen removed) may be used to effect miscible displacement.

The use of these fluids alone is expensive because of the amounts that are required for a full-scale project. Instead, a **slug** (or small amount) of the miscible fluid may be injected, followed by continuous injection of a cheaper fluid, such as water. A slug of alcohol, which is miscible with both water and oil, followed by water injection is less expensive than a continuous alcohol injection. This procedure can also be used with a carbon dioxide slug and subsequent water injection, or with a propane or LPG slug followed by low or medium pressure dry gas.

6. Polymer Injection

Polymers are long thread-like molecules which are flexible and resilient, much like rubber bands. Polymer injection may be used in some cases to increase oil recovery by increasing the sweep efficiency of the injected fluid.

This process is relatively complex. Simply, the polymer causes the injected fluid to reach a larger portion of the reservoir, thereby displacing more of the remaining oil.

Because polymers are extremely expensive, usually a slug of polymer is injected, followed by water injection, which pushes the polymer away from the injection wells and through the oil reservoir.

C. Fluid Injection Processes for Gas Reservoirs

The principal fluid injection processes used for gas reservoirs are **water injection** and **gas cycling**. These processes may be considered as secondary recovery; tertiary recovery has not been used for gas reservoirs.

The purpose for using injection methods for gas reservoirs differs from that for oil reservoirs. Instead of providing energy for producing or displacing the hydrocarbons, injection methods for gas reservoirs are generally employed to maintain pressure or increase condensate liquid recovery.

1. Water Injection

Water injection has been used to maintain the pressure of gas reservoirs in the same way that it has been used for oil reservoirs. Under certain circumstances, water injection has proved helpful in increasing the production from gas reservoirs. Many times, however, water injection can significantly reduce the recovery from a gas reservoir.

Water injection can be detrimental because some gas is trapped in the pore spaces of the rock by the water. This is the same reason that water influx in gas reservoirs may be damaging to the overall gas recovery. As much as one quarter to one third of the original gas in place may be trapped by water, decreasing the total recovery of gas.

For this reason, water injection into gas reservoirs is not a common practice. Where time limitations exist (water injection will recover gas faster because of the pressure maintenance), or where pressure maintenance is necessary to prevent ground subsidence, water injection may be used for gas reservoirs.

2. Gas Cycling

Gas cycling operations are often carried out in gas condensate reservoirs. The main purpose of gas cycling is to recover more of the condensate liquids than would be recovered otherwise. This process is important for retrograde condensate reservoirs and for water influx gas condensate reservoirs.

In gas cycling, a portion of the produced gas is reinjected, after condensate liquids have been removed or stripped from the gas. The removal process causes the gas to be deficient in liquids. Consequently, the gas has the capacity to hold more condensate liquids than it actually contains when reinjected. This cycled gas will pick up additional liquids that exist in the reservoir, and will push the richer gas towards the producing wells.

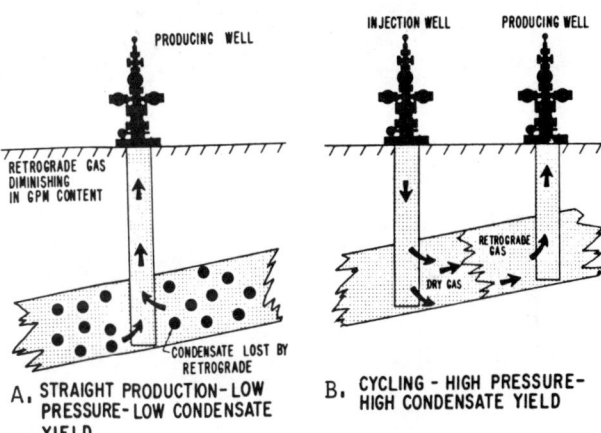

Figure 7.21. STRAIGHT PRODUCTION vs. CYCLING: NO WATER DRIVE.
(Figure courtesy of SPE of AIME)

Gas cycling also can be important for retrograde condensate reservoirs. As production occurs from a retrograde condensate reservoir, the pressure in the formation decreases and liquids are condensed out of the gas before it is produced (Figure 7.21A). Gas cycling recovers a large portion of these liquids, and at the same time, keeps the reservoir pressure up so fewer liquids are condensed out. By using this procedure, a larger proportion of the valuable condensate liquids may be recovered (Figure 7.21B).

A similar effect occurs during gas cycling of water influx condensate reservoirs. As the water moves into the portion of the reservoir originally containing gas, some of the gas is trapped by the water (Figure 7.22A). This condition is known as **trapped** or **residual gas saturation**. The liquid content of the trapped gas is then lost. Gas cycling operations in water influx reservoirs can be used either to prevent gas from being trapped or

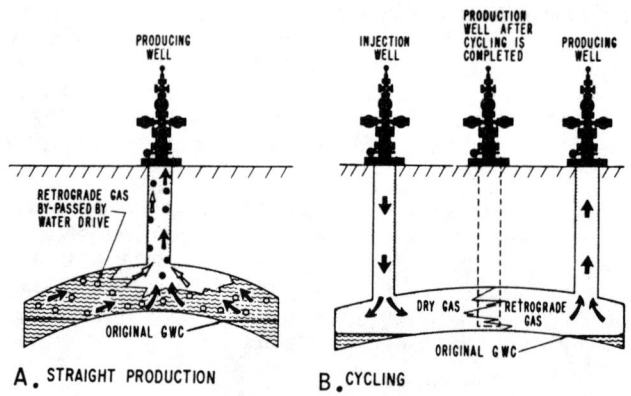

Figure 7.22. STRAIGHT PRODUCTION vs. CYCLING: WATER DRIVE.
(Figure courtesy of SPE of AIME)

to make the trapped gas deficient in liquids. Gas cycling can often recover a significant amount of liquids that would otherwise be lost (Figure 7.22B).

Condensate liquids from gas reservoirs are valuable. For example, in a wet gas containing 10 gpm (gallons per 1,000 cubic feet) of condensates, approximately half of the energy content of the wet gas is due to the liquids, and more than half of the economic value of the wet gas comes from the condensate liquids. A gas cycling operation recovering a greater percentage of the condensate liquids can maximize the amount of energy recovered and be economically attractive as well.

RECOVERY—GLOSSARY

API gravity—a measure of liquid density according to standards and procedures as adopted by the American Petroleum Institute, usually used to express the density of oil or other hydrocarbon liquids. It is derived from specific gravity according to the following equation:

$$\text{API gravity} = \frac{141.5}{\text{specific gravity}} - 131.5$$

API gravity is expressed in degrees, a specific gravity of 1.0 being equal to 10° API.

bottom water—water occurring in a producing formation below the oil or gas in that formation.

bubble point—the pressure at which solution gas will come out of the oil and become free gas.

combustion—the burning of some portion of the petroleum in a reservoir in order to increase the production of the reservoir. The process aids in recovery for a number of reasons: the heat makes the unburned oil more mobile and easier to produce; the injected air maintains pressure on the reservoir; and the gases produced by combustion drive the oil to the producing wells.

condensate—hydrocarbons that are gaseous under reservoir conditions, but which become liquid in passage through the well or at the surface.

distillation—the process of removing gas from liquids or solids by heating.

dry gas—natural gas that is produced without hydrocarbon liquids; also, gas that has been processed to remove such liquids.

dry gas reservoir—a petroleum reservoir containing no liquid hydrocarbons.

edge water—water occurring in a producing formation around the perimeter of the oil or gas in that formation.

energy balance—the amount of energy contained in a petroleum reservoir calculated as follows:

energy to drive fluids out of reservoir = natural energy + energy from fluid injection − energy withdrawn from the reservoir

enhanced recovery—the use of fluid injection processes, either secondary or tertiary, to recover additional amounts of oil.

entrained gas—gas suspended in bubbles in a stream of fluid, such as water or oil.

flue gas—gas generated by burning oil or other fossil fuels.

fluid injection—the injection of water, steam, and other fluids into a reservoir to increase the amount of energy available to recover additional oil or gas.

gas cap reservoir—a petroleum reservoir containing an accumulation of gas overlying the liquid hydrocarbons.

gas condensate reservoir—a petroleum reservoir in which condensate and gas exist in a single homogeneous phase. When production takes place and pressure drops below a certain point, condensate liquids form and are recovered at the surface.

gas injection—injection of natural gas into a producing reservoir for pressure maintenance.

gas-oil ratio—the number of cubic feet of gas produced with a barrel of oil.

hot oil—treatment of a well which produces viscous or paraffinic oil that is deposited in the well and restricts production. This is accomplished by the injection of hot, lighter oil from the surface to remove such deposits.

inert gas—normally inactive gas, especially non-combustible gas.

material balance—the amount of material contained in a petroleum reservoir calculated as follows:

fluid present in reservoir = original fluid present + natural influx + injected fluid − fluid withdrawn

miscibility—the ability of two or more liquids to mix in any proportion.

miscible flood—see *miscible fluid displacement*.

miscible fluid displacement—an enhanced recovery process in which a solvent is injected into a reservoir to reduce the interfacial forces between oil and water in the pore spaces, thereby displacing the oil from the reservoir rock.

moveable oil—oil which can be moved from the underground formation or trap into the wellbore and then pumped to the surface.

oil in place—the total amount of oil contained in an underground reservoir.

polymer—a long, flexible, resilient, thread-like molecule.

pressure maintenance—the injection of a fluid or fluids into an oil field to restore and/or maintain its original pressure.

primary recovery—the initial production of fluids from a reservoir using only available natural sources of energy to produce the oil or gas.

psi—pounds per square inch; a unit of measurement of pressure (or force per unit area).

recoverable oil—oil in place in the ground that can be produced.

reserves—the unproduced but recoverable oil or gas in place in an underground formation that has been proved by production.

reservoir pressure—the pressure in a reservoir.

residual saturation—the saturation of a fluid which remains in pore spaces after the partial displacement of that fluid by another fluid. For example, in a waterflood the residual oil saturation is the oil saturation behind the waterflood front which cannot be displaced by water.

retrograde condensation—the process in which certain hydrocarbons condense into a liquid instead of expanding into a gas when pressure is decreased.

secondary recovery—the use of waterflooding, gas injection, and other methods to recover additional amounts of oil beyond what could be recovered by natural reservoir energy.

service well—a well drilled or completed specifically to support production in an existing field.

slug—a small amount of some fluid injected into a reservoir to recover additional oil in place.

solution gas reservoir—an oil reservoir containing entrained gas (gas in solution with the oil) due to pressure on the reservoir.

specific gravity—the ratio of the weight of a given volume of substance at a given temperature to the weight of an equal volume of a standard substance at the same temperature.

standard cubic foot—the amount of gas present in a cubic foot of volume at some standard atmospheric pressure and temperature, usually 14.7 psia and 60°F.

steam injection—an enhanced recovery process that involves the injection of steam into a reservoir to heat the "thick" oil present to make it more mobile; the addition of the steam also provides pressure on the reservoir.

sweep efficiency—a measure of the percentage of an oil reservoir that is swept or contacted by an injected fluid displacing the oil.

tertiary recovery—the use of the injection of steam, carbon dioxide, polymers, micellar fluids, in situ combustion, and other methods to supplement the natural reservoir energy to recover additional amounts of oil.

undersaturated reservoir—an oil reservoir containing little entrained gas in the oil.

waterflooding—an enhanced recovery method in which water is injected into a petroleum reservoir to maintain the pressure in the reservoir and to displace the hydrocarbons out of the rock and into the producing wells.

water influx reservoir—a petroleum reservoir whose pressure is maintained naturally at a relatively high level due to the inflow and expansion of water around the bottom or on the perimeter of the reservoir.

wet gas—natural gas which contains liquid hydrocarbons.

RECOVERY—BIBLIOGRAPHY

Calhoun, John C., Jr.: **Fundamentals of Reservoir Engineering,** University of Oklahoma Press (Norman) 1960.

Clark, Norman J.: **Elements of Petroleum Reservoirs,** Society of Petroleum Engineers of AIME (Dallas) 1969.

Craft, B.C. and Murray F. Hawkins, Jr.: **Applied Petroleum Reservoir Engineering,** Prentice-Hall, Inc. (Englewood Cliffs, N.J.) 1959.

Craig, Forrest J., Jr.: **The Reservoir Engineering Aspects of Waterflooding,** Society of Petroleum Engineers of AIME Monograph, Vol. 3 (Dallas) 1971.

Frick, Thomas C. (ed.): **Petroleum Production Handbook,** Vol. 2, Society of Petroleum Engineers of AIME (Dallas) 1962.

Gulf Publishing Company: **Thermal Recovery Handbook #4** (Houston) 1969.

Gulf Publishing Company: **Waterflood Handbook #3** (Houston) 1969.

Oil and Gas Journal: **Thermal Recovery Handbook** (Tulsa) 1966.

PPC Books: **Enhanced Recovery Methods** (Tulsa) 1977.

Smith, Charles R.: **Mechanics of Secondary Oil Recovery,** Robert E. Krieger Publishing Company (New York) 1975.

Society of Petroleum Engineers of AIME: **Thermal Recovery Processes,** SPE of AIME Reprint Series #7 (Houston).

Society of Petroleum Engineers of AIME: **Thermal Recovery Techniques,** SPE of AIME Reprint Series #10 (Houston).

Society of Petroleum Engineers and Department of Energy: **Second Joint Symposium on Enhanced Oil Recovery,** Society of Petroleum Engineers, 1981.

ACKNOWLEDGEMENTS

Figures 7.10, 7.21, and 7.22 are taken from **Elements of Petroleum Reservoirs,** by Norman J. Clark, and are used with the permission of the Society of Petroleum Engineers of AIME. The Society of Petroleum Engineers of AIME owns the copyright to these drawings.

CHAPTER 8:
TRANSPORTATION

I. INTRODUCTION

Thus far in our discussion of the different aspects of the petroleum industry, we have concentrated on the processes involved in finding the petroleum and producing it on site. The next logical step is to consider how the crude oil and natural gas are transported from the oil fields to the refinery.

The transportation of petroleum is the transition point between the "upstream" and "downstream" functions of the industry. We are still dealing with crude petroleum, when we speak of transportation from producing fields to refineries. However, shipment of petroleum products from the refineries to their ultimate destinations can be considered a "downstream" function.

In this chapter, we will look at the history of the petroleum transportation industry and the methods of transporting petroleum today.

II. HISTORY

The petroleum transportation business began in earnest in America after the success of the Drake well in Titusville, Pennsylvania, in 1859. Initially, oil was transported by boats down Oil Creek, from Titusville to Oil City, then transferred to larger vessels for shipment to Pittsburgh, which soon became a major refinery and transportation center for the oil industry.

Transportation by water played a key role in the movement of crude oil and refined petroleum in the early history of the oil industry. Streams, creeks, and rivers provided a natural transportation network, connecting producing regions with processing facilities and urban areas. Barges, passenger boats, and rafts were used to carry crude oil to centralized shipping points, where steam boats waited to transport barrels of petroleum to refining and marketing centers.

The first overseas shipment of oil occurred in 1861, when three thousand barrels were shipped to London, England. In 1869, the "S.S. Charles" of Antwerp, Belgium, was the first vessel outfitted for bulk shipment of oil. The holds of the ship were converted to carry 59 tanks of oil, or the equivalent of 794 tons. Seventeen years later, the overseas transportation of petroleum entered a new era, as the first tanker set out on its maiden voyage.

Transportation of petroleum on inland waterways was not without its difficulties and problems. Numerous river boat accidents caused significant loss of oil and created pollution along the river. It soon became obvious that other transportation methods had to be found.

Railroad companies seized the opportunity, and began to construct new lines into the oil producing areas. At first, the railroads simply carried the barrels of oil on regular flatcars. The first true tank car was built in 1865, and consisted of two, round, vertical, wooden tanks (with a combined capacity of approximately 80 barrels) mounted on a flatcar (Figure 8.1). By that time, rail networks had been developed to ship crude oil from Pennsylvania into New York and other industrialized areas on the east coast.

Although the railroads and steam boats carried the oil to processing and marketing centers, a way had to be devised to transport the crude oil from the fields to water and railway shipping points. At first, the teamsters provided an answer to the problem by creating a system of boats, barges, and horse-drawn wagons. This setup proved adequate for a while, but soon the rising costs, congested traffic lanes, and hazardous conditions, demanded a safer and more economic method to move the oil.

To meet this need, the first pipeline for carrying oil in the United States was completed in 1865. Samuel Van Syckel, who organized the first company for transporting oil by pipeline, built a two-inch line from Pithole City to the Miller Farm station of the Oil Creek Railroad—a distance of about five miles.

The first pipelines in the United States were not well received by the teamsters, who earned their living by hauling oil. Construction was disrupted, individuals were harassed, and lines were torn up. Eventually, however, pipelines gained acceptance among people in the oil field, and grew to be a major part of the transportation sector of the early industry.

Railroads began to purchase and construct pipelines in order to secure a monopolistic hold on the transportation industry by controlling all the major facets of oil transportation. Private companies were forbidden by the railroads to build pipelines across railroad track. To counteract these measures, pipeline companies often would build their pipeline as far as the railroad track and install tanks on both sides of the track. The oil

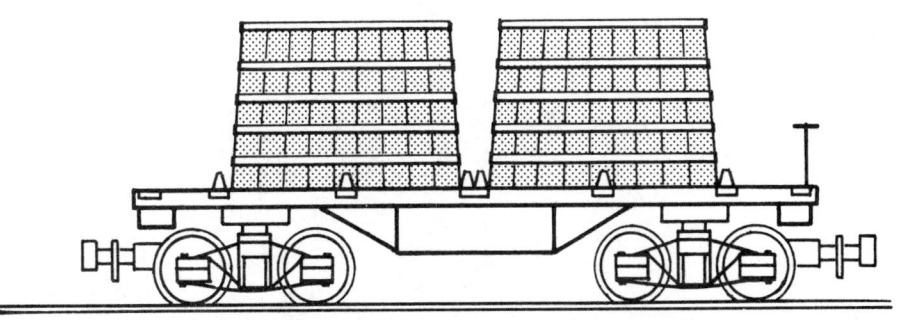

Figure 8.1. FIRST RAILROAD TANK CAR.

Figure 8.2. EARLY HORSE-DRAWN TANK WAGON.

would be pumped from the pipeline into the tanks on one side, emptied from the tanks and transported across the railroad track in horse-drawn tank wagons (Figure 8.2), transferred into the tanks on the other side, and then pumped into another pipeline.

Although this conflict with the railroads retarded the growth of pipelines in Pennsylvania for a while, the private companies ultimately won several legal battles in their efforts to cross railroad property. By 1872, the common carrier pipelines had been granted the privilege of eminent domain in acquiring rights-of-way. Thereafter, the use of pipelines for carrying oil from the fields to shipping points and refineries became a widespread practice.

Following the development of the eastern oil fields, the petroleum industry began to move westward into such areas as Texas, Oklahoma, and California. This westward movement precipitated a growth in the network of product lines and crude oil pipelines, that would eventually link all areas of the country. The expansion in the number of crude oil pipelines over the years in the U.S. is illustrated by Figure 8.3.

III. MODERN TRANSPORTATION METHODS

A. Introduction

Today, there are several methods used to transport oil and gas: **pipelines, tank trucks, railroad tank cars, barges,** and **tanker ships.** The determination as to which method is to be used depends on such factors as distance, product, area, and availability of suitable alternatives. For example, pipelines are very economical and can be used to cover great distances, but are limited as to route and destination. Tank trucks are only able to transport relatively small quantities, but provide the most flexibility and can be most economical over shorter distances. Railroad tank cars can be competitive for intermediate distances and sufficient quantities. Tankers are used to carry large volumes of petroleum across international waters to link exporting and importing nations.

Approximately 47 percent of the oil transported in the United States is moved by pipeline. There are more than 227,000 miles of crude oil and product lines, carrying approximately one billion tons annually. In addition, natural gas is shipped by a system of pipelines totaling more than one million miles.

Tank trucks, river barges, and tankers provide most of the remaining domestic oil transportation. There are more than 158,000 tank trucks hauling 29 percent of the oil in this country, while tankers and barges handle 22 percent of the total volume. Although there are 165,000 private railroad tank cars in use in the United States, only about one or two percent of our oil is currently transported by rail. This is in striking contrast to the early days, when the majority of oil in this country was moved by the railroads.

Internationally, ocean tankers are the most common method of transporting oil from one country to another. According to the latest available figures, there are approximately 5,140 tank ships (ocean-

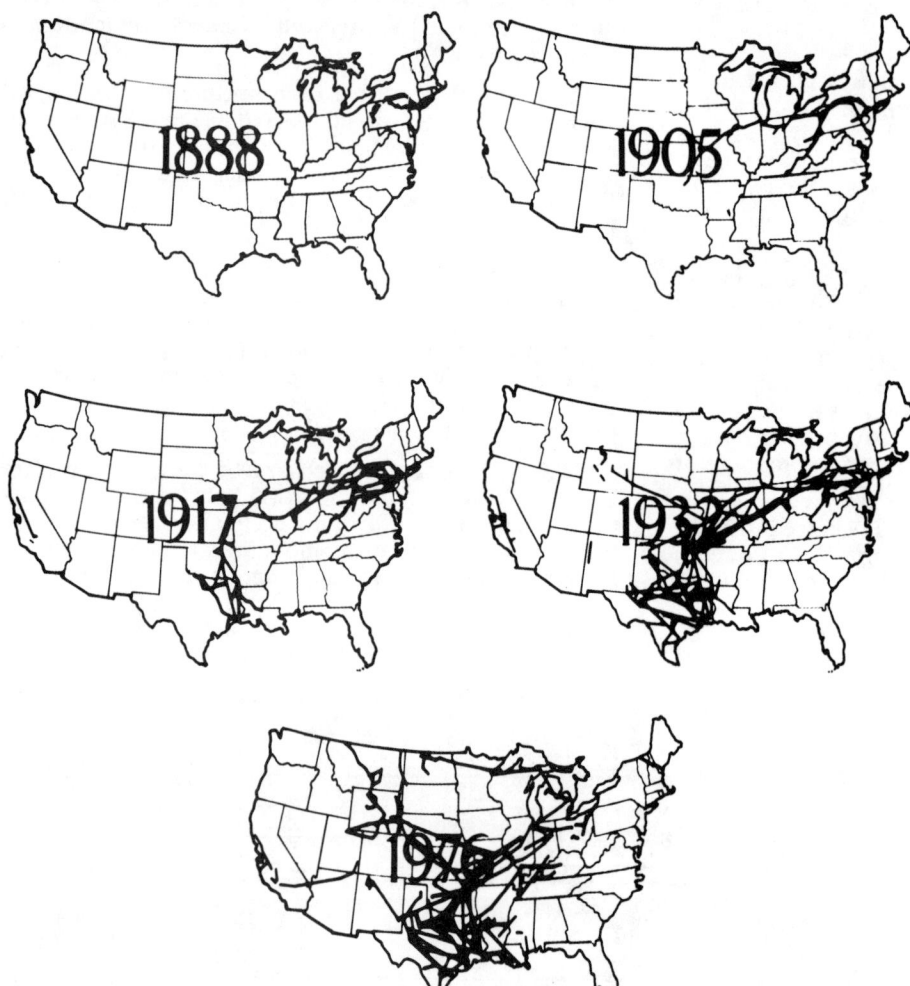

Figure 8.3. CRUDE OIL PIPELINES IN THE UNITED STATES.

going vessels of 2,000 gross tons and over) currently in use throughout the world, with the United States possessing 290 of these.

The following sections briefly describe these different methods of transporting petroleum.

B. Pipelines

In the late 1800s and early 1900s, pipelines were used exclusively on dry land. Today, because of advanced engineering techniques and improved materials, it is possible to lay pipelines in many different environments, such as swamps, shallow coastal waters, deserts, and forests. Pipelines now cross over mountain ranges and lie beneath the ocean floor. This versatility has made possible the transportation of petroleum through, and into, previously inaccessible areas and terrains.

Pipelines can be divided into two general classifications: **oil**, which are used for both crude and refined products, and **gas**, which carry natural gas. Except for a few engineering adjustments necessitated by the different characteristics of the fluids, construction methods are generally similar for the two types of pipelines. However, there are differences in the two systems: facilities at terminals and intermediate stations are different; gas pipelines utilize compressors and compressor stations to move the gas, while oil lines employ pumps and pumping stations; and gas lines use pressure control in their operating procedures instead of controlling the rate of flow, as happens in oil pipelines.

Although the discussion of pipelines in this section generally centers around oil pipelines, much of the information is applicable to gas pipelines, with the exception of those differences noted above.

1. Types of Pipelines

There are several different types of pipelines, each with their own distinct purpose and function (Figure 8.4). Lines connecting individual wells in an oil field to field storage tanks are generally called **flow lines**. They carry oil and gas to a central point where they can be treated, tested, and measured by equipment on the production site. After the petroleum has been treated, it is transported from the tank battery by intermediate lines known as **gathering lines**. Flow lines and gathering lines usually measure anywhere from two to twelve inches in diameter.

From the gathering lines, the oil is eventually transferred into **trunk lines**. The trunk lines are the large, long-distance pipelines that transport the oil to refineries, port, or central storage. During World War II, the "Big Inch" and "Little Big Inch" (24 inches and 20 inches in diameter, respectively) were exceptional lines for the time because of their size. Today, trunk lines can measure 48 inches in diameter and carry huge volumes of crude oil to various destinations. There are approximately 80,000 miles of crude oil trunk lines presently in the United States.

After the oil has been processed at the refinery, the resulting products are shipped in **product lines**. Product lines are the pipelines which transport gasoline, diesel fuel, jet fuel, etc., from the refinery to users or consumers. Throughout the United States today, a product pipeline network of over 80,000 miles distributes the various petroleum products directly to consuming centers.

2. Pipeline Construction

a. Onshore

The construction of pipelines on land has undergone many significant modifications over the years. In the early days, pipe was laboriously screwed together, one joint at a time, then lowered into hand-dug ditches and buried. Without adequate sealing procedures, these threaded and coupled pipes were subject to frequent leaks where they were joined together. Today, new methods and better materials have erased many of these earlier problems.

Before pipeline construction can begin, engineering studies are made to determine the best routes to be followed and the correct equipment to be used. Negotiations are conducted with landowners on the routing of the pipeline, and rights-of-way are obtained. After this preliminary work has been completed, the **spread** (the men and equipment used to construct the pipeline) is assembled, and construction can begin.

The first step in laying a pipeline is to clear the right-of-way, that is, to prepare the terrain and remove any obstacles in the path of the line. After that has been accomplished, an automatic ditching machine digs the trench in which the pipe will be laid. The ditch must be deep enough to prevent damage to the pipe from surface exposure or freezing.

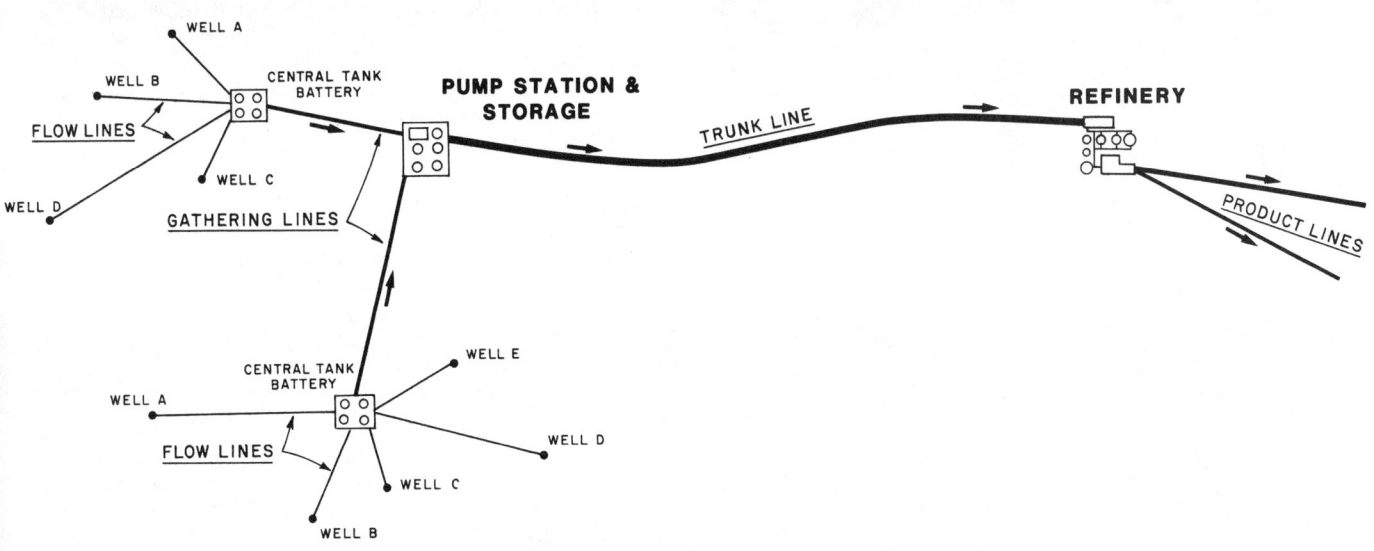

Figure 8.4. TYPES OF PIPELINES.

The pipe is first laid alongside the open trench. Sideboom tractors then lift the joints of pipe and hold them in position while they are welded together. High voltage electricity is applied to steel welding rods to melt the steel rods into the grooved ends of the pipe, forming a metal seal that joins the two lengths of pipe. This technique has greatly reduced the number of leaks experienced in pipelines.

After the welds are inspected, the pipe is cleaned, coated with tar, covered with fiberglass, and finally wrapped with asbestos felt, tar paper, or heavy kraft paper. The pipe is lowered into the ditch, and the trench is backfilled by bulldozers. Once these operations are completed, the right-of-way is regraded and restored as closely as possible to original conditions. (A cutaway diagram of a buried onshore pipeline is shown in Figure 8.5.)

It is often necessary for pipelines to cross obstacles such as rivers, railroad crossings, or highways. Boring machines have been developed that can tunnel out passageways beneath these obstacles, by drilling large horizontal holes which can be guided under the obstructions (Figure 8.6). The holes are lined with steel pipe to keep them from caving in. The pipeline is then inserted inside the lined passageway.

When large rivers have to be crossed, the pipeline can be routed on suspension bridges over the water (Figure 8.7). Suspension bridges also can be used to cross other natural obstacles such as canyons, where it would be impossible to lay the pipeline in the canyon, or to dig beneath it.

b. Offshore

In recent years, it has become necessary to develop methods for laying pipelines underwater in offshore areas where oil and gas fields have been developed. A **lay barge** is a typical vessel used for offshore pipeline construction (Figure 8.8). The joints of pipe are assembled and welded together on the barge, and slowly fed over the side of the barge onto the ocean floor. An anchor holds the lay barge in position as it works. As new pipe is added to the line, the barge is

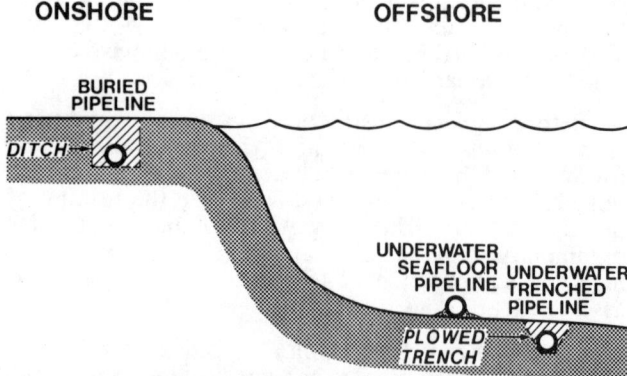

Figure 8.5. ONSHORE AND OFFSHORE PIPELINES.

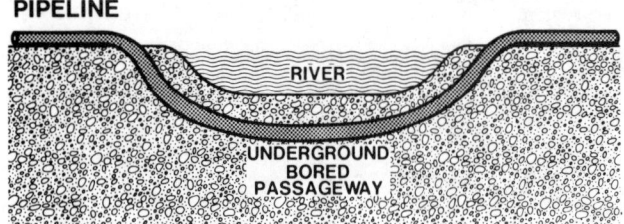

Figure 8.6. PIPELINE RIVER CROSSING: UNDER THE RIVER.

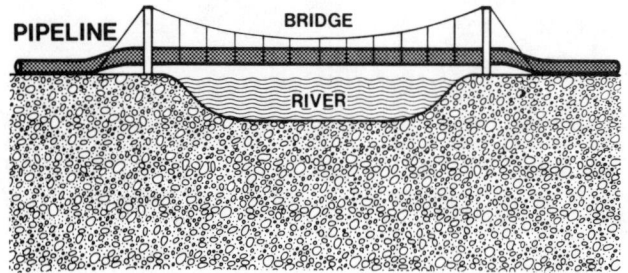

Figure 8.7. PIPELINE RIVER CROSSING: OVER THE RIVER.

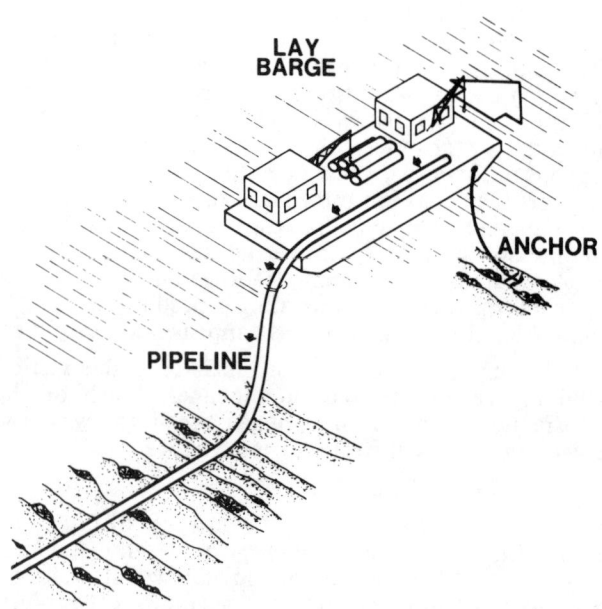

Figure 8.8. OFFSHORE PIPELINE CONSTRUCTION.

pulled forward on its anchor chains. Several different types of lay barges are used in the offshore industry to meet different situations. Although most of these vessels are used for laying pipeline in shallow water less than 300 feet deep, a number of large pipelines have been laid in the North Sea at depths of 500 to 600 feet. There have even been some test procedures conducted in the Mediterranean Sea at depths below 1,000 feet, but this is not yet a common practice.

Occasionally, the pipeline is **trenched,** so that it will lie below the mud line of the ocean floor (Figure 8.5). Trenching is required when the pipeline is to be protected from storms or fishing activities in the area, which might cause the pipeline to become entangled with trawler nets or anchors. The pipeline is normally lowered at least three feet below the mud line, but sometimes it must be submerged as much as ten feet in areas around platforms, or where there is considerable risk that the line might be damaged.

Common trenching methods use high pressure jets, powered by water pumped down from the surface, which scour the sea mud from under the pipeline, thus causing it to settle into a trench. To accomplish this, a trenching sled is mounted on top of the underwater pipeline and pulled along by a surface barge, which is

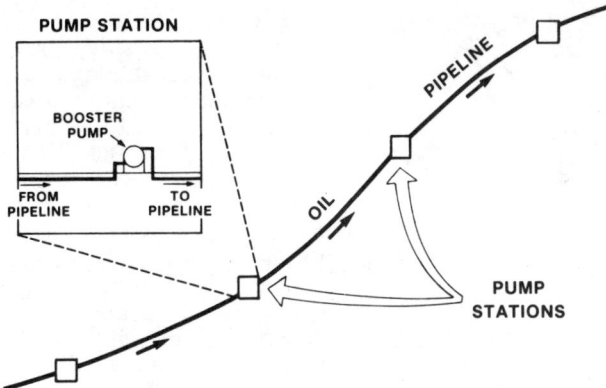

Figure 8.9. PUMP STATIONS.

equipped with pumps that provide high pressure water to specially-designed nozzles mounted in the sled. These nozzles create a trench by blowing the mud and soft ocean floor out from around the bottom and sides of the pipe. The pipe settles into this trench at the desired depth, away from possible hazards and interference.

3. Pump Stations

Since pipelines transport oil over great distances, it is necessary to construct pump stations to keep the oil flowing through the line (Figure 8.9). As the oil moves through a pipeline, friction causes the oil to slow down. Periodically, the oil must pass through **booster pumps,** which increase the pipeline pressure so the oil can continue to the next station. In former days, these pumping stations would be approximately 40 to 50 miles apart. Today, however, stations can be spaced as far as 80 to 150 miles, depending on the type of oil, the size of the pipe, and the terrain.

4. Batching

One of the options offered by pipeline transportation of oil is the ability to carry different petroleum products through the same line in rapid succession. Transportation of these shipments, or **batches,** can be carried out in a number of ways.

Years ago, slugs (or small quantities) of kerosene or water were used to separate the different shipments. Today, compatible products can be shipped one after another, without any intervening liquid separator. The points of commingling of the two different products are detected by recording instruments, electronic devices, or sampling.

One method of separating batches is illustrated in Figure 8.10. An inflatable rubber sphere, sometimes called a **pipeline pig,** is inserted between shipments in the pipeline. In the example used in the illustration, diesel fuel and gasoline are able to be shipped in the same line, because the sphere keeps them separate. Batching and batch separators have eliminated the expense of building several different pipelines to ship individual products.

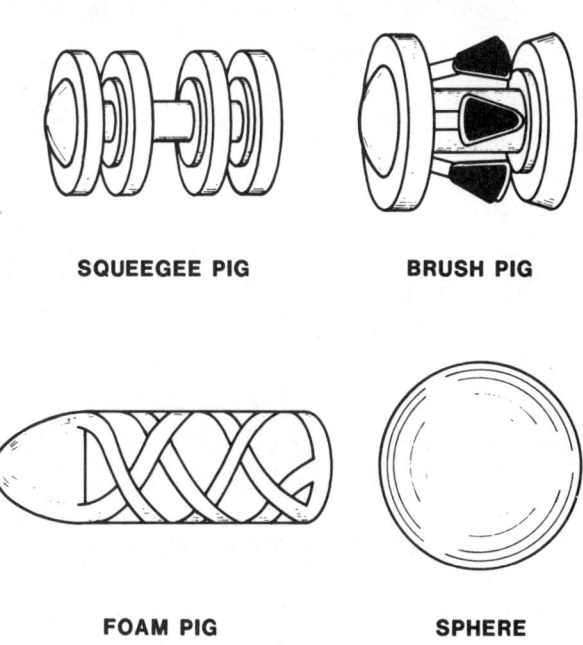

Figure 8.11. TYPES OF PIGS OR SCRAPERS.

5. Pigs

Certain devices, with the common classification of **pigs,** periodically are run inside the pipeline to perform different functions. Some of these pigs, also known as **scrapers** or **go-devils,** are pumped down the line in order to clean it. (Different types of pigs used for cleaning are shown in Figure 8.11). These scrapers are used to remove wax, dirt, and water accumulations from the line. Gauging pigs and electronic, or caliper, pigs also are used for detecting dents, buckles, or corrosion in the pipeline.

6. The Alaska Pipeline

The Alaskan Pipeline is an excellent example of the capability and flexibility of the modern oil pipeline. Oil from the large fields on the north slope of Alaska moves through a 48-inch diameter pipeline across three mountain ranges, 250 rivers and streams, and hundreds of miles of tundra and permafrost, a distance of 800 miles, to the Port of Valdez (Figure 8.12). The oil must pass through several pump stations, which in-

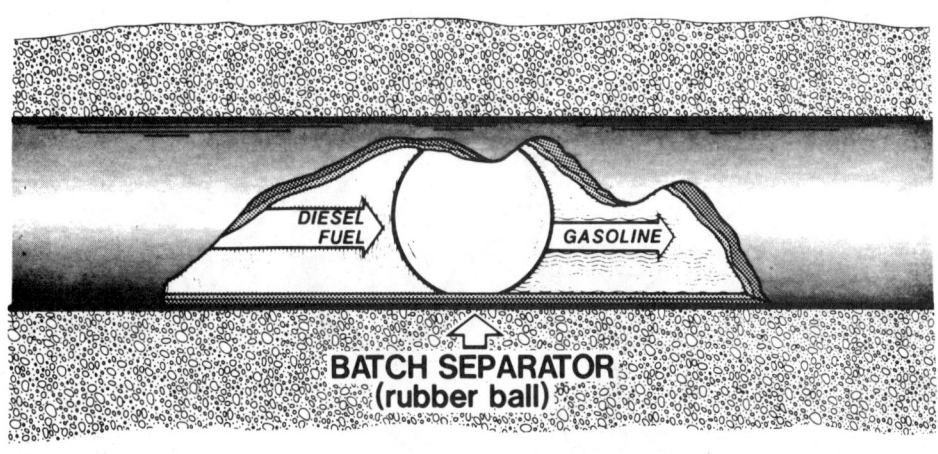

Figure 8.10. BATCH SEPARATION USING A RUBBER SPHERE.

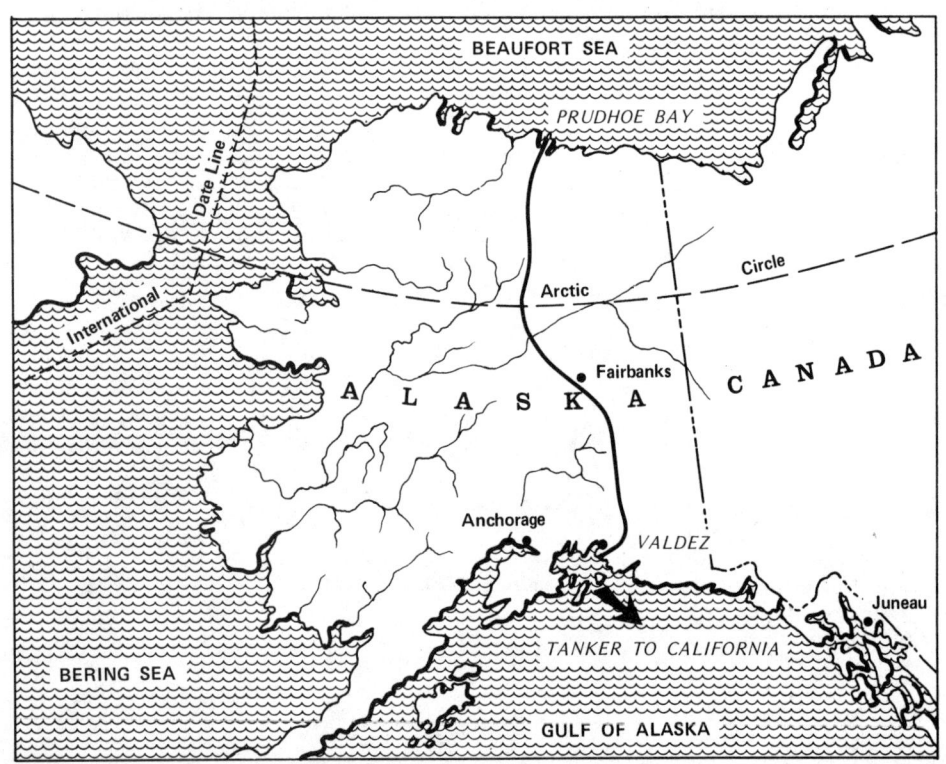

Figure 8.12. THE ALASKA PIPELINE.

crease the line pressure and move the oil to the next station.

The pipeline is positioned mainly above the ground on racks or carriages. If the pipeline were buried in the ground, the warm oil (135° F) flowing through the line would thaw the permafrost zone found in northern Alaska. If this were to happen, it would present a twofold problem. The thawing would undermine the pipeline's support in the frozen ground, allowing the pipeline to subside, and eventually causing a rupture of the line. Secondly, when the permafrost did freeze again, the pressure created would cause buckling in the earth, which could destroy the pipeline. In permafrost areas, the pipeline is supported high enough above the ground to allow the migration of the caribou.

The Alaska pipeline is considered one of the major engineering achievements of the last ten years. It cost more than 9 billion dollars to construct, and is currently pumping more than 1.7 million barrels of oil per day from the north slope to Valdez. The oil is carried from this port by tanker ship to the west coast, where it is distributed to refineries.

C. Tanker Ships

As the demand for petroleum around the world has increased, the use of tanker ships to transport great quantities of oil and liquefied natural gas has similarly expanded. The major oil producing and exporting nations rely heavily on tanker fleets to ship their products to consuming countries. Among the major areas which ship oil to other parts of the world are the Middle East, North Africa, the Carribean area of Latin America, and Southeast Asia. Parts of the world which receive this oil include Western Europe, the United States, and Japan. The principal routes that tankers use are indicated in Figure 8.13. Today's international tanker fleet consists of approximately 5,140 ships, transporting billions of barrels of oil each year.

Tankers can load oil from either onshore facilities, where oil has been brought from inland fields, or from offshore platforms. A common method for loading tankers, known as the **single point** (or **single buoy**) **mooring system**, is shown in Figure 8.14. In this system, oil is pumped from an offshore platform or from onshore facilities, through a pipeline on the ocean floor, to a marine riser, which is suspended at the surface by a large mooring buoy (usually 30 or more feet in diameter). The oil passes from the submarine pipeline into a flexible hose connected to the riser, up the riser, and through a floating hose to the ship.

By loading the tanker vessel from this mooring system, the tanker ship can be kept away from high risk areas, such as offshore platforms, while also remaining in deep water, which is often necessary for these large vessels. In areas where no deep water loading ports are available for the oil tankers to move in close to shore, oil can be pumped out of the tanker ships into shore tanks by using a similar system.

Tanker ships have grown tremendously in size over

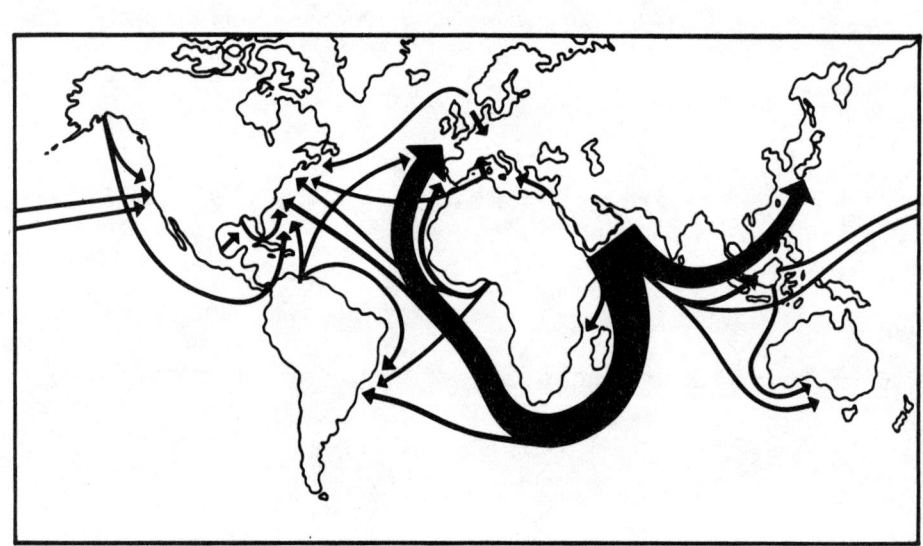

Figure 8.13. PRINCIPAL TANKER ROUTES.

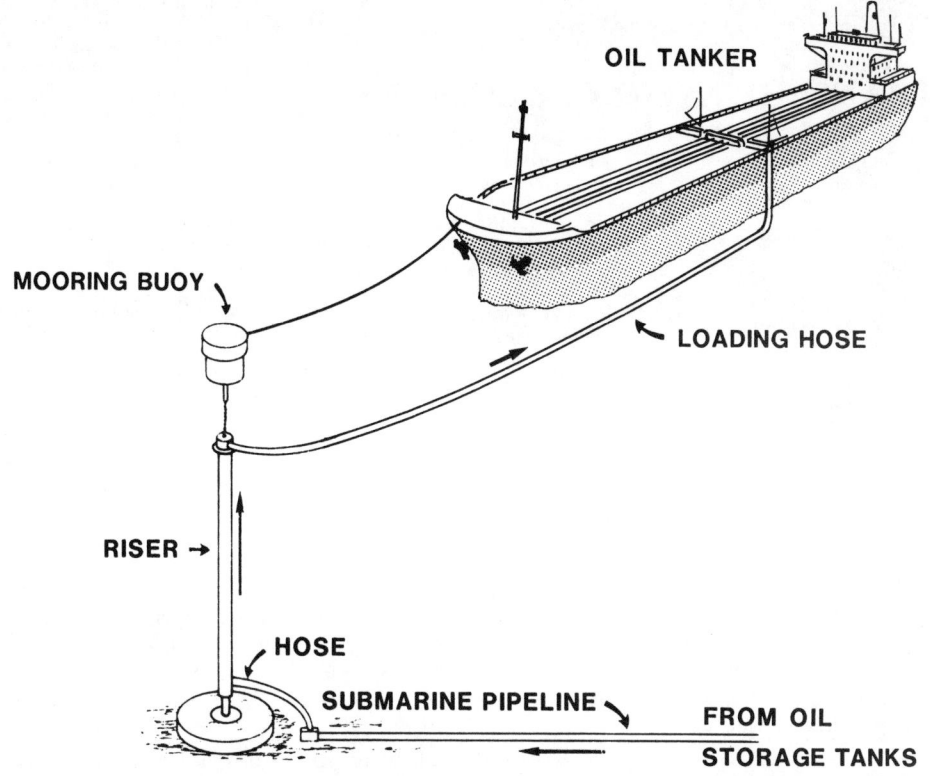

Figure 8.14. SINGLE POINT MOORING SYSTEM.

D. Barges

Our continent possesses one of the greatest inland water networks in the entire world, when one considers the Great Lakes and the numerous rivers and canals. At the very beginning of the oil industry in this country, riverboats of all types were used to transport oil from producing fields to refineries and marketing centers. Since many towns had been established along rivers, water transportation provided direct shipping to industrial and population areas.

Today, barges, varying greatly in size and use, still transport a great deal of oil in this country. Barges not only travel along inland rivers and canals, but also are used for shipping between coastal ports. Some of these barges are self-propelled, but most are either towed or pushed by tugboats. Load capacities range from 15,000 barrels for river and canal vessels, to 250,000 barrels for seagoing barges.

the last thirty years. Following World War II, tanker ships weighed approximately 30,000 tons. Today, **VLCCs (very large crude carriers)** with 160-320,000 DWT (dead weight tonnage) and **ULCCs (ultra large crude carriers)** with up to 540,000 DWT have been constructed to transport massive quantities of oil. Some of these super tankers have capacities exceeding 4 million barrels, and are often more than 1,000 feet long. These tankers are so large that presently there is only one port used in the United States having water deep enough to accommodate vessels of this size. Normally, smaller tankers or barges are used to shuttle the oil from the super tanker to a port. The single point mooring system also is used to unload these large tankers.

Tankers have been the subject of much controversy, because of the damage that would be done to the environment if an accident caused hundreds of thousands of barrels of oil to be spilled into the sea. Effort has been made to employ well-trained captains and crews on these vessels, and to use sophisticated communications and monitoring systems. Also, the presence of these large tankers eliminates the need for many smaller tankers, thus reducing the congestion of shipping traffic on the seas.

E. Tank Trucks

Tank trucks provide the capability of reaching remote producing areas and isolated consumers, that is not available with other means of transportation. Trucks are used to deliver much of the products processed by refineries to their final destinations, including factories, homes, farms, and service stations.

Because of the new alloys and materials used in their construction, modern tank trucks are built lighter and stronger than those of the past. Their load capacity has also increased, with some trucks now capable of hauling as much as 13,000 gallons of petroleum.

F. Railroad Tank Cars

Railroads were the primary means of transporting oil in the late 1800s, when the industry was still young. However, expansion of pipelining, increased use of inland waterways, and the versatility of tank trucks, have relegated the railroads to a relatively minor role in domestic petroleum transportation.

Improvements in design and construction have increased rail capabilities in recent years. The development of lighter cars, and cars with specialized features for transporting specific products, have opened up some new possibilities for rail shipment.

The following glossary, taken from **Introduction to the Oil Pipeline Industry**, is used with the permission of Petroleum Extension Service, the University of Texas at Austin (PETEX). We wish to express our thanks to the Petroleum Extension Service for this permission.

TRANSPORTATION—GLOSSARY

ACT—abbreviation for Automatic Custody Transfer; a metering system for automatically measuring and sampling oil or products at points of receipt or delivery.

ball—a neoprene spheroid that is run in the pipeline to clean the line, displace liquid hydrocarbons from natural gas pipelines, or to separate batches in liquid shipment.

barrel-mile—a unit of measurement of pipeline shipment of oil that signifies one barrel moved one mile.

batch—a tender or shipment of a single product that is handled through the line without mixing with preceding or following tenders.

batching—pumping shipments, tenders, or batches of a product through the line without mixing with other tenders.

Big Inch—colloquial term for a 24-inch crude-oil line constructed from Texas to the East Coast during World War II; a large-diameter pipeline.

booster station—a station used to increase the pressure of oil received through a main pipeline to transmit it to the next station or terminal.

centrifugal pump—a pump with an impeller or rotor, a rotor shaft, and a casing; it discharges fluid under pressure by centrifugal force.

centrifuge—a device used for separating BS&W and oil in routine tests; operates on the principle of differential gravities and centrifugal force.

commingle—the mixing of crude oil or oil products rather than moving as separate batches (see *batching*).

common carrier—any transportation system available for use by the public for transporting cargo; almost all interstate pipelines are common carriers.

deadwood—structural members within a tank whose volume is deducted when computing the capacity of a tank.

deliveryman—an employee who takes delivery of oil from a pipeline company at a terminal or junction.

dispatcher—the employee responsible for operation of the pipeline in accordance with a schedule of pipeline movements.

door-sheet—a plate at the base of a tank shell, which is removed when the tank is to be cleaned.

flange-up—to join two sections of pipe by means of flanges; also, to complete any assigned job or operation.

floating roof—a roof that rests on the surface of the oil contained in a tank rather than on structural members; it rises and falls with the level of the liquid in the tank.

float tank—a tank that is open to the main line at a station; oil from the main line may enter the tank and leave it as pumping rates in the line vary.

gauging—determining the liquid level of a tank so that its volume can be calculated.

grind-out—colloquial term for centrifuge test.

grind-out machine—a centrifuge.

holiday—a gap or void in the coating of a pipeline or in paint on a metal surface.

holiday detector—an electrical device used to detect weak places or holidays in pipeline and other coating.

hydraulic head—pressure exerted by, or imparted to, a column of fluid; usually expressed in feet or inches of water or other liquid; can be converted to psi.

ICC—Interstate Commerce Commission; a federal agency that has jurisdiction over oil pipelines engaged in interstate commerce.

inhibitor—a chemical used to inhibit or retard internal corrosion of pipelines.

joint movement—the shipment of a tender of oil through the facilities of two or more pipeline companies.

joint tariff—a rate sheet issued jointly by two or more companies, setting forth charges of moving oil over the facilities of each.

LACT station—Lease Automatic Custody Transfer station; an automated system for measuring and transferring oil from a lease, usually into a gathering system.

log book—book used to keep notes of current operating conditions and other useful information.

log sheet—daily report sheet on which operating data are entered by gaugers, dispatchers, and station engineers.

main line—a trunk pipeline.

manifold—an arrangement of piping and valves to provide interconnecting lengths between a number of pumps, tanks, and lines at a pump station.

microwave—ultra-shortwave radio communications system; signal waves in this system are focused to travel on a direct line-of-sight between sending and receiving equipment. All radio waves tend to travel in straight paths, but some are reflected back to earth, permitting communication between points not accessible by line-of-sight because of the curvature of the earth.

on-stream—a term to signify that a pump or pump station is operating normally.

on suction—a term indicating that a tank has been opened to the pump suction.

over and short station—a once-popular description of a pump station where one or more tanks could be floating on the line (see *floating tank*).

pig—a scraper; also called a go-devil.

positive-displacement pump—a pump that may use pistons, plungers, rotating vanes, or rotating gears to produce pumping pressure.

products cycle—the sequence or order in which a number of different products are batched through a pipeline.

products line—a pipeline used for the shipment of refined products.

prover—a device used to calibrate meters used in measuring oil.

reciprocating pump—a pump whose pumping effect is produced by reciprocating motion of pistons or plungers operating in cylinders.

right-of-way—a strip of land usually from 50 to 80 feet wide on which permission has been granted by land owners for the construction of a pipeline.

scheduler—the employee responsible for establishing pipeline oil movements based on the coordination of all shipper requirements.

scraper trap—special piping arranged to launch or receive a pipeline scraper.

Shepard's canes—an earth-resistivity meter used to measure the resistance of soil to the passage of electric current.

slop—collection of waste hydrocarbon liquids; in products pipelining, the commingled liquids consisting of contaminated mixtures that cannot be cut to any product.

sour crude oil—oil containing hydrogen sulfide or other sulfur compounds.

sphere—a neoprene ball that is run in a pipeline to clean it, to displace liquid hydrocarbons from natural gas pipelines, or to separate batches in liquid shipment (see *ball*).

strapping—measurement and calculation to determine the capacity of a tank and to provide tables for conversion of liquid level height to volume in barrels.

stringer bead—the foundation metal placed by a welder in making a pipeline weld; followed by two or more additional beads or passes to complete the weld.

supervisory control—the coordination of all facets of pipeline operations to insure proper handling of oil movements; begins before the physical movement of product and continues through record keeping of quality and quantity of movement.

sweet crude oil—oil containing little or no hydrogen sulfide or other sulfur compounds.

tank battery—a group of tanks to which crude oil flows from producing oil wells; also, a group of tanks at a tank farm.

tanker—see *tankship*.

tank farm—one or more tanks connected to a pipeline and a pump station by means of which oil is unloaded into tanks or withdrawn from them.

tankship—a seagoing vessel whose cargo space consists of a number of tanks; used for shipment of liquid cargo. Recently built vessels have capacities as high as 425,000 barrels and a speed of 15 knots.

tariff—a rate sheet of charges made by pipeline companies for moving oil.

tender—a shipment of oil presented by a shipper to a pipeline for movement.

terminal—a point to which oil is transported through pipelines; usually includes a tank farm and may include tanker loading facilities.

thief—a device that is lowered into a tank to take an oil sample at any desired depth; used to determine the BS&W content of the oil in a tank.

trunk line—a main pipeline.

welder—an employee who welds.

wind ring, or wind girder—a horizontal stiffening structural member installed near the top of a floating-roof tank to reinforce the tank wall against wind pressure.

TRANSPORTATION—BIBLIOGRAPHY

American Petroleum Institute: **Facts About Oil,** API (Washington, D.C.) no date.

Berger, Bill D. and Kenneth E. Anderson: **Modern Petroleum: A Basic Primer of the Industry,** PPC Books (Tulsa) 1978.

Petroleum Extension Service: **Fundamentals of Petroleum,** The University of Texas at Austin (Austin) 1979.

Petroleum Extension Service: **Introduction to the Oil Pipeline Industry,** 2nd Ed., The University of Texas at Austin (Austin) 1978.

Petroleum Extension Service: **A Primer of Offshore Operations,** The University of Texas at Austin (Austin) 1976.

The Science and Public Policy Program, University of Oklahoma: **Energy Alternatives: A Comparative Analysis,** U.S. Government Printing Office (Washington, D.C.) 1975.

CHAPTER 9:
ECONOMICS

I. INTRODUCTION

At each juncture in the search for and development of new oil and gas resources, the final determinant as to whether activities should proceed or stop is economics. Economic considerations receive particular attention in the exploration and production phases of the oil business. The explorationist uses economics as a screening tool to judge whether or not potential new resources are worth exploring and developing. Once a new field has been found, economic analysis determines the most profitable means of developing the resources. Decisions about the number of wells to be drilled in the new field, the type of equipment that will be used, and the ultimate direction of the overall operation are all based on economics.

The word "economics," as used here, refers to the profitability of a certain project. A project can be something as small as installing a new piece of equipment for a drilling operation, or as large as the overall exploration and development of a new field. The main factors used in petroleum economics include **cost estimations, production** and **revenue forecasts, cash flow determinations,** and **profitability analysis.**

To clarify our discussion, we have developed a hypothetical project called the Red Mesa Field. This example will be used to illustrate the overall project economics in the "exploration and production cycle." This term refers to a typical cycle involved in finding and developing a new petroleum reservoir. Each of the factors mentioned above will be applied to an economic analysis of the Red Mesa Field.

II. BASIC ECONOMICS

A. Cost Estimation

The first step in determining the economic value of a project or venture is generally a cost estimate. Cost estimates are important since the petroleum industry is known as a highly "capital intensive" industry. Considerable initial capital expenditure is required to begin new projects, particularly new exploration and drilling programs.

As shown in Figure 9.1, exploration and development expenditures by the petroleum industry have increased significantly in the last few years. A number of factors have contributed to the rise of these costs. The need to explore for oil in more hostile environments, such as offshore on the continental shelf or in the northern arctic regions of Alaska, has been a major factor in the escalation of these costs. Inflation and government regulations have also increased the cost of doing business in the petroleum industry.

Cost estimation, therefore, is a critical factor in making an economic evaluation of any project. Indeed, it is often the cost of a proposed project that determines whether or not that project ever leaves the drawing board to go into actual operation.

Cost items are generally classified as either **explora-**

Figure 9.1. ESTIMATED U.S. EXPLORATION AND DEVELOPMENT EXPENDITURES. (Source: The Chase Manhattan Bank, API: *Basic Petroleum Data Book***)**

tion or **development** costs. Many exploration projects never reach the development stage because commercial deposits of hydrocarbons are not found. These exploration costs must stand alone. In a successful project, development costs follow exploration costs, and the project must be evaluated at each stage.

Costs included in the **exploration** phase of this cycle may be categorized as follows:

1. Geological and Geophysical Expenditures

These include the costs of performing field geological and geophysical surveys, as well as the office work required to process the field data and interpret the results.

2. Land Department Expenditures

These are the costs of maintaining a group of people experienced in the following areas: obtaining mineral leases for drilling new exploration projects, scouting for available land, and following competitor activity in the area.

3. Acquiring Undeveloped Acreage

Mineral leases on undeveloped acreage are acquired either through bidding in public sales or through private negotiations with landowners. In either case, an initial acquisition cost is required to obtain a lease on these mineral rights.

4. Lease Rentals

Once a lease has been obtained, it is usually necessary to pay yearly rental fees to the landowner to maintain these mineral rights prior to beginning production.

5. Test Well Contributions

A company often holds leases in areas where other companies have similar leases and mutual exploration goals. In such cases, it is common for one company to contribute money to a test well which is being drilled by another company. This allows important information regarding the potential of the area to be available to both companies at a reduced cost.

6. Drilling and Equipping Exploratory Wells

Much of the cost of exploration activities is derived from expenditures required to drill and equip new exploratory wells. As shown in Figure 9.2, the costs for drilling wells have increased dramatically over the last few years. Drilling costs increase significantly as the depth of the well increases, or if the drilling has to be done in a more demanding environment, such as offshore. These factors must all be considered when estimating exploratory drilling costs.

If an exploratory program is successful, the **development** costs must then be taken into account. The following are the major cost items included in a development program:

1. Development Drilling Costs

After a field has been discovered and its size determined, development well drilling is necessary to create an adequate number of drainage points to efficiently produce the oil or gas from the new reservoir. This is often one of the major costs in developing a new field.

2. Well Completion Costs

After the wells have been drilled, the cost of completing the wells and equipping them for production must be determined and included in the economic evaluation.

3. Surface Facilities Cost

Surface equipment is necessary to operate an oil or gas field. This may include oil or gas treating facilities, pipelines, pumping equipment, roads, and buildings. If the field is offshore, major investments must be made to construct permanent platforms to house drilling and production facilities.

4. Operating Costs

In addition to the initial capital expenditure required to develop a new field, the cost of operating the field must also be considered. This operating figure includes labor, maintenance, fuel, taxes, and overhead costs. These are generally computed as annually recurring costs.

Exploration and development costs can be estimated from such sources as company or industry records on similar projects and from equipment catalogs published by manufacturers. Labor costs are generally estimated by contractors who have been selected to perform the actual work.

Other common sources of cost data are **cost correlations** such as the Nelson Cost Index (Figure 9.3). The Nelson Cost Index in Figure 9.3 is tabulated specifically for the construction costs of oil refineries, but other sources are available for other aspects of the industry. In this particular example, the December, 1975, value of the index was 594 and the December, 1980, value was 859. Using this correlation, it can be seen that refinery construction costs increased by a ratio of 859/594, or 1.45 to 1 during this five-year period. One could obtain a rough idea of the cost to build a refinery in 1980 by

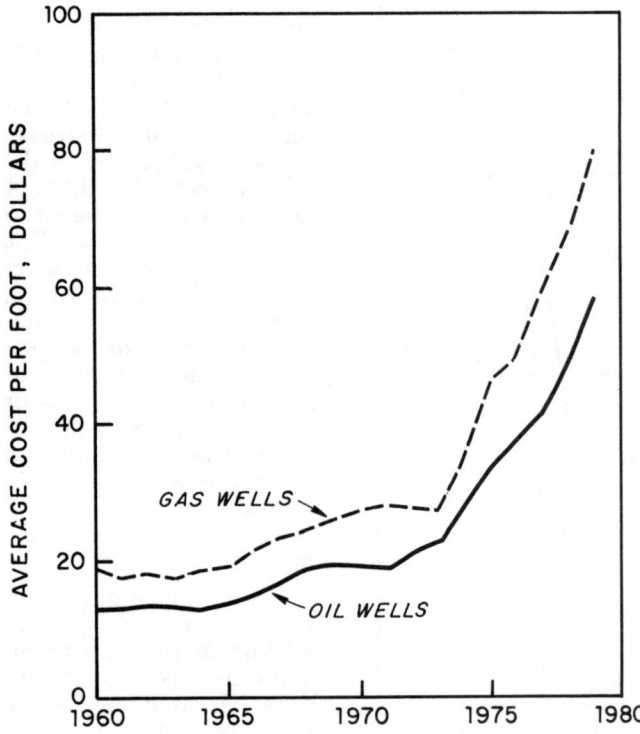

Figure 9.2. ESTIMATED DRILLING COSTS. (Source: "Joint Association Survey of the U.S. Oil and Gas Producing Industry" Annual, American Petroleum Institute, Independent Petroleum Association of America, Mid-Continent Oil and Gas Association)

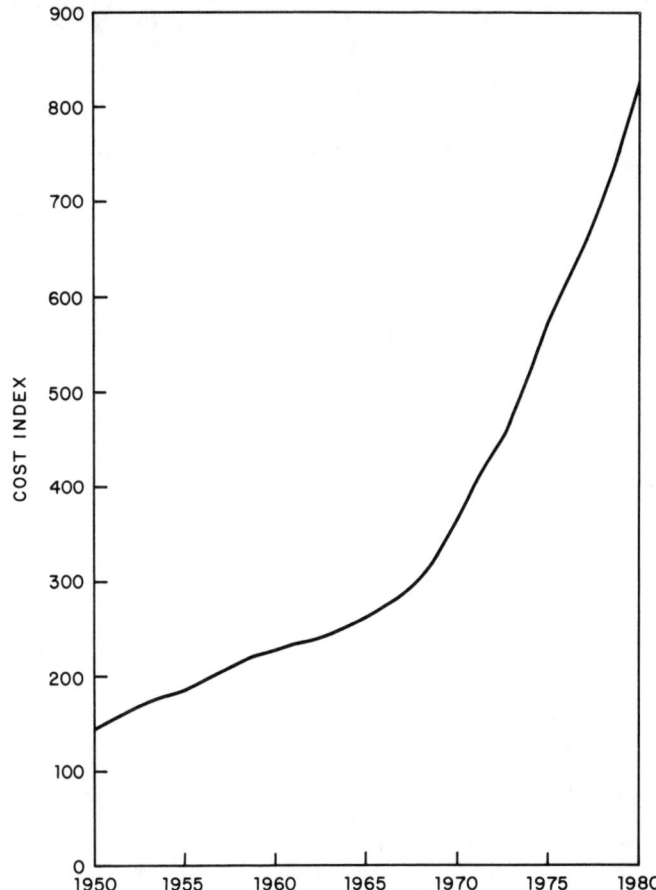

Figure 9.3. NELSON COST INDEX: REFINERY CONSTRUCTION.

multiplying the 1975 cost by this particular value. Similar cost correlations are often developed for other types of projects in the petroleum industry.

B. Reserves and Production Forecasts

One of the primary goals of an economic evaluation of a project is to predict the expected yearly income from the project. The first step in analyzing a new field is to estimate the petroleum reserves present in the field and the field's potential production rates, once the reservoir is fully developed. A field's expected size and performance can be determined using engineering and geological data.

Reserve forecasts involve estimating how much oil and gas is present in the new reservoir, and how much can be withdrawn or produced. These forecasts are used to determine the ultimate value of the field and at what rates it can be produced over time.

Production forecasts are used to determine the expected cash flow from a project. The amount of oil or gas which can be produced each year can be predicted by studying the history of similar wells in the area. Over a period of time, a well's production rate tends to decline; this must be considered when making an economic evaluation and projection of a project. Information about a field's past production can be used to predict its future cash income. (This will be illustrated in the Red Mesa Field example.)

C. Revenue and Cash Flow Analysis

Once reserve and production forecasts have been made, revenue and cash flow information can be obtained. This is the final step in preparing for a profitability analysis of the project.

Revenue simply means the value received for the products sold, in this case the oil and gas. **Gross revenue** is the total value of the oil or gas, which is the amount of oil or gas sold (the production) multiplied by its market price. **Net revenue** generally means the amount of revenue remaining after royalties.

The term **cash flow** refers to amounts of money which are either spent or earned as a function of time, usually on an annual basis. Figure 9.4 illustrates the cash flow concept in a bar graph.

The annual cash flow for an oil property is determined by deducting all operating expenses and taxes (other than income taxes) from net revenues to arrive at a net yearly operating profit or loss for the project. This value is often referred to as the **net operating income,** and includes all revenues and expenses from the project, except federal income taxes. (This, too, will be illustrated in the Red Mesa Field example that follows.)

D. Example: Red Mesa Field

1. Background

The Red Mesa Field is used here as a hypothetical project to illustrate various economic principles discussed in this chapter. Consider the field to be a small oil field in the western United States, in an area where oil is found in a sandstone formation called the Mesa Verde sand. The Red Mesa Field consists of a small anticlinal structural trap. The oil is found at a depth of approximately 8,000 feet. Similar oil fields have been found in the same general area. The geological department of a small oil company becomes interested in the area,

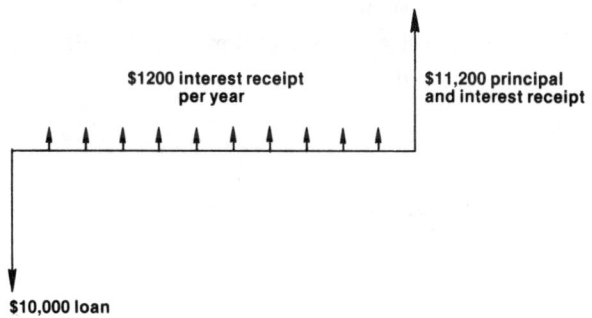

Diagram of Lender's Cash Flow

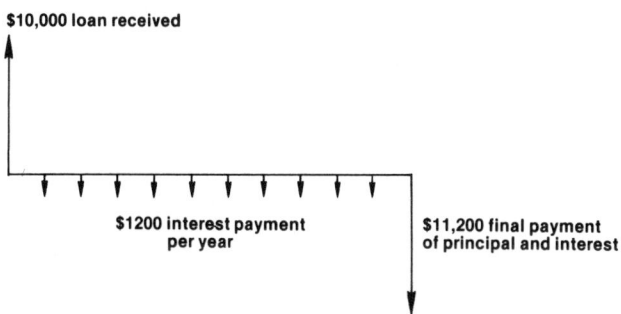

Diagram of Borrower's Cash Flow

Figure 9.4. CASH FLOW.

and conducts a geological study to determine whether any undrilled reservoirs exist. As a result, the Red Mesa anticline is identified and acreage is leased.

The company contributes money to another company, which is planning to drill a test well on land adjacent to the property which the former company has leased overlying the Red Mesa anticline. By contributing to the cost of drilling this well, the company obtains information which confirms the original geological study. The company decides to drill its own exploratory well to determine if the Red Mesa anticline exists, and if it contains oil. The exploratory well shows that commercial quantities of oil are present in an anticlinal trap. The Red Mesa anticline is now ready for development into a producing oil field.

Development drilling follows the successful completion of the first exploratory well. In the course of defining the field, ten producing wells (including the original well) are drilled, and four wells are drilled that encounter only water. These "dry holes" are plugged, except for one well. This well is used for subsurface disposal of water produced with the oil. Total oil in place is estimated at six million barrels. Very little gas is present in the reservoir. Within two years of the initial discovery of the field, all the wells have been drilled and production equipment has been installed.

2. Costs

The costs associated with discovering, developing, and operating the Red Mesa Field are as follows:

a. Exploration Costs

(1) Geological studies	$ 50,000
(2) Lease acquisition cost	150,000
(3) Lease costs	
(a) Straight rental	year/50,000
(b) Royalty payment on production	15%
(4) Test well contribution	100,000
(5) Drill initial exploratory well	400,000
TOTAL (excluding rental)	$700,000

Note: In (3) above, lease rentals must be paid as long as there is no oil production. After oil is produced, lease rentals are paid to the landowner as a percentage of gross oil revenues (called royalty); in this case, the amount is 15 percent.

b. Development Costs

(1) Drill 13 development wells @ $400,000 each	$5,200,000
(2) Production equipment for 10 wells	1,000,000
(3) Surface facilities (treating, water disposal, storage, shipping, etc.)	800,000
TOTAL	$7,000,000

c. Operating Costs

(1) Labor	year/$150,000
(2) Maintenance	200,000
(3) Fuel, electricity	100,000
(4) Overhead	50,000
TOTAL	year/$500,000

3. Reserve and Production Forecasts

As stated previously, geological information revealed that the Red Mesa Field contains approximately six million barrels of oil. Only a portion of this amount, however, can be recovered. By examining historical production data from similar fields nearby and analyzing the initial performance of the new wells in the Red Mesa Field, the company determines that approximately one third of the oil in place, or two million barrels of oil, can be recovered. A yearly production forecast is also developed, before and after royalty, as shown in Columns 1 and 2 of Table 9.1. Note that oil production is highest in the first year of production and declines steadily each year thereafter. This is caused by the decreasing amount of natural reservoir energy available to force the oil into the wells. For the sake of brevity, only the first ten years of production are shown, although the field could have a life of 25 years or longer.

Figures 9.5 and 9.6 show the projected Red Mesa Field oil production rate versus time. Both Figures represent the same projections, but they are plotted on different types of graph paper. The decline curve in Figure 9.5 is plotted on rectangular (or Cartesian) coordinate paper which can be used in any number of simple graphing applications. The curve in Figure 9.6 is plotted on what is known as semi-logarithmic paper (semi-

Table 9.1. RED MESA FIELD: CASH FLOW ANALYSIS

YEAR	(1) OIL PRODUCTION, BBLS GROSS	(2) NET AFTER 15% ROYALTY	(3) NET OIL REVENUE @ $35/Bbl	(4) OPERATING COST	(5) STATE TAXES @ 10%	(6) WINDFALL PROFIT TAXES	(7) NET OPERATING INCOME	(8) EXP. & DEV. CAPITAL INVESTMENT	(9) BEFORE-TAX NET CASH FLOW	(10) FEDERAL INCOME TAX	(11) AFTER-TAX NET CASH FLOW
1	None—Under Exploration		0	0	0	0	0	750,000	-750,000	-230,000	-520,000
2	None—Under Development		0	0	0	0	0	7,050,000	-7,050,000	-1,840,000	-5,210,000
3	300,000	255,000	8,925,000	500,000	893,000	1,209,000	6,323,000	0	6,323,000	2,625,000	3,698,000
4	255,000	217,000	7,586,000	500,000	759,000	1,007,000	5,321,000	0	5,321,000	2,206,000	3,114,000
5	217,000	184,000	6,456,000	500,000	646,000	839,000	4,471,000	0	4,471,000	1,852,000	2,619,000
6	184,000	156,000	5,474,000	500,000	547,000	696,000	3,731,000	0	3,731,000	1,542,000	2,188,000
7	157,000	133,000	4,671,000	500,000	467,000	581,000	3,123,000	0	3,123,000	1,288,000	1,835,000
8	133,000	113,000	3,957,000	500,000	396,000	480,000	2,581,000	0	2,581,000	1,062,000	1,519,000
9	113,000	96,000	3,362,000	500,000	336,000	398,000	2,128,000	0	2,128,000	872,000	1,256,000
10	96,000	82,000	2,856,000	500,000	286,000	329,000	1,741,000	0	1,741,000	710,000	1,031,000
11	82,000	70,000	2,439,000	500,000	244,000	221,000	1,475,000	0	1,475,000	601,000	874,000
12	69,000	59,000	2,053,000	500,000	205,000	100,000	1,248,000	0	1,248,000	509,000	739,000
TOTAL	1,606,000	1,365,000	47,799,000	5,000,000	4,778,000	5,859,000	32,141,000	7,800,000	24,341,000	11,197,000	13,144,000

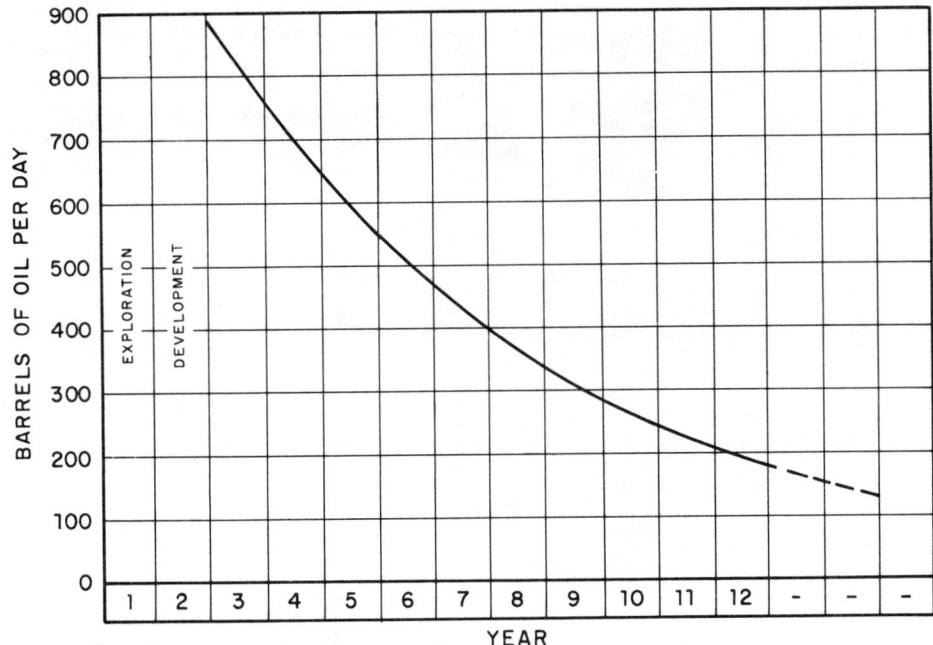

Figure 9.5. PROJECTED FUTURE PRODUCTION OF RED MESA FIELD.
Plotted on rectangular coordinate paper.

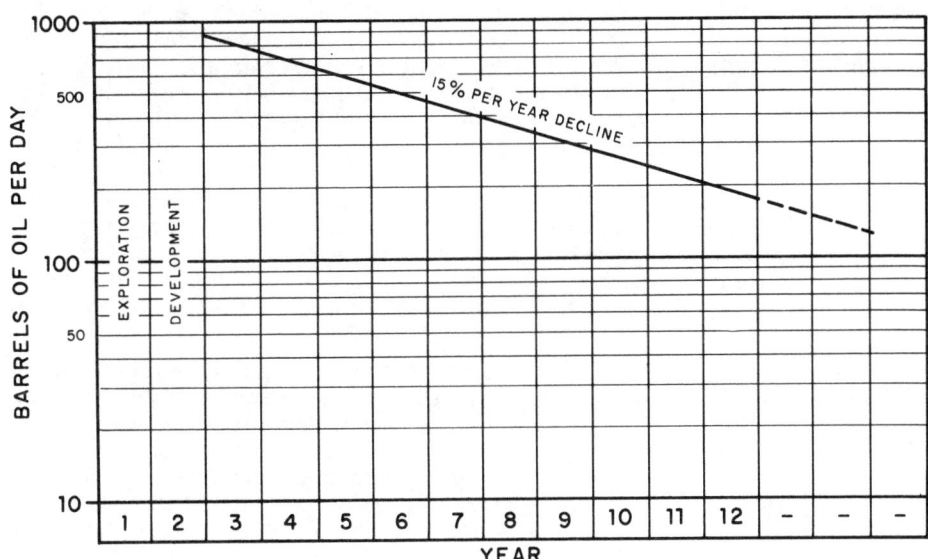

Figure 9.6. PROJECTED FUTURE PRODUCTION OF RED MESA FIELD.
Plotted on semi-logarithmic paper.

the amount of oil which actually belongs to the oil company. This is calculated by deducting royalty oil as shown in Column 2 of Table 9.1. **Net oil after royalty** is then multiplied by **oil price** to determine **net oil revenue** by year (Column 3). For the sake of simplicity in our Red Mesa Field example, oil price is assumed to be constant at $35/Bbl for the ten years of producing life analyzed here. To determine **net operating income** (Column 7 of Table 9.1), annual **operating costs, state production** or **severance taxes**, and the **windfall profits tax** (Columns 4, 5, and 6) are deducted from **net oil revenue**. For this example, it is assumed that state taxes are equivalent to 10% of **net oil revenue** in each year. In many oil and gas producing areas, there is a combination of state and local taxes, some of which may be based on the value of production, and some of which may be based on other factors, such as the assessed value of the property. The **windfall profit taxes** (Column 6) have been determined using an appropriate set of assumptions about the Red Mesa Field. The **windfall profits tax** is, in fact, too complicated to address within the scope of our discussion here. However, because of its importance in the economics of future United States oil producing operations, it has been included in the Red Mesa Field example.

After the **net operating income** has been determined (Column 7 of Table 9.1), the **before-tax cash flow** (Column 9) can be calculated by deducting the **exploration and development capital investments** presented earlier and shown in Column 8. **Federal income taxes** (Column 10) are deducted from Column 9 to determine the company's **after-tax net cash flow** (Column 11). Since federal income tax law is highly complex, the **income tax** shown in Column 10 has been determined using the simplest possible set of assumptions. The complexities of income taxation, like windfall profit taxes, are outside the scope of our discussion here. Again for the sake of simplicity, state income taxes which might also apply are not considered.

To illustrate how the gross oil revenues from the Red Mesa Field are distributed, Figure 9.7 has been constructed. This Figure shows that gross oil revenues over the 12-year period will be about 57 million dollars.

log). Note that for the estimated decline rate of 15% per year, the graph of production is a straight line. This and other similar types of paper are ordinarily used when projecting future oil or gas production, because production for each year is usually some percentage of the production for the prior year.

4. Cash Flow Analysis

A production forecast for the Red Mesa Field makes it possible to determine the project's anticipated cash flow. In this example, we will assume that oil production does not start until the end of two years, when all the wells have been drilled and all production equipment has been installed.

The first step is to determine the **net production**, or

Fifteen percent of gross revenue, or approximately 8.4 million dollars, goes to the mineral owner in the form of royalties. About 39% of gross revenue (in this case, 21.8 million dollars) is paid out in taxes. Thus, after royalty and taxes, the company has only 46% of gross revenues left for capital investment, operating costs, and net company profits after-tax. As shown in Figure 9.7 for the Red Mesa Field, the total capital investment and operating costs required to discover, develop, and produce the oil, amounts to less than one-fourth of the gross revenue. The company's net profit (or net cash flow) after-tax is also less than one-fourth of gross revenue.

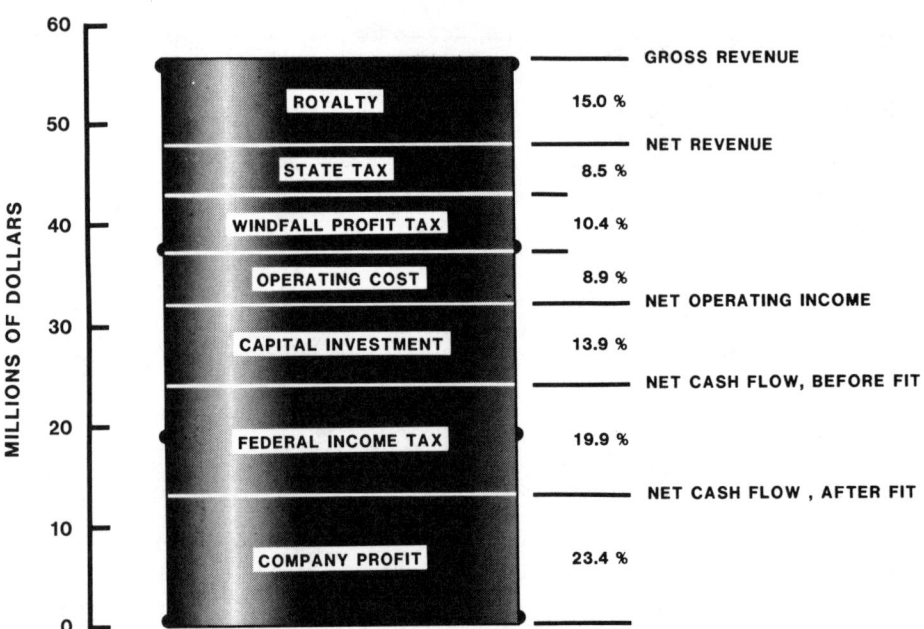

Figure 9.7. DISTRIBUTION OF OIL REVENUES FROM RED MESA FIELD.

The cash flow presented on Table 9.1 and Figure 9.7 are representative only of the Red Mesa Field example. The magnitude and distribution of revenues and costs for actual petroleum industry projects will, in fact, vary widely due to the nature of the project, the company or companies involved, and the economic conditions prevailing at the time.

III. PROFITABILITY ANALYSIS

A. Introduction

Profitability analysis is used to determine the value of a project or property for a given set of conditions. This information is necessary when buying or selling the property or comparing it to other projects for investment purposes. In the Red Mesa Field example, the oil company would develop a profitability analysis prior to undertaking full field development. The results of such an analysis would be compared with similar analyses of other possible projects to select the best investment opportunities.

Three common measures of profitability which are used to evaluate oil and gas properties are discussed in this section. **Payout, present worth,** and **annual rate-of-return** will be applied to our Red Mesa Field example. There are a number of other methods for determining profitability, but they are generally more complicated than these three and are often used only for specialized purposes. Only the basic principles behind payout, present worth, and annual rate-of-return need be considered in this section.

B. Payout

Payout, one of the simplest profitability yardsticks, refers to the amount of time required to recover the capital investments made on a project. This may be calculated either before or after income tax. Payout is generally measured from the time of initial capital investment in a project.

Using the Red Mesa Field example, before-tax and after-tax payouts are calculated by tabulating cumulative before-tax and after-tax net cash flows over the first few years of the project life (Table 9.2).

The annual before-tax and after-tax net cash flows in Columns 1 and 3 of Table 9.2 are from Columns 9 and 11, respectively, of Table 9.1. As shown on Table 9.1, the negative cash flows during the first and second year of the project are due to exploration and development capital investments made before production starts at the beginning of the third year. To get the cumulative before- and after-tax cash flows in Columns 2 and 4 in Table 9.2, each year's cash flow is added to the balance existing at the end of the previous year. When the cumulative net cash flow, either before-tax or after-tax, reaches zero, "payout" has occurred. For the Red Mesa Field, Table 9.2 indicates that payout occurs during the fourth year of project life. The actual payout times are 3.28 years before-tax and 3.65 years after-tax.

If, in the example, the beginning of production at the start of the third year was used as a reference date, payout would occur in less than two years, both before- and after-tax. Another alternative would be to measure payout from the beginning of field development in the second year, when most of the capital investment occurs. If we assumed that the exploration phase of the project was already complete, the beginning of development would probably be the best reference date for payout. In that case, only the capital investments for development would be considered in cumulative cash flows.

Table 9.2. RED MESA FIELD: NET CASH FLOWS

	BEFORE-TAX NET CASH FLOWS		AFTER-TAX NET CASH FLOWS	
	(1)	(2)	(3)	(4)
YEAR	ANNUAL	CUMULATIVE	ANNUAL	CUMULATIVE
1	-$ 750,000	-$ 750,000	-$ 520,000	-$ 520,000
2	- 7,050,000	- 7,800,000	- 5,210,000	- 5,730,000
3	6,323,000	- 1,477,000	3,698,000	- 2,032,000
4	5,321,000	3,844,000	3,114,000	1,082,000
5	4,471,000	8,315,000	2,619,000	3,702,000

C. Present Value

Present value or net present value is another widely used profitability indicator. The present value of a project is found by taking all future net cash flows, either before-tax or after-tax, and "discounting" these amounts back to the reference date of the evaluation.

To understand present value, it is necessary to understand the **time value** of money. For example, a certain sum of money, when deposited in a savings account, grows because of the interest it earns while on deposit. The value of the initial deposit is not fixed, but rather is a function of the interest rate and the length of time the money is on deposit. One dollar invested at an annual interest rate of 8 percent would be worth $1.08 after one year. Conversely, $1.08 to be received one year from now has a present value of only $1.00 at the same interest rate of 8 percent. This illustrates how a sum of money to be received in the future is discounted to get its present value. Any future sums of money or cash flows can be expressed in terms of their present value.

Present value is determined by using **discount factors** which are, in effect, the inverse of interest that might be applied to our example savings account. In that example, the discount factor at the eight percent interest rate can be calculated by taking the ratio of the present value to the future value:

$$\text{One Year, 8\% Discount Factor} = \frac{\$1.00}{\$1.08} = 0.9259$$

This discount factor can be used to determine the present value of any sum of money or cash flow to be received one year from the present. To illustrate this, suppose that you will receive $5,000 one year from now. The present value of this amount would be:

$$\$5,000 \times 0.9259 = \$4,629.63$$

If you deposited $4,629.63 in a savings account at 8 percent annual interest, you would have $5,000 after one year.

The equation for calculating discount, or present worth, factors at the end of any given year is:

$$\text{Discount Factor} = \frac{1}{(1+i)^n}$$

where i = interest, and n = number of years.

Using this equation, we obtain the same discount factor we calculated in the previous example:

$$\frac{1}{(1+0.08)} = 0.9259$$

Since cash flow projections are normally made on an annual basis, we may wish to consider that the cash flow for any given year is received or paid out in a lump sum at the middle of that year. The discount factor to be used for any year then would be:

$$\frac{1}{(1+i)^{n-.5}}$$

This type of discounting is best suited and most often used for oil and gas cash flow analyses because costs and production revenue will occur during each month over the year. It would be misleading in our analysis to assume that a company had to wait until the end of each year to receive its cash flows. Using the discount factor equations, a table of discount factors can be developed for any interest rate over any number of years.

Using the last formula above, the present value of the Red Mesa Field can be determined based on the cash flows presented in Table 9.1. If a 15 percent discount rate is used, the present value of the Red Mesa Field after-tax would be calculated as shown in Table 9.3.

As seen from Table 9.3, the present value (or discounted value) of the Red Mesa Field project is considerably less than the total of after-tax net cash flows. It is important to note that large project investments cause large negative net cash flows in the first two years, while the discount factor is still close to one. At the same time, net cash flows received in the later years of the evaluation are worth much less in terms of their present value. For example, the present value of each dollar received in the twelfth year is about $.20 when discounted at 15 percent.

In determining present value, proper selection of the reference date is important. In our example here, the beginning of exploration activities was chosen as a reference date. In practice, the start of the first year of development, which involves significantly greater capital expenditures, might be used as a reference date. Using that assumption, the discounted net cash flows, both before- and after-tax, would increase. The present value of the Red Mesa Field would also increase somewhat if the full project life were considered.

The significance of the present value (or discounted net cash flow) for the Red Mesa Field may be explained this way. Assuming that the cost estimates and income projections used in the foregoing analysis were correct, the oil company would earn, after-tax, the equivalent of 15 percent interest on the money they actually invested plus future profits equivalent to $4,496,000 received at the start of the project. Although this sounds attractive, the element of risk is not considered. In the case of the Red Mesa Field, the exploration program was successful in finding oil in commercial quantities. There are many unsuccessful exploration programs, the cost of which must be borne by profits from other projects which are successful. Even after the discovery of commercial quantities of oil or gas, there is still a risk that the actual production from a project may be significantly less than predicted, or that the costs of developing and producing

Table 9.3. RED MESA FIELD: PRESENT VALUE

YEAR	(1) MID-YEAR 15% DISCOUNT FACTOR	(2) AFTER-TAX NET CASH FLOW	(3) DISCOUNTED AFTER-TAX NET CASH FLOW	(4) CUMULATIVE DISCOUNTED AFTER-TAX NET CASH FLOW
1	.9325	-520,000	-484,900	-484,900
2	.8109	-5,210,000	-4,224,700	-4,709,600
3	.7051	3,698,000	2,607,500	-2,102,100
4	.6131	3,114,000	1,909,300	-192,800
5	.5332	2,619,000	1,396,400	1,203,600
6	.4636	2,188,000	1,014,400	2,218,000
7	.4031	1,835,000	739,800	2,957,800
8	.3506	1,519,000	532,500	3,490,300
9	.3048	1,256,000	382,900	3,873,200
10	.2651	1,031,000	273,300	4,146,500
11	.2305	874,000	201,400	4,347,900
12	.2004	739,000	148,100	4,496,000
TOTAL		13,144,000	4,496,000	

the property may be higher. Putting money into the bank is a safe investment relative to putting the same amount of money into the Red Mesa Field or any other oil and gas project. Thus, oil companies seek projects which are calculated to yield a greater present value profit than could be expected by putting available investment capital in the bank.

The interest or discount rate at which the present value of projects is determined is based on the interest rate a company could earn at any given time by putting its money in the bank or in some other low-risk investment.

D. Annual Rate-of-Return

Profitability can also be measured in terms of annual rate-of-return. There are a number of different methods by which rate-of-return is determined; however, the most common is called the **discounted cash flow rate-of-return** (DCFROR). In determining annual rate-of-return (or interest rate), a procedure similar to that used for determining present value is used. However, to get rate-of-return on a project, it is not necessary to assume a discount rate as it is to get that project's present value. Instead, the actual annual rate-of-return is determined by discounting net future cash flows from the project at different discount rates. As successively higher discount rates are used, the net present value of a project becomes smaller and smaller. When a discount rate is reached which causes the net present value of a project to be zero, that discount rate is known as the annual rate-of-return. Because this procedure takes many repetitive calculations, it is usually done with a computer.

For the Red Mesa Field, the annual rate-of-return (in this case, the DCFROR) was determined to be 57 percent before-tax and 43 percent after-tax. This tells us that to obtain equivalent future revenues by depositing our investment money in a bank, we would have to receive 57 percent interest before-tax.

E. Summary

Each of the three measures of profitability discussed in this section are useful in evaluating a project and comparing it to other projects. **Payout** measures the time it will take for a company to recover its investment in a project. Shorter payout times will be best for the company's capital budget. The sooner a company can recover its money from one project, the sooner it will be able to invest it in another. Payout alone, however, has the disadvantage of not revealing anything about the value of revenues to be received after payout. A company might have two projects requiring about the same investment, both of which payout in four years. However, the projected cash flows for one indicate a profitable life of six years, while the other project is expected to be profitable for twenty years. If payout was considered as the only investment yardstick, the much longer profitable life of the second project would have been overlooked.

The **present value** is an indicator of the value of a project compared to the return-on-investment that a company could obtain by putting its money in the bank or some other low-risk alternative. Present value reflects the fact that current investments and revenues to be received in the near future are more important to a company than revenues to be received in the distant future. If, for example, a company has $50,000,000 to invest over the next two years and has ten different projects to choose from, it might select the combination of projects which is expected to give it the greatest total present value.

Annual rate-of-return is determined using the same mathematical methods that are used to determine present value. However, annual rate-of-return by itself tells nothing about the size of the project. Rate-of-return will tell whether a given project is an attractive investment opportunity relative to the interest a company might receive by putting its money in the bank.

To arrive at the best investment decisions, a company must consider all of the important measures of profitability. It is apparent from the foregoing analysis of the Red Mesa Field that it presents an attractive investment opportunity, but this is without any consideration of risk factors. As mentioned in the chapter on Exploration, only one out of fifty exploratory wells in the United States today encounters significant oil accumulations. Many failures occur, and in an overall economic analysis, the cost of these failures must be considered. In the oil and gas industry, profits from fields like Red Mesa would pay for many unsuccessful exploration efforts and for other fields that were far less profitable.

ECONOMICS—BIBLIOGRAPHY

Anderson, Truman E.: **Oil Program Investments,** PPC Books (Tulsa) 1972.

Campbell, John M.: **Oil Property Evaluation,** 3rd Ed., Prentice-Hall, Inc. (Englewood Cliffs, N.J.) 1965.

Frick, Thomas C. (ed.): **Petroleum Production Handbook,** Vol. 2, McGraw-Hill Book Company (New York) 1962.

Grant, Eugene L. and W. Grant Ireson: **Principles of Engineering Economy,** 4th Ed., The Ronald Press Company (New York) 1960.

Megill, Robert E.: **An Introduction to Exploration Economics,** PPC Books (Tulsa) 1971.

Megill, Robert E.: **An Introduction to Risk Analysis,** PPC Books (Tulsa) 1977.

Society of Petroleum Engineers of AIME: **Oil and Gas Property Evaluation and Reserve Estimates,** Petroleum Transactions Reprint Series #3 (Dallas) 1970.

CHAPTER 10:
GOVERNMENT

I. INTRODUCTION

Government, at both the national and local levels, is involved with the petroleum industry in a number of different ways. Federal agencies and state commissions exercise regulatory control, provide statistical information, and establish operational guidelines in their relationship with the oil industry. Gas and oil companies must obtain permits, provide periodic reports, and conform to legislative demands on pricing, production, and taxation in order to comply with numerous governmental regulations.

In the first part of this chapter, we will look at some of the federal agencies that are directly involved with the oil and gas industry. The function, scope, and purpose of each agency is outlined briefly, together with its type of relationship to the petroleum industry. Much of the information in this section is taken directly from the **United States Government Manual: 1980-1981.** The second half of the chapter deals with state energy agencies, their programs, and required reports.

The reader should remember that none of the material presented in this chapter is carved in granite. Governmental organization and structure can undergo frequent and significant changes. Regulations and legislation can be extended, modified, or reversed, depending on the mood of the country, the needs of the times, and the philosophical bent of the administration in power. For example, much of the information on the structure of federal agencies is different from what it was four years ago, and undoubtedly could change again during the coming years. The purpose of this chapter is to give an overview of the current situation, and a general idea of the extent of government involvement with the petroleum industry.

II. FEDERAL AGENCIES

A. The Department of Energy

The Department of Energy was established by the Department of Energy Organization Act, approved August 4, 1977, and effective October 1, 1977. The act consolidated the major federal energy functions into one Cabinet-level Department, transferring to DOE all the responsibilities of the following: the Energy Research and Development Administration, the Federal Energy Administration, and the Federal Power Commission. Also transferred to DOE were certain functions of the Interstate Commerce Commission and the Departments of Commerce, Housing and Urban Development, the Navy, and the Interior.

The Department of Energy provides the framework for a comprehensive and balanced national energy plan through the coordination and administration of the energy functions of the federal government. The Department is responsible for the research, development, and demonstration of energy technology; the marketing of federal power; energy conservation; the nuclear weapons program; regulation of energy production and use; pricing and allocation; and a central energy data collection and analysis program.

The Department of Energy includes a number of programs related to the areas of resources and energy research and development:

- The Office of Energy Research monitors the physical and energy research and development programs of the Department, and manages the basic energy sciences program.

- The Assistant Secretary for Nuclear Energy is responsible for the planning, development, and execution of DOE programs for nuclear reactor research and development, and for the management of nuclear energy.

- The Assistant Secretary for Fossil Energy is responsible for research, development, and demonstration programs involving fossil fuels—coal, petroleum, and gas. The nature of these programs includes applied research, technology development, engineering development, and technology demonstration related to fossil fuel extraction, recovery, supply, conversion, or utilization.

- The Assistant Secretary for Resource Applications is involved with developing, managing, and directing policies and programs to increase domestic supplies of petroleum, natural gas, coal, and uranium; to reduce regulatory and financial constraints on resource development and utilization; and to demonstrate and encourage the commercial use of developed energy technologies.

- The Assistant Secretary for Conservation and Solar Energy formulates and directs conservation and solar commercialization programs designed to improve energy efficiency and system utilization, and reduce energy consumption in the transportation industry, public and private buildings, and agricultural and industrial process heating.

- The Assistant Secretary for Environment is responsible for ensuring that the implementation of all departmental programs is consistent with environmental and safety laws, regulations, and policies.

1. Economic Regulatory Administration

The Economic Regulatory Administration administers the Department's regulatory programs, other than those assigned to the Federal Energy Regulatory Commission. These functions include oil pricing, allocation, and import programs, which affect the prices and distribution of crude oil, petroleum products, and natural gas liquids.

The Administration ensures compliance with

existing regulations and carries out new regulatory programs as assigned. ERA also administers other regulatory programs, including conversion of oil and gas-fired utility and industrial facilities to coal, natural gas curtailment priorities and emergency allocations, regional coordination of electric power system planning and reliability of bulk power supply, and emergency and contingency planning.

2. Energy Information Administration

The Energy Information Administration is responsible for the timely and accurate collection, processing, and publication of data on energy reserves, the financial status of energy-producing companies, production, demand, consumption, and other areas.

Analyses of data to assist government and non-government users in understanding energy trends are also performed. Specifically, analyses are prepared on complex, long-term energy trends and the impacts of energy trends on regional and industrial sectors. Special purpose analyses are prepared involving competition within energy industries, the capital/financial structure of energy companies, and interfuel substitution. Extensive field audits are conducted to ensure the validity and accuracy of pertinent data.

3. Federal Energy Regulatory Commission

The Federal Energy Regulatory Commission is an independent, five-member commission within the Department that has retained many of the functions of the Federal Power Commission, such as setting of rates and charges for the transportation and sale of natural gas and for the transmission and sale of electricity, and the licensing of hydroelectric power projects. In addition, the authority to establish rates or charges for the transportation of oil by pipeline, as well as the valuation of such pipelines, has been assigned to this Commission from the Interstate Commerce Commission.

B. The Department of the Interior

The Department of the Interior was created on March 3, 1849, and incorporated the following: the General Land Office, the Office of Indian Affairs, the Pension Office, and the Patent Office. Over the period of its existence, different functions have been added and removed, so its particular role has evolved from general housekeeper for the federal government to that of custodian of the nation's natural resources.

As the nation's principal conservation agency, the Department of the Interior has responsibility for most of our nationally owned public lands and natural resources. This includes advocating the wisest use of our land and water resources, protecting our fish and wildlife, preserving the environmental and cultural values of our national parks and historical places, and providing for the enjoyment of life through outdoor recreation. The Department assesses our mineral resources and works to assure that their development is in the best interests of all our people.

The Department provides several programs involved with the development and uses of our natural resources and minerals:

- The Assistant Secretary for Energy and Minerals is responsible for programs associated with mineral policy, data, and analysis; surface mining reclamation and enforcement functions; topographic, geologic, and mineral resource matters; metallurgical and mining research and development; development and coordination of ocean mineral resource affairs; earth seismic research; remote sensing activities; water resource evaluation and analysis; and emergency preparedness and natural disaster minerals functions.
- The Assistant Secretary for Land and Water Resources is responsible for programs associated with land use and water planning; public land management, including mineral leasing; development and management of water resource projects and facilities; water resources research, including saline water conversion; and emergency preparedness and natural disaster water resources functions.

The Department of the Interior will play a significant role in the development of our country's oil and gas resources in the coming years, since federal holdings, both onshore and offshore, contain substantial reserves of petroleum. Policies related to the leasing of federal lands will have to be established, which will provide for both the environmental integrity of our natural resources and the energy demands of the various sectors within our country.

1. Bureau of Mines

The Bureau of Mines is primarily a research and factfinding agency. Its goal is to help ensure that the nation has adequate mineral supplies for security and other needs. Applied and basic research are conducted to develop the technology for the extraction, processing, use, and recycling of the nation's mineral resources at a reasonable cost without harm to the environment or the workers involved. Typical areas of research are mine health and safety, recycling of solid wastes, abatement of pollution and land damage caused by mineral extraction and processing operations, and development of ways to use domestic low-grade ores as alternative sources of critical minerals that must currently be imported.

The Bureau also collects, compiles, analyzes, and publishes statistical and economic information on all phases of mineral resource development, including exploration, production, shipments, demand, stocks, prices, imports, and exports. Special studies are frequently made on subjects of particular national interest, such as the effects of potential economic, technologic, or legal developments on resource availability. The effects of policy alternatives on mineral supply and demand are also analyzed.

2. U.S. Geological Survey

The broad objectives of the Geological Survey are to perform surveys, investigations, and research covering topography, geology, and the mineral and water resources of the United States; classify land as to mineral characteristics and water and power resources; enforce departmental regulations applicable to oil, gas, and other mining leases, permits, licenses, development contracts, and gas storage contracts; and publish and disseminate data relative to the foregoing activities.

The Conservation Division classifies federal lands as to their value for leaseable minerals or for reservoir and waterpower sites; evaluates federal and Outer Continental Shelf lands for tract selection, tract evaluation, and reserve inventory purposes in aid of mineral leasing and subsequent operations; supervises the operations of private industry on geothermal, oil shale, mining, and oil and gas leases on public domain, acquired, Indian, and Outer Continental Shelf lands, to ensure maximum utilization and to prevent waste of the mineral resources, and to ensure the protection of the environment and to prevent pollution; assures the public a fair market return for the disposition of its mineral resources; establishes maximum rates of production for producing wells on the Outer Continental Shelf; maintains production accounts and collects royalties; prepares and publishes maps and reports of mineral and water resources investigations on federal lands; and provides advice and services to other federal agencies on the disposition of public lands and mineral resources.

The Geologic Division conducts highly diversified research programs to increase understanding and to aid in management of the mineral and energy potential of the land area of the U.S. and of the adjacent continental margins. These programs provide basic information on the character, magnitude, location, and distribution of mineral, energy, and land resources, as well as on the principles and processes involved in their formation. This information also provides a basis for many critical decisions and actions relating to land use, urban planning and development, construction practices, environmental and health problems, and earthquake, volcanic, and other natural hazards. Special programs include the investigation and evaluation of geothermal resources, the maintenance of seismic and geomagnetic observatories as part of an earthquake hazards reduction program, offshore oil and gas resource appraisal, onshore oil and gas investigations, and mineral land assessments.

The National Mapping Program includes the preparation, publication, and revision of the several map series that include topographic maps at several standard scales, photo-image maps, land-use-land cover maps, and other special map products. Area of coverage includes the United States and its outlying areas, and Antarctica.

The Water Resources Division provides the hydrologic information and understanding needed for the optimum utilization and management of the nation's water resources by: 1) evaluating the quantity and quality of the nation's water resources; 2) appraising the characteristics of surface and ground water; 3) conducting research in hydraulics and hydrology; 4) disseminating the water data; 5) coordinating the activities of federal agencies in the acquisition of water data; and 6) providing scientific and technical assistance to other federal, state, and local agencies.

3. Bureau of Land Management

The Bureau is responsible for the total management of 417 million acres of public lands. These lands are located primarily in the Far West and Alaska, however, some scattered parcels exist in other states. In addition to minerals management responsibilities on the public lands and the Outer Continental Shelf, BLM is also responsible for subsurface resource management of an additional 169 million acres where mineral rights have been reserved to the federal government.

Resources managed by BLM include timber, minerals, oil and gas, geothermal energy, wildlife habitat, endangered plant and animal species, rangeland vegetation, recreation and cultural values, wild and scenic rivers, designated conservation and wilderness areas, and open space. BLM programs provide for the protection, orderly development, and use of the public lands and resources under principles of multiple use and sustained yield. Land use plans are developed with public involvement to provide orderly use and development, while maintaining and enhancing the quality of the environment. Under certain conditions, the Bureau also makes land available through sale to individuals, organizations, local governments, and other federal agencies when such transfer is in the public interest.

The Bureau has responsibility to issue rights-of-way, in certain instances, for crossing federal lands under other agencies' jurisdiction. It also has general enforcement authority. In addition, the BLM is responsible for the survey of federal lands, and establishes and maintains public land records and records of mining claims.

C. Environmental Protection Agency

This federal agency has tried, since its establishment on December 2, 1970, to develop programs that would answer this fundamental question: how does the United States permit desired economic growth and at the same time preserve a liveable environment? The growth of technology and the spread of its by-products —chemicals in the air and water, refuse, noise—have aroused concern over the dangers of pollution to public health and well-being. Environmental damage has become widespread, precipitating the need for national action. As a result, the EPA was formed to control and abate pollution in the areas of air, water, solid waste, toxic substances, noise, and radiation.

The Environmental Protection Agency was created to permit coordinated and effective governmental action on behalf of the environment. EPA endeavors to reduce and control pollution systematically, by proper integration of a variety of research, monitoring, standard setting, and enforcement activities. As a complement to its other activities, EPA coordinates and supports research and antipollution activities by state and local governments, private and public groups, individuals, and educational institutions.

In its air programs, the agency conducts research, measures pollutants, develops standards, and works with state and local agencies. The solid waste management program supports demonstration projects to convert trash to fuel for heating and electric power generation. The pesticides program is engaged in an effort to help certify workers so that they can apply chemicals safely on farms and ranches. The program on toxic substances is also designed to develop strategies to control toxic substances and to establish criteria for assessing the impact of chemicals on people and the environment.

The National Environmental Policy Act, passed January 1, 1970, is the law which requires that, for any proposed energy projects, an Environmental Impact Statement (EIS) must be prepared and submitted for review by the EPA. This statement analyzes the environmental effects that may result from the project, and makes that analysis available for public review. The intention of the EPA for these statements is twofold: to offer important input into policy and decision making, and to change the practice of permitting development irrespective of environmental consequences.

III. STATE OIL AND GAS COMMISSIONS AND REPORTS

A. Introduction

Almost all 50 states have a state agency which establishes policies for oil and gas operations, from drilling to transporting to refining. The size, complexity, and title of the agencies vary, depending on how widespread the petroleum or natural gas industry in a state may be. The jurisdiction, in most of the states, for regulating oil field operations to prevent waste and pollution lies with an Oil and Gas Conservation Commission. These commissions, made up of a supervisor and a board of one to ten members, establish a set of rules and regulations that serve as guidelines for state oil and gas operations. A booklet containing these regulations is available in each state for public distribution.

One of the ways in which a state commission regulates oil and gas operations is by requiring operators to file for various permits. Although the exact number, purpose, and deadline schedules of these permits are not the same in all states, there are some common forms used in a majority of states. Several of these forms are described briefly in this section. By no means is this a comprehensive list, but the material presented here does provide a sampling of the types of forms used, and the information required on these forms.

The following three sections are divided into the general categories or groupings under which individual forms can be included. Within each section, there is a brief description of some of the forms and reports that fall under the general heading of the section.

B. Drilling and Re-entering Operations

1. Designation of Agent or Operator

Before any oil or gas operations can begin in a given state, all lease owners, producers, operators, transporters, refiners, and gasoline plant operators must file this form. The purpose of this form is to designate an operator or agent (acting on behalf of the operator, lease owner, producer, etc.) who has full authority and is responsible for complying with any applicable state regulations. The agent is the person whom the state regulatory agency (or a representative thereof) will contact when serving written or oral instructions about the regulations and rules of practice and procedure for oil and gas operations within the state. The designation of an agent does not relieve the lessee, or lease owner, of responsibility for complying with the terms of the lease or the relevant state regulations and guidelines.

The agent can, and does, sign any papers or reports required by the state regulatory agency. All changes of address and any termination of an agent/operator's authority must immediately be reported in writing to the supervisor of the state agency. In the latter case, a new agent/operator must be designated immediately. This is usually done by means of another form, known as a "Notice of Change of Operator."

2. Organization Report

Every person acting as a principal or agent, or who is independently engaged in oil and gas operations, shall file under oath a statement giving the following information:

a. Name of the company or individual under which the business is being operated or conducted

b. Address of the company, organization, or individual

c. Form of the organization; in the case of corporations, the law under which the corporation was chartered

d. Purpose of the organization (the specific aspect engaged in)

e. Name and address of any trustees, officers, or directors

If any changes occur in the above information as it is originally filed, an amended report, also under oath, must be sent immediately to the state commission.

3. Application for Permit to Drill, Deepen, Re-enter, or Operate

This form must be filed with the state Oil and Gas Conservation Committee, usually accompanied by an appropriate fee, before any of the listed activities can begin. The state will then issue a permit authorizing the commencement of the designated activity. (In many states, a survey plan map is also required, which shows the location of the proposed well in reference to the nearest boundary lines of an established public survey.)

In most states, activities must begin within a certain number of days after the commission has given its approval and issued the permit. If unforeseen circumstances should arise that delay operations, an appeal can be made to the commission to obtain an extension of this deadline.

There is usually no fee exacted for drilling a stratigraphic test hole (a hole that is drilled to gather data on the rock formations), but a state permit is still required.

4. Bonding Requirements

Except for those cases in which an approved bond is filed by the owner in accordance with federal or Indian lease requirements, and such bond is presented to the state Oil and Gas Commission, most states require either an individual bond for each well drilled, or a blanket bond covering all the wells. Both of these types of bonds are accompanied by appropriate fees, which are determined by the number of wells to be drilled, or in some cases, by the depths of the wells to be drilled. Naturally, the individual bond carries a smaller fee, since it is only in effect for one well, while the blanket bond necessitates a larger fee, because it is established for all the wells drilled.

Bonds remain in effect from the purchase date until the particular well (or wells) is plugged and abandoned. The state commission must approve the plugging and surface restoration. The liability under any bond usually cannot be terminated without written approval of the state commission.

5. Sundry Notices

If an owner or operator plans to carry out operations that involve a change in original plans, or that are not covered by a specific individual state form, notice must be given on a "Sundry Notices" form to the commission supervisor, and approval must be obtained before operations can begin. The "Sundry Notices" form usually lists about ten different possible situations that can be reported.

Some of the operations reported originally on a "Sundry Notices" form are later reported in more detail on a specific form. For example, on a recompletion project, an operator files a "Sundry Notices" form for prior commission approval to recomplete, then files a detailed "Recompletion Report" after the actual operation is finished.

6. Application to Construct/Use an Earthen Pit

Many oil and gas operations (drilling, testing, completion, and producing) require the use of one or more earthen pits on the location. These pits usually hold such substances as produced water or oil, drilling mud and cuttings, or other waste liquids from drilling operations. A permit must be obtained to construct such pits to insure that no waste liquids contaminate potable surface or subsurface water supplies.

If a pit is not dug in a tight soil, such as clay or hardpan, or is not adequately lined with material that will prevent seepage, damage may occur to water supplies underlying the pit. Porous soil, such as gravel or sand, easily allows such seepage. Most states also require that, when water is to be stored in the pit, no surface accumulations, such as oil or other liquid hydrocarbons, be permitted on the water. Should these accumulations appear, they must be cleaned up immediately.

Sometimes, detailed information is required to supplement the application to construct the pit. This information can include:

a. Legal location of the proposed pit

b. A schematic sketch of the complete operating system used in conjunction with the proposed pit(s), and a USGS topo map large enough to indicate the location of surrounding surface drainage

c. A plan showing the size, depth, and proposed lining of the pit

d. A detailed log describing the nature of the soil in which the pit is to be constructed, and the strata existing between the bottom of the pit and the top of the first fresh water source below the pit

e. A statement of the proposed final disposition of the water, the anticipated amount of water to be produced, and all relevant calculations

f. A chemical analysis of the water to be stored, of the fresh water below the pit, and of all other fresh water strata which might be affected

When a drilling operation is finished, these pits are usually drained and filled with earth, and the surface is restored as much as possible. However, these pits are used sometimes by landowners for storage purposes when there is no public health hazard in terms of water contamination, and if commission approval has been granted.

C. Testing and Completion Operations

1. Completion or Recompletion Report

Most states require the owner or operator to submit this report to the state commission within thirty days after a completion or recompletion operation. In many of these states, a wireline log or driller's log evaluating the well must accompany the report. Upon request, well logs that are submitted in certain states can be marked "confidential," and held from public inspection and filing for a period of time.

The "Completion Report" contains a designation of the type of well: exploratory or wildcat, production test or core hole, oil or gas, producing or dry hole. In addition, the exact operation performed on the well is noted, such as drill, workover, deepen, plug back, change in producing interval, etc. The spud date, total depth, and records on the casing, liner, and tubing, are other items reported on this particular form. Initial production or test data can also be recorded.

2. Notice of Intent to Plug or Plug and Abandon

An owner or operator cannot let any well or hole drilled for oil, gas, saltwater disposal, injection, or exploratory operations, remain unplugged after the well or hole is no longer being used for the purpose for which it was drilled or converted. Plugging a well prevents the migration of fluids from their original formations, thus protecting the surrounding fresh water supplies. The plugging material usually consists of a mud-laden liquid, cement, a combination of these two, or a mechanical plugging device.

Before a well can be plugged, an operator must receive approval from the state commission for the following: the method of plugging, the time and date the procedure is to begin, and the plugging material to be used. Specific plugging methods can vary for cased and uncased holes as to the requirements for weight and viscosity of the cement, and the length and placement of the plugs. These specifications also differ from state to state. A representative from the state commission will usually be on hand to witness the operation. Frequently, a minimal fee may be assessed with the plugging permit.

After the well or hole has been plugged, a permanent monument must be constructed. This monument is commonly a piece of pipe about four inches in diameter and ten feet long. A little over half of the pipe is embedded in the top of the cement plug, or welded to the surface casing, while the remainder protrudes above the surface. (The length and diameter of this pipe can vary from state to state.) This marker pipe must indicate the name of the well, the location, the total depth, and the elevation.

When a landowner wishes to use a well as a fresh water well instead of having it plugged, and if it has been determined that this can be safely done, the well does not have to be filled above the required sealing plug which is set below the fresh water level. In these instances, the landowner must provide written authority for such use, and must secure approval from the state water division for a water well permit. The landowner assumes the responsibility for plugging the well, in accordance with state regulations, once it is abandoned as a water well.

Except in the cases of water wells, or when the abandonment will only be temporary, a "Notice of Intent to Abandon" is normally filed at about the same time as the "Notice of Intent to Plug." Some states will even combine these two notices into one form. A well may be maintained as temporarily abandoned as long as the hole is cased or left in such a manner that fluids cannot migrate from their original formations. The manner in which the well is to be maintained has to be reported to the state agency. Bonding requirements are kept in force until the well is permanently abandoned.

3. Plugging Report

This report is usually filed within a specified number of days after the plugging procedure is completed. The report gives the final details of the method and material used for plugging the well, and the date the operation was completed. In some cases, a "Subsequent Report of Abandonment" is also filed at the same time as the "Plugging Report."

4. Report of Subsurface Injections

This report provides the state commission information on injection operations which will affect underground formations. The purpose of the report is to assure the commission that an injection or disposal well will be completed, equipped, operated, and maintained in a manner that will prevent pollution or contamination of fresh water sources, or damage to sources of oil and gas, while confining the injected fluids to the approved interval or intervals.

A report is filed on a periodic basis that covers these major operations: injection for disposal and injection for purposes of enhanced recovery. The important details on this form include the type of fluid injected, its source, the purpose of the injection operation, the formation injected, cumulative injection totals (in mcf, bbls, or gals), and a description of the types of chemicals, if any, that are used to treat the injected fluids. Some states require that fluids used in fracturing, acidizing, or other treatment operations also be reported on this form. Specific state requirements regarding information about the tubing and casing can vary.

5. Application to Dispose of Salt Water by Injection

While the "Report of Subsurface Injections" may include information on wells used for the disposal of salt water or brine, some states require a separate form for this operation. States usually prefer that the water be injected back into its original formation. Again, specifics on the condition of the well's tubing and casing are different from state to state.

Some of the items of information that are to be included on such an application are:

a. Location plat of the water disposal well(s), which also shows a location history of all other holes or wells of any type within a certain radius of the disposal well(s)

b. Name, description, and depth of the formation into which the water is to be injected; the original source of the water or brine, and a chemical analysis; the minimum and maximum daily amounts of water to be injected

c. Description of the existent casing in the disposal well(s), and the proposed method of testing this casing to prevent fresh water or producing formation damage

D. Production Operations

1. Operator's Monthly Production Report

A report of all oil, gas, and water production is made in most states on a monthly basis to the state commission. In some instances, each product is reported on a separate form ("Monthly Gas Report," "Monthly Report of Oil Production," "Monthly Report of Water Production"). The "Operator's Monthly Report" presents a summary of the number of wells producing for the month, the amount of oil or gas produced, and the disposition of the product. The deadline for filing this form is different for each state, but two common dates are the first and twenty-fifth of each month.

2. Gas-Oil Ratio Report

Either on a regular monthly basis, or as the state requests, gas-oil ratio tests must be run on producing wells, and the results reported to the state commission. These tests are taken at a specified pressure and temperature, and are used to compare the amounts of gas and oil produced (SCF/bbl). The gas-oil ratio information can be used to determine the gas allowables for a given well (the amount of gas allowed to be produced per barrel of oil per time period).

3. Gasoline Plant Report

Operators of a gasoline plant, cycling plant, or any other plant in which gasoline, butane, propane, condensate, kerosene, or other liquid products are extracted from gas, must file a monthly plant report. The principal categories included on this report are product intake volumes, disposition of unprocessed gas and residue, and details of sale or other disposition of plant products (receipt, deliveries, fare, and stock).

4. Transporter's and Storer's Reports

Transporters of oil or condensate must furnish, usually for each calendar month, a report describing the stock of oil on hand, and the movements of oil by pipeline, water transport, tank trucks, or other conveyance, except railroads. These reports deal with the transportation of oil in the following situations: from leases to storers or refiners, between transporters within the state, between storers within the state, between refiners within the state, and between storers and refiners within the state.

Each storer of oil must furnish, for the same time period as the "Transporter's Report," complete data concerning the storage of oil within the state. Both of these reports must be filed on, or before, a specified date in the succeeding month.

5. Refiner's Report

Each refiner of oil or condensate within a state must regularly furnish a report containing information on the oil, condensate, and products involved in the refining operations for a certain period of time. Again, if this is a monthly report, the report is to be filed by a specific date in the succeeding month. The report includes information on the following: stocks at the beginning and end of the reporting period, products manufactured, feedstocks, deliveries of product, and any plant losses which might occur.

GOVERNMENT—BIBLIOGRAPHY

Bureau of Land Management and U.S. Geological Survey: **Surface Operating Standards for Oil and Gas Exploration and Development,** Department of the Interior (Washington, D.C.) 1978.

Office of the Federal Register: **Code of Federal Regulations: Title 10, Energy, Parts 200-1599,** U.S. Government Printing Office (Washington, D.C.) 1981.

Office of the Federal Register: **Code of Federal Regulations: Title 18, Conservation of Power and Water Resources, Parts 0-399,** U.S. Government Printing Office (Washington, D.C.) 1980.

Office of the Federal Register: **Code of Federal Regulations: Title 30, Mineral Resources,** U.S. Government Printing Office (Washington, D.C.) 1980.

Office of the Federal Register: **Code of Federal Regulations: Title 40, Protection of Environment,** U.S. Government Printing Office (Washington, D.C.) 1980.

Office of the Federal Register: **Code of Federal Regulations: Title 43, Public Lands: Interior,** U.S. Government Printing Office (Washington, D.C.) 1980.

Office of the Federal Register: **United States Government Manual: 1980-1981,** U.S. Government Printing Office (Washington, D.C.) 1980.

CHAPTER 11:
ENERGY OUTLOOK

I. INTRODUCTION

Two critical issues facing the United States are the present status of its energy supplies and the future outlook of its energy resources. The assessment of these topics is interrelated with the country's economic stability and national security. Contemporary society is based to a great extent on the assumption that the world has an abundant supply of cheap energy. In the past, these supplies have been readily available, and have contributed to the technological progress of the nation. However, as consumption has increased and supplies have dwindled, the energy picture has changed. Costs to find and develop energy resources have escalated, while demand on existent supplies continues to climb.

Following the Arab oil embargo in October, 1973, the United States became aware of the importance of stable energy supplies, especially resources which are free from foreign interference or control. Supply shortages, both during the oil embargo and the following period of steadily declining crude oil production in the United States, led to the belief that our country is experiencing an "energy crisis." Opinions vary as to the exact nature of this "crisis," and the up-and-down nature of the supply of oil in this country over the last few years has clouded the issue and fueled the controversy.

Whatever the "current" situation might be at any one time regarding the supply and availability of petroleum products, the fact remains that petroleum is a finite resource that will one day be exhausted. Enhanced recovery processes and new discoveries can provide short-term remedies, but they are not final solutions. We must work at creating a long-range solution to the present energy problems—a program that will incorporate the prudent development and use of current resources with the evolution of alternative sources of energy which can adequately supply future needs.

This chapter will concentrate on describing current trends and forecasts for the supply of energy resources in the United States, the implications of the present energy situation, and some of the alternate sources of energy being considered and developed. The complexity of the subject and the limited scope of this text preclude a detailed analysis of this issue, but pertinent observations and possible developments will be included in this chapter.

II. PRESENT ENERGY SITUATION

A. U.S. Energy Consumption

There has been a particularly rapid growth in energy consumption throughout the world during the last 500 years. The patterns of this increase and the changing areas of concentration are indicated in Figure 11.1. Growing energy consumption in the United States has reflected this historical pattern, and can be directly related to the development of technologies which support our advanced lifestyle.

The United States presently consumes approximately 29 percent of all the energy used in the world. In comparison, Europe uses 21 percent, the Sino-Soviet world 28 percent, and Japan 7 percent. From 1960 to 1979, United States energy consumption has grown by an average of about 3 percent a year.

The consumption of energy can be divided among four sectors in our society: **residential-commercial, industrial, transportation,** and **electrical generation.** The trends in the amount of

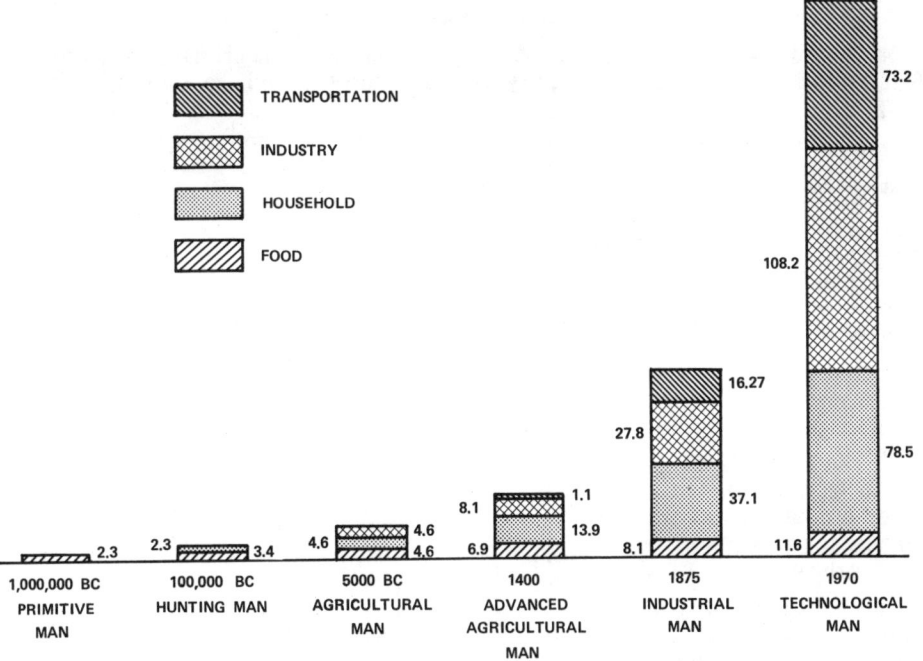

Figure 11.1. HISTORY OF MANKIND'S ENERGY CONSUMPTION. Units = watt hour. (Source: *Scientific American*, "The Flow of Energy in an Industrial Society")

Table 11.1. ENERGY USE BY SECTOR
(in quadrillions of BTUs)

	Residential-Commercial	Industrial	Transportation	Electrical Generation
1950	8.1 (24.1%)	12.7 (37.8%)	8.8 (26.2%)	4.0 (11.9%)
1955	9.3 (23.7%)	14.7 (37.5%)	10.2 (26.0%)	5.0 (12.8%)
1960	11.9 (27.0%)	15.8 (35.8%)	10.5 (23.8%)	5.9 (13.4%)
1965	14.2 (26.8%)	18.7 (35.3%)	12.3 (23.2%)	7.8 (14.7%)
1970	17.8 (26.7%)	21.4 (32.0%)	16.1 (24.1%)	11.5 (17.2%)
1975	17.8 (25.2%)	20.2 (28.6%)	18.2 (25.7%)	14.5 (20.5%)
1979	19.1 (24.2%)	22.6 (28.7%)	19.8 (25.1%)	17.3 (22.0%)

(Source: U.S. Department of Energy)

energy used by each sector over the last thirty years are shown in Table 11.1. These statistics reflect an early rise, followed by an eventual decline, in the share of energy used by the residential-commercial segment, a steady decline in the percentage used by the industrial sector, and an initial decline, followed by a subsequent return to near previous levels, by the transportation sector. The rise in the use of electricity in our society during the thirty year period is reflected in the substantial increase in energy used for electrical generation.

The Exxon Company has made projections about the future consumption of energy in this country by these individual sectors, as shown in Table 11.2. Instead of being calculated as a separate quantity, the amount of energy consumed in electrical generation has been allocated proportionately to the other individual sectors. Exxon introduces an additional category, **nonenergy**, to account for the consumption of gas, oil, and coal as raw materials or feedstock, rather than fuel. Residential-commercial and transportation uses of energy are projected to increase at slower rates in the future, due to greater efficiency in design for homes and automobiles. The demand for energy in the industrial sector is anticipated to increase, as expansion offsets conservation.

Table 11.2. PROJECTED U.S. ENERGY DEMAND BY CONSUMING SECTOR
(in quadrillions of BTUs)

	Residential-Commercial	Industrial	Transportation	Nonenergy
1980	25.5 (33.2%)	25.7 (33.5%)	19.0 (24.8%)	6.5 (8.5%)
1990	27.1 (33.1%)	28.8 (35.1%)	18.6 (22.7%)	7.5 (9.1%)
2000	27.8 (30.2%)	36.9 (40.0%)	19.0 (20.6%)	8.5 (9.2%)

(Source: **Energy Outlook: 1980-2000**, Exxon Company, USA)

B. U.S. Energy Demand

The major sources of energy in the United States at the present time are **oil, gas, coal, nuclear, hydro,** and **geothermal**. Table 11.3 indicates the amount of each energy source that was used in 1980, and the growth rate in the consumption of each resource for the year.

Although Exxon has projected a decline in the use of oil and gas in this country during the next twenty years, these two resources will still account for 52% of the fuel consumed in the year 2000. The use of coal will grow most significantly, both in direct consumption and for use in the production of synthetic fuels. Nuclear

Table 11.3. 1980 U.S. ENERGY CONSUMPTION BY FUEL SOURCE

Source	Use (quadrillion BTUs)	% of Total*	Annual Growth Rate
Coal	15.66	20.6%	+3.5%
Oil	34.25	45.0%	−8.2%
Gas	20.39	26.8%	−1.1%
Nuclear	2.73	3.6%	−0.8%
Hydro/Geothermal	3.11	4.1%	−1.8%

*sum does not equal 100% due to rounding of numbers
(Source: U.S. Department of Energy, Energy Information Administration: Monthly Energy Review; from **Basic Petroleum Data Book**, API)

Table 11.4. PROJECTED U.S. ENERGY CONSUMPTION BY FUEL SOURCE
(in quadrillions of BTUs)

	1980*	1990	2000
Coal	15.8 (21%)	20.3 (25%)	28.6 (31%)
Oil	34.2 (44%)	32.8 (40%)	31.0 (33%)
Gas	20.7 (27%)	17.8 (21%)	17.2 (19%)
Nuclear	2.8 (4%)	6.5 (8%)	9.9 (11%)
Hydro/Geo/Solar	3.2 (4%)	4.7 (6%)	5.5 (6%)

*some 1980 figures are still preliminary, and therefore are not exactly equivalent in Tables 11.3 and 11.4
(Source: **Energy Outlook: 1980-2000**, Exxon Company, USA)

energy's role will depend greatly on regulatory activity and the progress made in the construction of new facilities. Nevertheless, an increase in the use of nuclear power is expected by the end of the century. Solar, hydro, and geothermal sources are forecast as still playing minor roles twenty years from now. Table 11.4 illustrates these projections for the consumption of energy sources for the next two decades.

C. U.S. Energy Supply

As seen in Table 11.3, the primary energy sources currently in this country are oil, gas, and coal, which

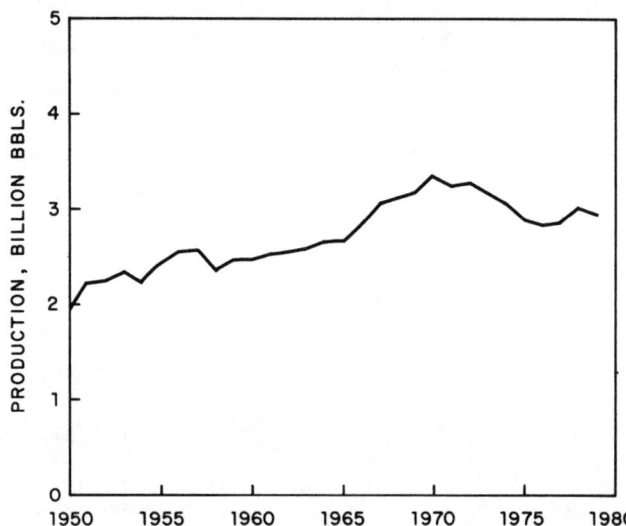

Figure 11.2. U.S. OIL PRODUCTION. (Source: American Petroleum Institute, Committee on Reserves and Productive Capacity)

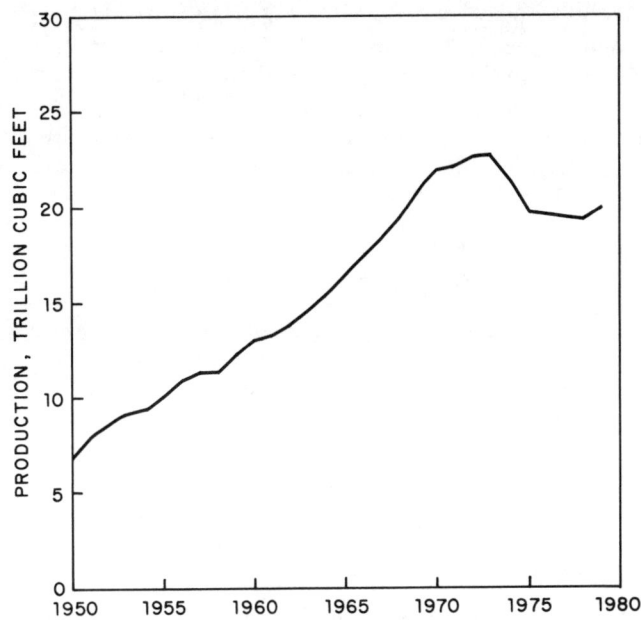

Figure 11.3. U.S. GAS PRODUCTION. (Source: American Gas Association, Committee on Natural Gas Reserves; from API: *Basic Petroleum Data Book*)

together provide over 92% of the energy consumed domestically. (Coal will be discussed in more detail in the section on alternate sources of energy in this chapter.) Production of crude oil and natural gas has fluctuated somewhat the last five years, but the overall trend during the past decade has been one of decline, after years of steady increases (Figures 11.2 and 11.3). Recent reversals and levelling off in the pattern of decline have been stimulated by such factors as increased drilling activity and favorable market prices, but it is still too soon to predict a definite trend. The slowing of the decline will persist if drilling continues to escalate, new exploration areas prove productive, and production from existing fields is maximized by well workovers and enhanced recovery projects.

The one constant over the last ten years (since the inclusion of the Alaskan reserves) has been the steady

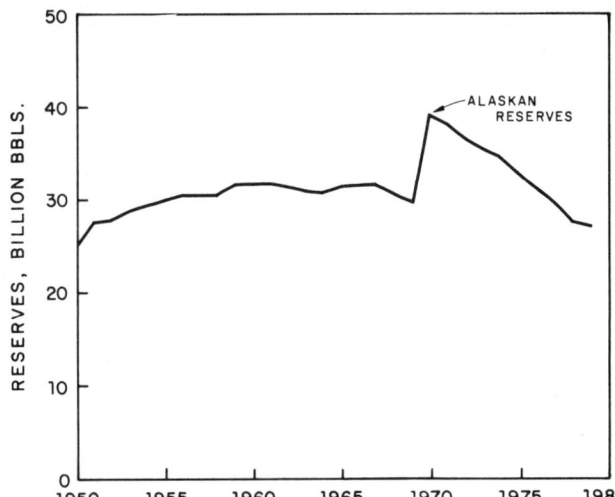

Figure 11.4. U.S. OIL RESERVES. (Source: American Petroleum Institute, Committee on Reserves and Productive Capacity)

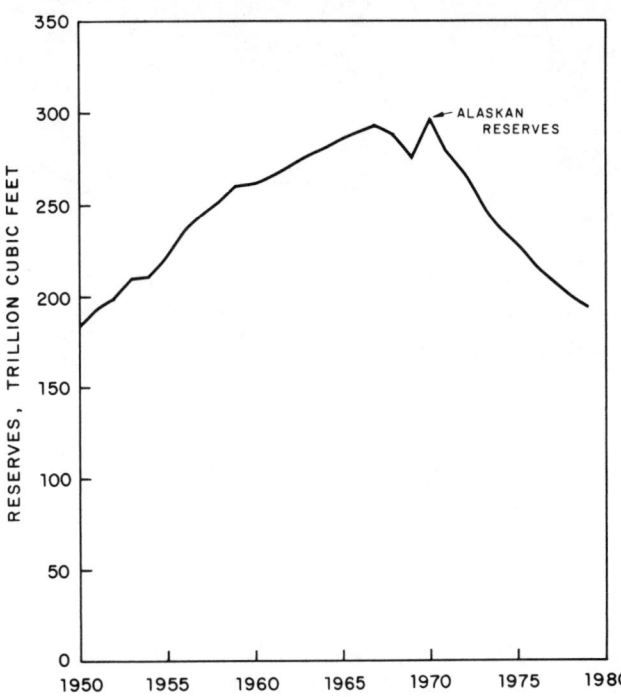

Figure 11.5. U.S. GAS RESERVES. (Source: American Gas Association, Committee on Natural Gas Reserves; from API: *Basic Petroleum Data Book*)

reduction of U.S. crude oil and natural gas reserves (Figures 11.4 and 11.5). If foreign imports and future discoveries are not taken into account, existing domestic oil and gas reserves would last approximately ten years, at the current rate of consumption.

The decrease in our petroleum reserves over the last decade, coupled with a rise in the demand for petroleum

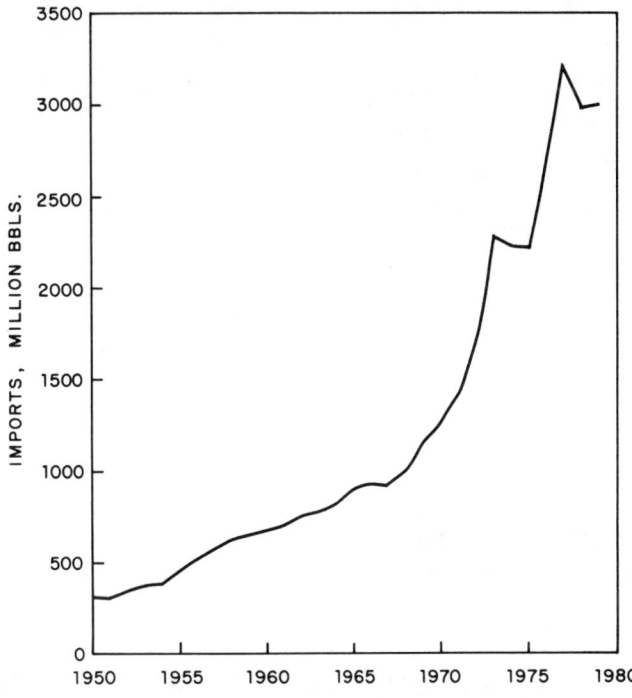

Figure 11.6. U.S. PETROLEUM IMPORTS. (Source: U.S. Energy Information Administration; from API: *Basic Petroleum Data Book*)

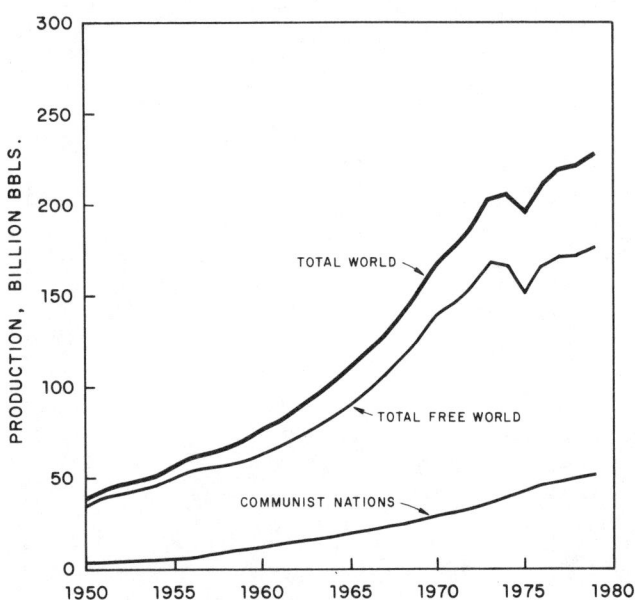

Figure 11.7. WORLD OIL PRODUCTION. (Source: U.S. Bureau of Mines; from API: *Basic Petroleum Data Book*)

The following excerpt from **World Petroleum Availability: 1980-2000,** published recently by the Office of Technology Assessment, summarizes trends in world oil supply and production over the next two decades.

"In spite of the importance of the issue, uncertainties about world supplies of oil from conventional sources during the next two decades are surprisingly large. Nevertheless, it is highly likely that there will be little or no increase in world production of oil from conventional sources. Prudent planning should consider this possibility.

"Oil production in the industrialized non-Communist world could begin to decline by the early 1980s. While it may be physically possible to increase world production of oil significantly (perhaps 33%) by the 1990s, substantial increases are unlikely because the nations capable of contributing to such an increase have no financial or political incentive to do so and because a number of practical problems would arise if a significant increase in production were attempted.

"Enough is known about world oil supplies to make a few specific observations:

"(1) Assuming political stability in the major exporters, non-Communist world oil supply is likely to range between 45-60 MBD in 1985 and 40-60 MBD in 2000 (compared to 52 MBD in 1979). (Throughout this paper oil production will be given in millions of [42 gallon] barrels per day [MBD]. Unless otherwise specified, oil production will include liquids associated with production of natural gas.) **The sizes of potential increases in Saudi Arabia, Mexico, and Iraq and of the decrease in the United States** account for a major portion of the variation in production possibilities (10 MBD or approximately 50% of the variations in the year 2000).

"(2) As a group, the non-Communist industrialized countries will experience no significant increase in production. In fact, production in these countries may decrease by as much as 50% by the year 2000.

"(3) In the short term, U.S. production may decline from its current level of 10.2 MBD to a level of 7.2-8.5 MBD in 1985. Production in the year 2000 may range between 4-7 MBD. The high estimate for the year 2000 (7 MBD) depends on both the annual addition of 1 billion barrels to proven reserves and the extensive use of enhanced recovery techniques.

"(4) OPEC production during the next 20 years will not differ significantly from its current level of 31 MBD. Any increases in the production rate will be strongly dependent upon Arab OPEC producers. Except for Iran, only Saudi Arabia, Kuwait, and United Arab Emirates have the reserves and Iraq the estimated potential to increase production rates. Substantial dependence on Arab OPEC (the Persian Gulf region) is likely to continue with its obvious implications for foreign policy. In particular, Saudi Arabia retains its central position in OPEC. Although the Saudis have reserves which could permit an increase in production capacity to 16 MBD, they have announced their intentions to not exceed a maximum sustainable capacity of 12 MBD, and plan to reach this level of capacity in 1987 at the earliest.

products, has increased this country's dependence on foreign imports (Figure 11.6). In 1977, imports supplied 48% of the nation's demand for petroleum. During the last two years, however, rising prices, conservation efforts, and a resolve to become energy self-sufficient, have contributed to a reduction in import levels. In 1980 alone, total imports are estimated to have declined 19 percent from the previous year. Despite this recent trend, oil imports are projected to have a significant impact on the oil supply in the United States for the remainder of this century, and will continue to affect the nation's economy and national policies.

Because of the dwindling supplies of oil and gas in the U.S., concentrated efforts must be made in the following areas: maximizing production in existing oil and gas fields through enhanced recovery projects; discovering new reserves, especially in untested areas offshore and in Alaska; developing alternate sources of energy; and conserving our present resources.

D. Worldwide Trends in Oil and Gas Production and Supply

Trends in worldwide oil and gas production must be considered if the domestic situation is to be placed in the proper perspective. Figure 11.7 outlines world oil production over the last thirty years. Note that production figures for non-Communist nations climbed steeply in the 1960s and early 1970s, but now have begun to level off. In fact, early production estimates for 1980 indicate a decline of 7 percent for non-Communist countries, as a result of various political, military, and economic factors, in particular the conflict in the Middle East. Communist areas still showed an increase in production for 1980.

Because of complicated international factors, it is difficult to predict when world oil production will peak. Most experts estimate that it will probably occur in the 1980s. When this happens, crude oil shortages in many parts of the world will become frequent, causing major changes in world energy consumption patterns and trends. Rapid development of non-petroleum energy supplies and conservation programs will be essential.

"(5) Although production in the non-OPEC less developed countries (LDCs) will increase above its current level (principally as a result of increases in Mexican production), much, if not all, of the increases in LDC oil supply will be offset by increases in LDC demand.

"(6) The Communist countries may cease being a net exporter of oil to the free world by the early 1980s as a result of declines in Soviet production. The increased pressure caused by the entry of the Eastern European countries (currently more than 80% dependent on the Soviet Union for their imported source of oil) and conceivably the Soviet Union itself as buyers on the world oil market has serious implications both for price and foreign policy. While the Soviet Union has potentially promising areas for increases in oil production, most of the petroleum "frontier" Soviet areas lie either north of the Arctic Circle or in deep water where drilling is expensive and slow; development is at least a decade away. Many technical difficulties have already been encountered and it appears that the Soviets have not accumulated enough technology or experience to develop these areas rapidly. The Office of Technology Assessment has a study underway of the contributions which U.S. and other Western technologies could make to Soviet oil production.

"(7) In addition to Mexico, some new large discoveries are possible outside the Middle East, but there appears little possibility that the Middle East oil fields will be duplicated elsewhere. All promising areas for large discoveries outside the Middle East are either in the Arctic or involve areas of territorial dispute (the Malvinas basin off the Falkland Islands and the South China Sea). Even under optimistic discovery assumptions, it is unlikely that substantial production from new sources will occur in this decade. Because of the remoteness of most prospective areas, there would be a considerable delay between the time of discovery and significant production from any new discoveries in these areas.

"(8) Major additions to the world's known oil supplies are likely to result from additional recovery in known fields rather than new field discoveries. These new additions are not expected to alter the dominance of the Middle East since over half of the new additions are expected to be in the Middle East. Moreover, the world distribution of ultimately recoverable oil is not believed to differ significantly from the known distribution today.

"(9) There is some speculation regarding the petroleum potential of the deep ocean areas and Antarctica. The technology and the price of oil are not sufficient now to encourage active exploration and development of either of these regions. Future development of these areas may also necessitate new international agreements."

E. New Oil Potential

Several new geographical areas remain to be explored in North America, which could contain undiscovered oil and gas fields. These include offshore areas of the Atlantic coast, California, and the Gulf of Mexico, as well as interior regions and offshore areas of Alaska. The Arctic regions of Canada, such as the Beaufort Sea, also hold considerable promise for future exploration. However, these areas, if successfully exploited, will only help replace existing reserves which are nearing depletion. They are not expected to increase oil production over the long run. In addition, many of these areas are made up of physically hostile environments that are difficult and costly to explore and develop.

It is often asked if large new oil and gas fields could be found in existing petroleum provinces within the United States. In answering this question, it is interesting to consider the history of the Permian Basin in west Texas and southeast New Mexico. This is one of the largest petroleum provinces in the United States, with recoveries expected to exceed 23 billion barrels of oil.

From the first discovery in the area in 1921 to 1948, 74 percent of the province's reserves were discovered. Also during this period, only 14 percent of the area's exploration wells were drilled. In the succeeding 25 years, the industry drilled six times as many exploration wells, and discovered only one-third the amount of oil. This example illustrates the present problem for U.S. oil production: existing areas have been extensively explored, and further large supplies of new, undiscovered oil cannot be expected.

Synthetic liquids, converted from coal and oil shale, possess the potential to supplement conventional oil production. Exxon projects that these fuels could account for 21 to 34 percent of our country's oil supply by the year 2000. However, there are many issues still to be resolved, and policy decisions to be made, in the areas of leasing, water sources, environment, and socioeconomic impact, before this type of production can be realized. Since there is considerable lead time involved in constructing facilities and establishing programs, action must be taken soon in these areas, if synthetic fuels are to be a viable source of energy in this century.

During the last few years, there have been a number of estimates made by various groups and organizations regarding the ultimate petroleum resources available in the United States. These estimates often vary widely due to differences in methodology and the premises upon which the particular study is based. However, figures from a recent geological estimate of recoverable petroleum in this country, by the U.S. Geological Service, may help put the present situation in perspective.

According to this study, the amount of undiscovered recoverable petroleum resources in the U.S. ranges from 50 to 127 billion barrels of oil, and 322 to 655 trillion cubic feet of gas. These figures do not include present reserves (identified resources recoverable with current technology and economic conditions), but refer only to undiscovered deposits that are economically producible, or to identified resources that are not currently recoverable. (It should be remembered that revisions are constantly being made in these estimates as changes occur in geological information, economic conditions, and technological development.)

These projections indicate that the undiscovered recoverable quantities of petroleum are approximately one half of the amount identified and produced to date. To put it another way, two-thirds of our ultimately recoverable petroleum resources have already been either produced or identified. At the current rate of consumption, petroleum resources in this country would be exhausted early in the twenty-first century.

F. Implications of the Present Situation

Speculation varies as to the impact the present

energy situation will have on the future of the United States. Technical, sociological, economic, and political factors will influence the way existing U.S. energy resources are developed and used, and what effect they will have on society in general. Many segments of society seem certain to be affected.

Prices. Prices for petroleum products, such as gasoline and heating fuel, will no doubt continue to escalate in the future. This will certainly affect every American's budget and style of living. Inflation and national economic stability are influenced by oil prices, and continued increases in these prices may produce long-term detrimental effects on the economy.

Supply. Immediate supply tends to fluctuate as demand and economic climate change. The extended outlook, however, indicates a decline in the availability of certain petroleum products as resources are consumed. This has already happened in certain parts of the country, where natural gas connections have been unavailable for new housing.

Conservation. As a result of increasing prices and decreasing supply, conservation of our energy resources has become imperative. The residential, industrial, and transportation sectors of our society face increasing pressures to develop new products and improved operational procedures that will conserve energy. Fuel-efficient cars, better insulated homes, and more efficient planning and location of urban residential areas and transportation systems will be necessary.

National security. The dependence of the United States on foreign oil imports, which carry the possibility of abrupt increases in prices or curtailment of production, presents serious implications for the national security of our country. Interruptions of imports from future oil embargoes, or conflicts within oil producing countries, could result in disruptions in everyday life for Americans, the loss of jobs, economic instability, and a decline in the nation's ability to function as a world power.

Alternate energy sources. Development and use of alternate energy sources is certain to become more important as petroleum resources are diminished. The transition to other fuels will probably result in higher prices for energy than is now exacted by present resources. The following section will examine briefly some of the alternatives that are presently available, together with comments on their real potential at this point in time.

III. ALTERNATIVE SOURCES OF ENERGY

A. Coal

Coal is the lithified remains of ancient plant life that existed in swamps millions of years ago. As decay and decomposition occurred, geological processes, such as burial, pressure, and heat, were at work transforming these remains into the various kinds, or **grades**, of coal. The ranks of coal, in order of increasing hardness, carbon content, and energy production, are **lignite, subbituminous, bituminous,** and **anthracite.**

Coal is found in layers (also called **beds** or **seams**) between beds of sedimentary rock. These seams can vary greatly in thickness (from one inch to more than one hundred feet), depth, and area.

Coal is the largest single energy resource in the United States today, making up 90 percent of all conventional energy sources now available. The coal resources in this country could supply twice the energy of the entire oil reserves in the Middle East. Large deposits of coal exist in the Appalachians, the Midwest, the Rocky Mountains, and Alaska. The United States possesses about one-fifth of the entire world's coal supply. (The Soviet Union has approximately one-half.)

Coal resources in the U.S. are estimated at 3.9 trillion tons, with 2.2 trillion tons being estimated undiscovered resources, and 1.7 trillion tons identified resources. Of the identified coal resources, 438 billion tons are considered technically and economically minable, but only half of that amount (or 219 billion tons) is currently recoverable.

Whereas coal was the principal energy source in the late 1800s and early 1900s, and supplied approximately half of the energy demand in the U.S. as late as the second World War, it now accounts for only 21 percent of national fuel consumption. Coal, however, is again being viewed as the principal energy source to replace crude oil and natural gas in the immediate future. Coal has provided a growing share of the total energy for this country over the last few years, and could nearly equal oil by the year 2000 in the amount being consumed. The primary uses of coal for the next twenty years would be in direct consumption as a fuel, as a raw material for conversion into synthetic fuels, and possibly in underground conversion into gas.

The two major methods for producing coal are **surface mining** and **underground mining. Surface mining** is used to produce coal near the surface, generally less than 200 feet deep. Overlying sediments are removed so the exposed bed of coal can be mined. **Area surface mining** takes place on flat ground, while **contour mining** is used in hillside operations (Figure 11.8). Companies now make every effort to reclaim and restore the land after surface mining operations.

Underground mining, which utilizes a network of tunnels beneath the surface, is necessary when the coalbed is too deep for surface mining. The types of underground mining are classified by the kind of opening that is used: **shaft** (vertical), **drift** (level), and **slope** (inclined) (Figure 11.9).

One of the major problems to be solved, if coal is to make further gains as a significant fuel source, is the control of environmentally unacceptable emissions. Research is being conducted, and processes

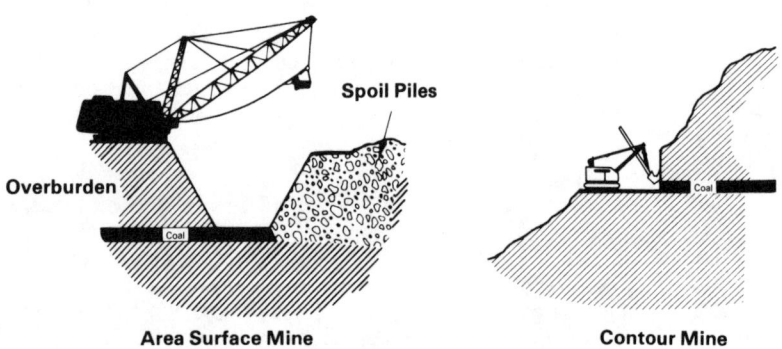

Figure 11.8. SURFACE COAL MINING METHODS. (Source: U.S. Energy Information Administration: *Coal Data: A Reference*)

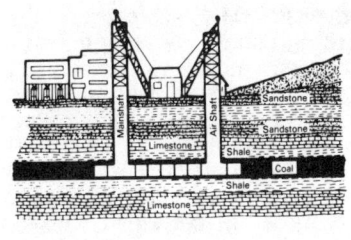 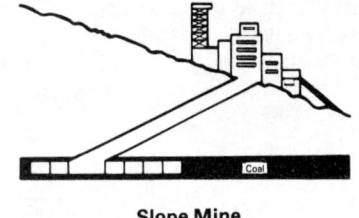 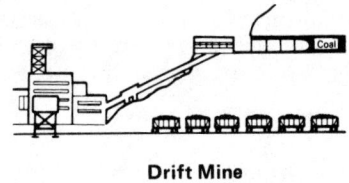

Shaft Mine   Slope Mine   Drift Mine

Figure 11.9. UNDERGROUND COAL MINING METHODS. (Source: U.S. Energy Information Administration: *Coal Data: A Reference*)

are being developed, to control the soot and sulfur that pour into the air from coal-burning operations. Progress in this area will facilitate an even greater use of this plentiful resource.

B. Oil Shale

The substance **marlstone,** commonly called oil shale, contains a bituminous, solid material known as **kerogen.** By heating the shale, the large hydrocarbon molecules can be broken down, thereby releasing the kerogen as both a gas and a heavy oil.

Oil shale is usually found in a layer or series of layers, called **zones,** located between other layers of sedimentary rock. Oil shale resources are classified by their average oil yields: **high grade** oil shale averages 30 or more gallons of oil per ton of shale, while **low grade** shale averages 10 to 30 gallons per ton.

Geologists have estimated that more than 2 trillion barrels of oil are contained in U.S. oil shale deposits. The world's largest deposits of high grade oil shale exist in Colorado, Wyoming, and Utah. These deposits were laid down in two large lakes about 50 million years ago, and are part of the Green River formation (Figure 11.10). The richest oil shale is found in a layer called the Mahogany Zone, which is normally 50 to 100 feet in thickness. Large quantities of oil shale also exist in Alaska and Montana. In addition, deposits of bituminous or oil shales are found in many of the Appalachian and Midwest states.

The oil shale is converted to oil by a process known as **retorting.** In this process, oil shale is ground into small pieces, and then heated to a temperature of 900° to 1000°F. This heating causes the kerogen in the shale to turn into oil vapors, which are then condensed into oil. A carbonaceous material is left as residue at the end of the retorting process.

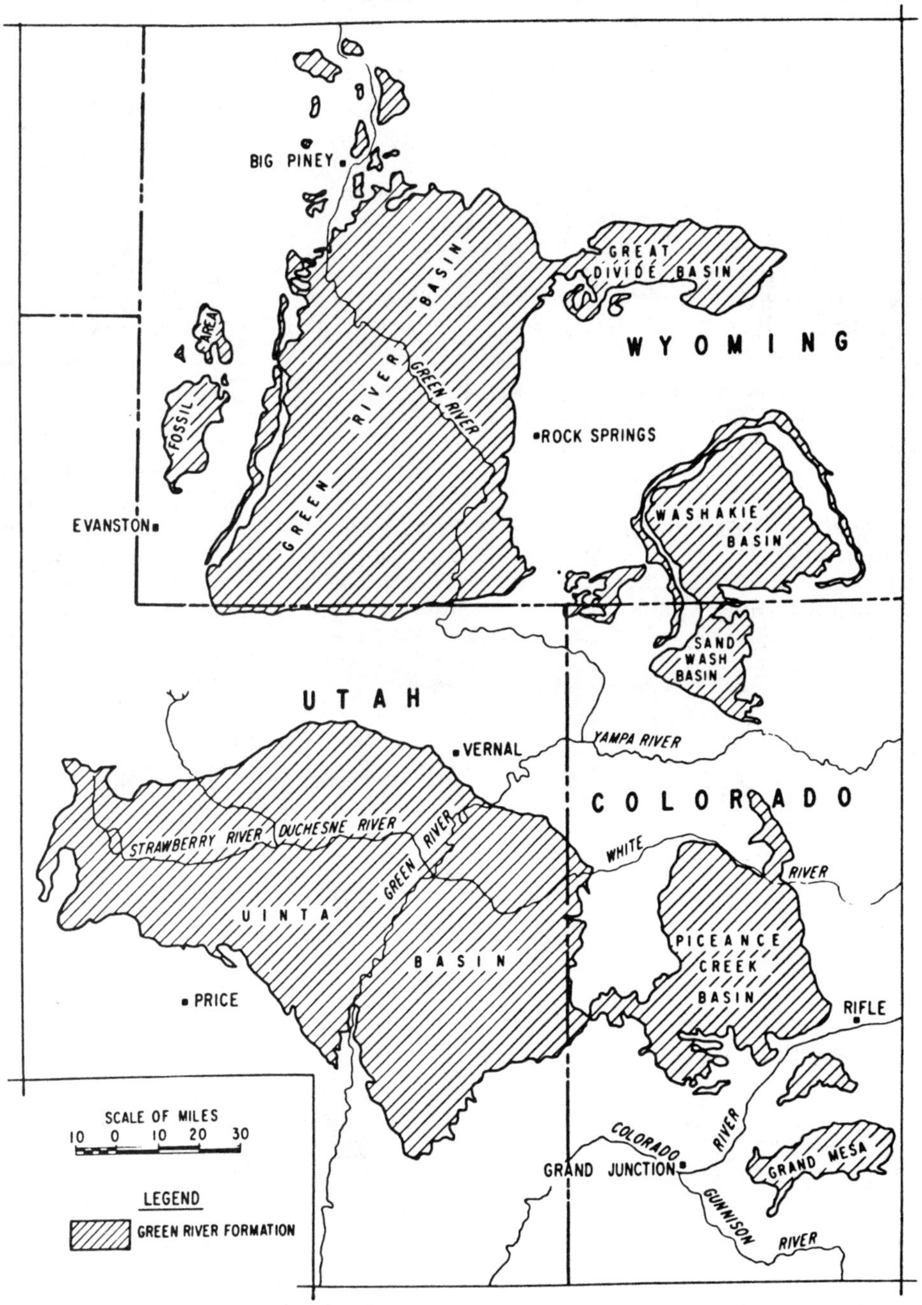

Figure 11.10. OIL SHALE DEPOSITS IN COLORADO, UTAH, AND WYOMING. (Figure courtesy of SPE of AIME)

Since the oil must be extracted from such large volumes of rock, the difficulty with oil shale has been to find ways of economically producing the resource. Several methods have been devised to accomplish this immense task of recovering shale oil, including **underground mining, surface mining,** and **in situ processing.**

Subsurface mining, often at depths of 1,500 feet, is used to remove oil shale from the ground to surface retorts. This method has been demonstrated by the U.S. Bureau of Mines as a feasible way of recovering oil shale. An underground mine is usually constructed by the "room and pillar" system, in which large areas (rooms) are excavated, with only pillars of oil shale left to support the roof of these areas. After the oil shale is retorted on the surface, the residual products are returned to the mine for disposal.

Oil shale that is removed by **surface mining** is retorted in the same manner as in underground mining. One of the primary disadvantages of **open pit,** or surface, mining is the environmental damage caused by the removal of such large quantities of surface rock. Large-scale reclamation and restoration projects are necessary after this type of oil shale recovery.

The major problems encountered in mining oil shale are: (1) the necessity of physically transporting large quantities of rock to surface retorts, and (2) the disposal of waste materials left over from the retorting process. These problems are both overcome by **in situ processing.** The term "in situ" implies that the oil shale is processed in place, without removing it from the ground.

There are several ways to accomplish this. One of the currently more successful techniques has been developed by Occidental Oil Shale, Inc. As shown in Figure 11.11, a large underground column of rock is drilled with shot holes, and reduced to rubble by the use of explosives. A fire is then started in the top of the column and allowed to burn to the bottom, over a period of time. As the fire progresses, the high temperature retorts the oil shale. The produced oil runs to the bottom of the column, where it is collected in a sump and pumped to the surface. This process also involves mining out working rooms above and below the column, as well as access and service tunnels to support the operation.

C. Tar Sands

Tar sands are sandstones saturated with highly viscous, heavy oil, which is immovable under normal circumstances. Tar sands have the same origin as oil sands, but the loss of the volatile fractions of the petroleum reservoir to the atmosphere has left only the heavy oil and asphalt fractions in the sandstone. This is caused by erosion exposing the reservoir beds on the surface, with the accompanying changes in temperature, pressure, and bacterial and chemical action. There is a fine line of distinction between heavy oil deposits and tar sands, since it is often difficult to determine whether a substance is a very thick liquid, a semi-solid, or a solid.

The Athabasca tar sands, found in Alberta, Canada, are the largest and best known deposits of tar sands. There are also over 10 billion barrels of oil contained in the Uinta Basin, the Tar Sand Triangle, and the Circle Cliffs deposits in Utah. Much of this oil is not easily recoverable, however, since it is either buried too deep in the ground, or has such low concentrations that it is uneconomical to produce.

Two general processes, **in situ recovery** and **surface mining,** are used to recover the oil. **In situ recovery** methods concentrate on improving the mobility of the oil in the formation, thus enabling the oil to flow into the wells and be produced at the surface. In many respects, in situ methods used to recover tar sand oil are similar to the thermal recovery methods used for heavy oil in California and other areas.

Several in situ recovery processes have been tested and are in use in Canada. One of these consists of conventional steam soak and fire-flooding techniques, similar to those described in the chapter on Recovery. Several, more exotic, thermal techniques involving the injection of hot alkaline solutions into the tar sands are also used. The alkaline solution helps neutralize acidic compounds often found in oils of this type, and thereby reduces corrosion problems in the production wells.

Tar sands can also be mined using conventional **surface mining** operations. Surface retorting is used to extract the oil by refining the tar sand in special containers. The refined oil is then ready to be marketed.

D. Nuclear Power

The United States reached a milestone in 1980, when more electricity was produced with nuclear energy than with oil. Although coal still remains the major source of electrical generation, nuclear energy has now begun to replace petroleum in at least one area of our energy demands.

Nuclear fission is the process of splitting certain heavy

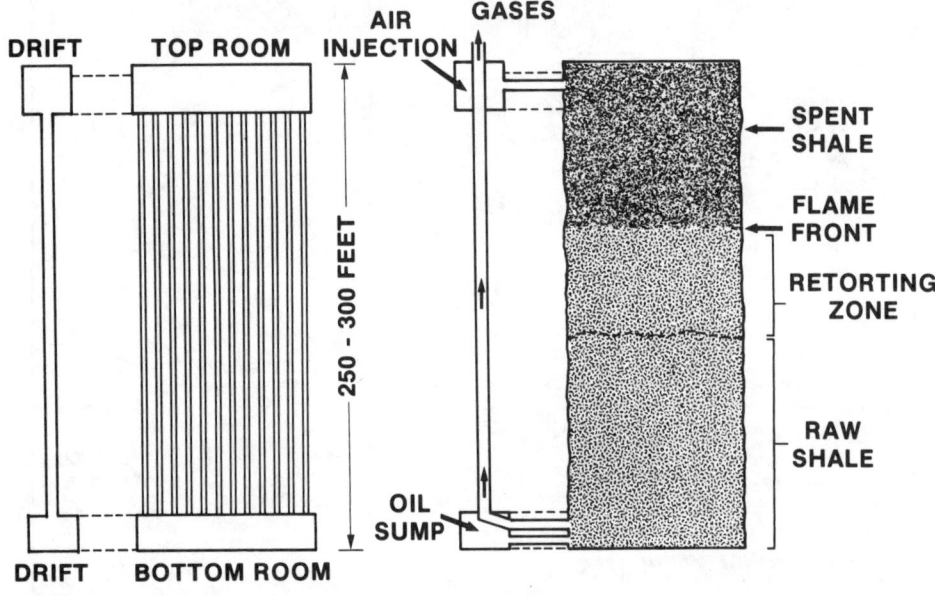

Figure 11.11. IN SITU RETORTING PROCESS.

atoms into two dissimilar atoms, thereby releasing energy and one or several neutrons. These neutrons can react with other atoms, causing them to fission, and thus creating a chain reaction. In a **nuclear reactor,** the controlled chain reaction creates heat, which can be converted into electrical energy.

The resource base for conventional fission reactors is the element **uranium,** which occurs in nature as a compound. Almost all of the uranium mined in the United States exists as **uraninite,** or **uranium oxide.** Most commercial uranium reserves in this country are found in ancient stream bed deposits, where the uranium was originally precipitated out of the ground water by chemical reaction (Figure 11.12).

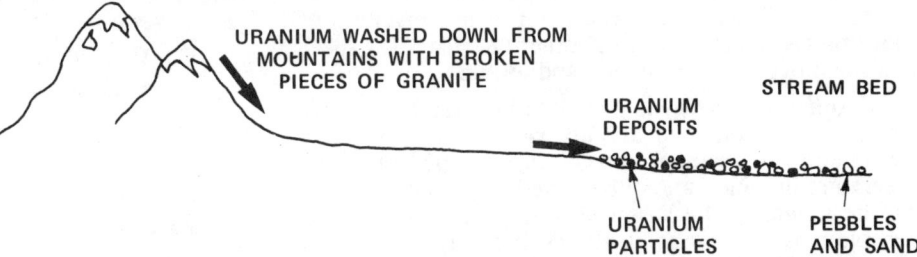

Figure 11.12. URANIUM DEPOSITION.

Uranium probably had its origin in igneous rocks. Granite, for example, normally contains about four grams of uranium (less than one-hundredth of a pound) per ton of rock. This may seem small, but the weathering of each cubic mile of granite releases almost 50,000 tons of uranium for transportation.

Most of the uranium reserves in the United States are found in Wyoming, Utah, Colorado, New Mexico, and Arizona. Canada also has large amounts of uranium ores.

There has been some controversy over the estimated amount of uranium reserves in this country. Whatever the case, there is some concern over the supply available, which in part has led to the development of the **breeder reactor.** Breeder reactors produce **plutonium,** which may be used to fuel other breeder reactors, thus reducing the amount of uranium required annually for each reactor.

By the end of this century, nuclear energy could account for almost 30 percent of electrical generation in this country. However, nuclear energy remains controversial, and its future contribution is difficult to predict with any certainty. Environmental and safety concerns, security measures for nuclear plants, uncertainty about the supply of uranium—all of these elements have clouded the future development of this energy source. If these issues can be settled satisfactorily, nuclear power constitutes a valuable resource for this country's energy needs.

Nuclear fusion is also being intensely investigated as an alternative source of energy. This process consists of controlling the fusion of atoms in a reactor to generate energy. Nuclear fusion is an attractive alternative both from an environmental standpoint, and from the availability of fuel sources. Initially, fusion reactors are expected to use **deuterium** and **tritium,** heavy isotopes of hydrogen, as fuel sources, both of which are available in plentiful quantities. Environmentally, fusion reactors also are expected to have less serious fuel handling problems and lower radioactive inventories.

If fusion could be made to work, it would supply almost unlimited energy. All electrical power would be generated by this process; it could produce hydrogen and other synthetic fuels to replace petroleum; and hydrocarbons could then be used as raw materials for manufacturing various products. At the present time, however, no way has been developed that will produce more energy than is consumed in the fusion-generating process. Nuclear fusion is not currently projected as a source of commercial energy in this century.

E. Solar Power

The sun is a tremendous source of energy. In just the transformation of a portion of its mass from hydrogen to helium, through the fusion of hydrogen nuclei, the sun radiates 70,000 horsepower of energy per square yard of surface. In three days time, the earth intercepts solar energy equivalent to all the energy stored in our fossil fuel reserves. If it could be efficiently converted, the energy reaching 5,000 square miles of the desert Southwest could theoretically supply the needs of the entire country.

Obviously, there are problems involved with solar energy, since such an unlimited resource is not currently being used more extensively. Because solar energy is not constant, energy needs to be stored, or a backup system must be available, to provide power at night or when the sun is concealed. In addition, solar radiation has a low density, requiring large areas of land assigned to energy collection. Finally, since solar energy sources operate at relatively low efficiencies, there is considerable cost involved in collecting, storing, and transforming the energy.

In spite of the current disadvantages, this alternative energy source has a distinct advantage over all the others: it is not damaging to the environment and is virtually inexhaustible in supply. For this reason, the study and development of solar energy is receiving considerable support and attention.

The technical capacity to use solar energy for space heating is now well developed. A number of new homes and buildings are being equipped with solar heating and cooling systems, which promise to be even more common in the future. Unfortunately, these systems are still expensive. Solar energy will become commercially feasible as new technology lowers the cost to levels comparable to that of conventional sources.

Solar radiation is also the primary resource responsible for other possible sources of energy: **wind, organic fuels** or **biomass, direct radiation,** and **ocean thermal energy conversion.** Each of these is being considered and studied, but problem areas preclude any major impact by any of these sources in the immediate future.

F. Geothermal Energy

The core of the earth, consisting of magma (or molten rock) generates temperatures approaching 8,000° to 10,000° F. This geothermal energy penetrates the earth's crust in some areas, in the form of volcanoes, geysers, and hot springs. This natural energy has been converted in some areas to provide electrical power or heat. For example, steam from the earth at Geysers, California, generates approximately half of San Francisco's electricity. Geothermal steam for space heating is produced in Oregon and Idaho.

Geothermal energy is available in three forms: **dry steam, wet steam,** and **dry hot rock** formations. **Dry steam** and **wet steam** are produced from hydrothermal reservoirs in which an aquifer overlies the heat source, or the magma. The steam that is produced by these reservoirs is run through turbines to generate electricity (Figure 11.13). In **dry hot rock reservoirs,** no permeable aquifer exists over the heat source; consequently, water must be injected into the system.

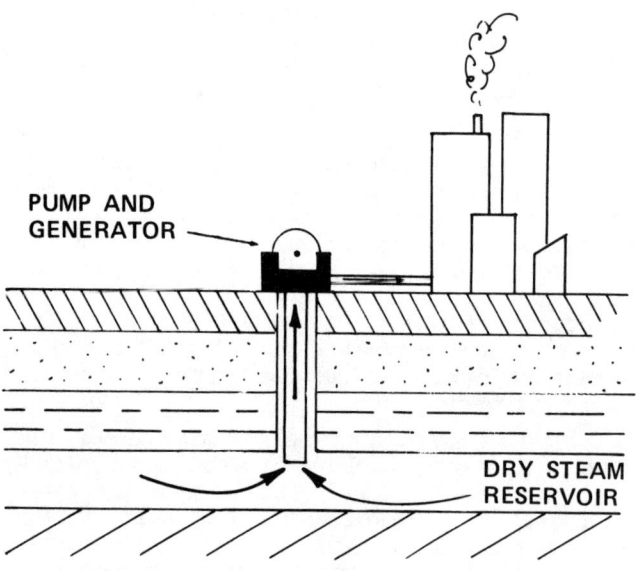

Figure 11.13. GEOTHERMAL ENERGY.

There are some problems associated with geothermal energy. Noise pollution can be intense because of the high-pitched sound steam makes as it rushes from the earth. Hydrogen sulfide and methane are also emitted into the air with the steam, and pose environmental problems. The saturation of geothermal water with salt and other minerals also causes problems in waste water disposal and corrosion of equipment. Finally, geothermal energy is relegated to localized areas, since it cannot be transported in its original form, but must first be converted to electricity if it is to be used elsewhere.

G. Hydro Power

Water power has been used to generate electricity in this country since the 1880s. Water exists as a hydroelectric source when an adequate flow rate is combined with a suitable elevation difference between the surface of the water storage and the outlet of the turbine discharge (the minimum is usually twenty feet). Because hydroelectric resources are renewable, they are calculated as annual rates for producing power, rather than as fixed quantities of a depletable fuel.

The components of a hydroelectric system consist of a **water source,** a **storage reservoir,** a **pipe transport system,** and a **turbine-generator complex.** There is also a **pumped storage** component that can elevate water back into the reservoir for use during peak demand periods.

Hydroelectric energy resources have not grown substantially for a number of reasons: sites for dams are limited, capital costs have escalated, and environmental concerns have been raised over changed land use and impact on wildlife from dam construction projects. It seems unlikely that the share of U.S. energy supplied by hydroelectric power will increase during this century.

H. Waste Conversion

The potential exists to convert dry, organic, solid waste materials from dumps and waste disposal systems into liquid, gaseous, and solid fuels. Each year we generate almost a billion tons of solid wastes in this country, in the form of animal manure, urban refuse, and agricultural waste. Three processes are already available to convert such substances into fuel: **hydrogenation** (the addition of hydrogen to an organic molecule for conversion to oil), **pyrolysis** (the chemical decomposition of waste without oxidation), and **bioconversion** (the conversion of organic wastes into methane).

The advantages of waste conversion are: a solution to waste disposal problems, a reduction in landfill requirements, and production of products low in sulfur. The major drawback, at the present time, is cost—including the expense involved in collecting and preparing the waste material, and the costs of any necessary elements to be used in the waste conversion processes.

ENERGY OUTLOOK—APPENDIX

"Worldwide Oil and Gas at a Glance," taken from the December 31, 1984, issue of **Oil and Gas Journal,** is used with the permission of the **Oil and Gas Journal.**

COUNTRY	ESTIMATED PROVED RESERVES 1-1-1985 Oil (1,000 bbl)	Gas (10⁹ cu ft)	OIL PRODUCTION Producing wells** July 1, 84	Estimated 1984 (1,000 b/d)	% change from 1983	No. of ref.	REFINING Capacity (b/cd) Jan. 1, 1985 Crude	Thermal operations	Catalytic cracking	Catalytic reforming
ASIA-PACIFIC										
Australia	1,430,900	17,850	545	481.0	+ 15.3	10	696,850	...	175,500	170,020
Bangladesh	...	7,000	...	...	...	1	31,200	...	...	1,650
Brunei	1,400,000	7,300	649	160.0	+ 3.2	1	9,400	...	...	...
Burma	28,000	170	445	30.0	...	2	26,300	1,700	...	...
China, Taiwan	(c)5,700	540	78	2.6	− 7.1	2	542,510	3,933	22,600	56,400
Guam	...	...	...	...	...	1	43,700	...	...	...
India	3,500,000	15,000	1,560	543.0	+ 26.3	12	704,752	107,030	86,900	26,560
Indonesia	8,650,000	40,000	4,840	1,332.0	− 3.8	6	630,500	98,500	...	60,700
Japan	56,000	720	348	6.4	− 7.2	45	4,813,000	80,520	422,650	537,550
Korea, South	...	...	...	...	...	6	776,000	43,990	...	36,550
Malaysia	3,000,000	50,000	368	462.0	+ 25.9	4	205,000	...	...	20,600
New Zealand	(c)155,000	5,438	27	18.0	+ 20.0	1	53,000	...	...	18,000
Papua New Guinea	(c)50,000	500	...	...	...	...	...	...	...	...
Okinawa	...	...	...	...	...	3	153,000	...	...	11,500
Pakistan	82,000	15,760	38	18.0	+ 50.0	3	128,750	...	...	5,600
Philippines	16,300	12	10	12.0	− 14.3	3	286,000	...	16,100	39,400
Singapore	...	...	...	...	...	5	1,072,000	133,000	...	50,000
Sri Lanka	...	...	...	...	...	1	50,000	12,500	...	3,750
Thailand	156,000	5,900	84	19.0	+171.4	3	171,600	9,600	8,000	24,600
Total Asia-Pacific	18,529,900	166,190	8,992	3,084.0	+ 8.5	109	10,393,562	490,773	731,750	1,062,880
WESTERN EUROPE										
Austria	118,500	530	1,243	23.0	− 4.2	1	273,000	18,000	27,000	33,000
Belgium	...	...	...	...	...	5	692,500	51,600	100,200	92,420
Cyprus	...	...	...	...	...	1	16,000	...	...	3,400
Denmark	400,000	3,500	32	45.0	+ 4.7	3	176,000	72,500	...	29,700
Finland	...	...	...	...	...	2	299,000	45,500	37,500	48,000
France	148,000	1,500	550	35.0	...	17	2,386,000	119,000	279,950	310,000
Germany, West	299,000	5,957	3,136	79.0	− 2.5	25	2,171,600	331,800	179,700	385,850
Greece	51,000	3,440	...	27.0	+ 12.5	4	390,000	19,520	23,000	30,300
Ireland	...	800	...	...	...	1	56,000	...	...	11,000
Italy-Sicily	800,000	4,360	140	45.0	...	23	3,095,200	309,630	279,500	340,400
Netherlands	310,000	68,482	633	61.0	+ 15.1	7	1,498,500	174,900	122,500	171,900
Norway	8,300,000	89,000	171	688.0	+ 12.8	4	244,200	67,000	...	29,000
Portugal	...	...	...	...	...	3	289,922	...	10,096	38,169
Spain	115,000	700	42	45.0	− 22.4	10	1,493,000	129,400	103,500	179,700
Sweden	...	...	...	...	...	6	439,000	69,000	24,000	73,500
Switzerland	...	4	...	...	...	2	127,000	24,000	...	25,000
Turkey	294,000	600	440	41.0	− 6.8	4	460,475	...	38,600	45,057
United Kingdom	13,590,000	27,800	535	2,452.0	+ 7.0	16	2,007,700	205,000	356,500	349,500
Total Western Europe	24,425,500	206,673	6,922	3,541.0	+ 7.0	130	15,654,622	1,636,850	1,543,446	2,150,839
MIDDLE EAST										
Abu Dhabi	30,500,000	20,750	254	750.0	− 3.4	3	185,250	...	...	12,334
Bahrain	170,000	7,260	257	41.0	...	1	250,000	19,000	36,000	17,100
Divided (Neutral) Zone	5,420,000	8,290	464	420.0	+ 7.1	...	...	...	...	...
Dubai	1,440,000	4,320	141	324.0	− 3.6	...	...	...	...	...
Iran	48,500,000	478,600	530	2,166.0	− 10.7	4	530,000	80,800	...	63,845
Iraq	44,500,000	28,800	290	1,218.0	+ 36.4	8	318,500	...	...	43,500
Israel	750	8	10	0.1	...	2	170,000	62,000	18,000	24,500
Jordan	...	...	...	...	...	1	100,000	...	4,410	8,640
Kuwait	90,000,000	32,500	530	925.0	+ 6.6	5	668,600	...	...	51,000
Lebanon	...	...	...	...	...	1	17,000	...	...	3,100
Oman	3,500,000	7,377	555	404.0	+ 7.7	1	50,000	...	...	10,000
Qatar	3,350,000	7,377	555	404.0	+ 33.9	1	56,058	...	...	10,463
Ras al Khaimah	(c)100,000	400	2	5.8	...	...	...	...	...	...
Saudi Arabia	169,000,000	123,270	555	4,545.0	− 7.1	4	840,000	...	10,000	38,000
Sharjah	450,000	6,500	18	62.0	+ 77.1	...	...	...	...	...
South Yemen (Aden)	...	...	...	...	...	1	178,000	...	...	12,000
Syria	1,450,000	1,280	802	161.0	− 4.2	2	228,790	41,199	...	19,348
Total Middle East	398,380,750	869,355	4,582	11,416.9	− 0.7	38	4,052,673	202,999	107,010	358,887

EDITOR'S NOTE: All reserves figures except those for the U.S.S.R. (and gas for Canada) are reported as proved reserves recoverable with present technology and prices. U.S.S.R. figures are "explored reserves" which include proved, probable, and some possible. Canadian gas figure, under criteria adopted by Canadian Petroleum Association in 1980, includes proved and some probable. (c)Condensate. *U.S. reserves numbers based on Energy Information Administration estimates. †Estimates based on capacity as of Jan. 1, 1984, plus known 1984 closures and expansions. ††Includes Albania, Bulgaria, Cuba, Czechoslovakia, East Germany, Hungary, Mongolia, North Korea, Poland, Romania, Viet Nam, and Yugoslavia. **Does not include shut in, injection, or service wells.

COUNTRY	ESTIMATED PROVED RESERVES 1-1-1985		OIL PRODUCTION			REFINING Capacity (b/cd) Jan. 1, 1985				
	Oil (1,000 bbl)	Gas (10⁹ cu ft)	Producing wells** July 1, 84	Estimated 1984 (1,000 b/d)	% change from 1983	No. of ref.	Crude	Thermal operations	Catalytic cracking	Catalytic reforming
AFRICA										
Algeria	9,000,000	109,100	1,090	608.0	− 9.9	4	464,700	...	...	55,600
Angola-Cabinda	1,800,000	1,650	291	207.0	+ 16.9	1	32,100	...	...	1,900
Benin	100,000	...	6	6.9	+ 23.2	...	...	...	...	...
Cameroon	550,000	4,150	166	125.0	+ 9.6	1	43,000	...	...	6,500
Congo	480,000	2,118	246	118.0	+ 35.6	1	21,000	...	...	2,000
Egypt	3,200,000	7,000	617	790.0	+ 14.7	6	369,330	...	...	10,800
Ethiopia	...	...	...	...	...	1	14,600	...	...	1,898
Gabon	510,000	550	270	150.0	...	1	20,000	7,200	...	1,400
Ghana	3,800	4	3	0.6	− 50.0	1	28,000	...	...	5,550
Ivory Coast	108,000	3,000	19	22.0	− 8.3	2	90,073	...	...	14,600
Kenya	...	...	...	...	...	1	95,000	...	...	9,000
Liberia	...	...	...	...	...	1	15,000	...	...	2,000
Libya	21,100,000	21,200	905	1,090.0	+ 1.3	3	329,800	...	...	14,182
Madagascar	...	...	...	...	...	1	16,350	...	...	2,600
Morocco	250	...	...	0.2	...	2	79,703	...	...	13,058
Mozambique	...	...	...	...	...	1	17,000	...	...	2,500
Nigeria	16,650,000	35,600	1,177	1,414.0	+ 13.9	3	250,400	...	44,100	38,100
Senegal	...	...	...	...	...	1	29,800	...	...	2,870
Sierra Leone	...	...	...	...	...	1	10,000	...	...	...
South Africa	114,500	350	...	...	...	5	388,500	59,500	80,000	62,500
Sudan	300,000	...	...	...	...	1	23,800	...	...	1,700
Tanzania	...	200	...	...	...	1	13,500	...	...	3,000
Togo	...	...	...	...	...	1	20,000	...	...	4,400
Tunisia	1,514,000	2,224	131	114.0	− 5.0	1	34,000	...	...	3,300
Zaire	110,000	30	16	27.0	+ 12.5	1	17,000	...	...	3,000
Zambia	...	...	...	...	...	1	24,500	...	...	5,600
Total Africa	**55,540,550**	**187,176**	**4,937**	**4,672.7**	**+ 6.6**	**43**	**2,447,156**	**66,700**	**124,100**	**268,058**
WESTERN HEMISPHERE										
Argentina	2,266,000	24,628	8,189	467.0	− 2.5	11	678,400	121,600	102,500	37,700
Bahamas	...	...	...	...	...	1	350,000	...	...	...
Barbados	600	...	61	1.7	+ 54.5	1	3,000	...	...	...
Bolivia	158,000	4,270	352	20.0	...	3	46,930	...	...	14,158
Brazil	1,976,000	2,840	3,386	437.0	+ 32.8	13	1,305,100	16,200	315,400	21,600
Canada	7,075,000	92,300	39,146	1,430.0	− 1.4	27	1,868,534	79,700	401,767	349,338
Chile	736,000	2,360	266	38.0	− 2.6	2	141,000	8,400	34,000	9,300
Colombia	624,000	3,786	2,509	165.0	+ 8.6	4	211,000	61,000	91,000	6,000
Costa Rica	...	...	...	...	...	1	15,600	...	...	1,500
Dominican Republic	...	...	...	...	...	2	44,000	...	...	8,300
Ecuador	1,400,000	3,000	...	254.0	+ 8.1	3	82,282	11,550	11,550	2,513
El Salvador	...	...	...	...	...	1	16,300	...	...	2,800
Guatemala	500,000	30	8	5.2	− 23.5	1	16,000	...	...	3,000
Honduras	...	...	...	...	...	1	14,000	...	...	1,800
Jamaica	...	...	...	...	...	1	35,500	...	...	3,540
Martinique	...	...	...	...	...	1	12,800	...	...	3,000
Mexico	48,600,000	77,000	...	2,743.0	+ 2.1	9	1,269,000	82,000	297,000	163,800
Netherlands Antilles	...	...	...	...	...	2	740,000	295,000	44,000	15,000
Nicaragua	...	...	...	...	...	1	15,000	...	...	3,000
Panama	...	...	...	...	...	1	100,000	...	...	7,500
Paraguay	...	...	...	...	...	1	7,500	...	...	...
Peru	670,100	1,100	4,083	201.0	+ 17.5	6	175,560	...	21,670	1,680
Puerto Rico	...	...	...	...	...	2	121,000	...	12,000	5,800
Suriname	...	...	12	0.5	...	...	...	...	...	...
Trinidad & Tobago	540,000	10,550	3,145	169.0	+ 5.6	2	320,000	...	25,000	27,000
Uruguay	...	...	...	...	...	1	45,000	...	5,000	3,000
Venezuela	25,845,000	55,367	13,512	1,724.0	− 2.5	7	1,224,200	41,600	179,400	6,000
Virgin Islands	...	...	...	...	...	1	545,000	80,000	...	125,000
United States	*27,300,000	*198,000	615,500	8,750.0	+ 0.7	†202	†15,400,000	†1,611,705	†4,671,000	†3,476,520
Total Western Hemisphere	**117,690,700**	**475,231**	**690,169**	**16,405.4**	**+ 1.4**	**308**	**24,857,706**	**2,368,755**	**6,211,287**	**4,298,849**
Total non-Communist	**614,567,400**	**1,904,625**	**715,602**	**39,120.0**	**+ 2.4**	**628**	**57,405,719**	**4,766,077**	**8,717,593**	**8,139,513**
COMMUNIST AREAS										
China	19,100,000	30,900	NA	2,250.0	+ 6.1	20	2,150,000	NA	NA	NA
U.S.S.R.	63,000,000	1,450,000	110,000	12,230.0	− 0.8	38	12,200,000	NA	NA	NA
Other††	2,000,000	16,500	NA	490.0	...	45	3,150,000	NA	NA	NA
Total Communist	**84,100,000**	**1,497,400**	**110,000**	**14,970.0**	**+ 0.2**	**103**	**17,500,000**	**NA**	**NA**	**NA**
TOTAL WORLD	**698,667,400**	**3,402,025**	**825,602**	**54,090.0**	**+ 1.8**	**731**	**74,905,719**	...	...	...

ENERGY OUTLOOK—BIBLIOGRAPHY

Abelson, Phillip H.: **Energy: Use, Conservation and Supply,** American Association for the Advancement of Science, 1974.

American Institute of Architects: **Energy,** AIA Energy Notebook, 1975.

Brinkworth, B.J.: **Solar Energy for Man,** Halstead Press, John Wiley and Sons, Inc. (New York) 1972.

Central Intelligence Agency: **The International Energy Situation, Outlook to 1985,** Superintendent of Documents (Washington, D.C.) 1977.

Clark, Wilson: **Energy for Survival,** Anchor Press/Doubleday (New York) 1975.

Committee on Energy and Natural Resources: **The Geopolitics of Oil,** U.S. Government Printing Office (Washington, D.C.) 1980.

Cook, Glenn L.: **Oil Shale: An Impending Energy Source,** SPE Paper 3861, Society of Petroleum Engineers of AIME (Dallas) 1972.

Department of Commerce: **1981 U.S. Industrial Outlook,** U.S. Government Printing Office (Washington, D.C.) 1981.

Department of Energy: **Coal Data: A Reference,** U.S. Government Printing Office (Washington, D.C.) 1980.

Department of the Interior: **Geological Estimates of Undiscovered Recoverable Oil and Gas Resources in the United States,** Geological Survey Circular 725 (Washington, D.C.) 1975.

Exxon Company, U.S.A.: **Developing the Many Sources of Energy,** Public Affairs Department (Houston) no date.

Exxon Company, U.S.A.: **Energy Outlook: 1980-2000,** Public Affairs Department (Houston) 1980.

Hammond, A.L., W.D. Metz, and T.H. Maugh, II: **Energy and the Future,** American Association for the Advancement of Science, 1973.

Kreider, J.F. and F. Kreith: **Solar Heating and Cooling,** Hemisphere Publishing Corporation, McGraw-Hill (New York) 1975.

Kruger, Paul and Carol Otte (eds.): **Geothermal Energy,** Stanford University Press, 1973.

Office of Technology Assessment: **Application of Solar Technology to Today's Energy Needs,** 2 Vols., U.S. Government Printing Office (Washington, D.C.) 1978.

Office of Technology Assessment: **World Petroleum Availability: 1980-2000,** U.S. Government Printing Office (Washington, D.C.) 1980.

The President's Commission on Coal: **Coal Data Book,** U.S. Government Printing Office (Washington, D.C.) 1980.

Ray, Dr. Dixy Lee: **The Nation's Energy Future,** United States Atomic Energy Commission, 1973.

The Science and Public Policy Program, University of Oklahoma: **Energy Alternatives: A Comparative Analysis,** U.S. Government Printing Office (Washington, D.C.) 1975.

Scientific American: **Energy and Power,** W.H. Freeman and Company (San Francisco) 1975.

Subcommittee on Synthetic Fuels: **Synthetic Fuels,** U.S. Government Printing Office (Washington, D.C.) 1979.

Voegeli, Henry and J.J. Tarrant: **Survival 2001,** Van Nostrand Reinhold Company, 1975.

ACKNOWLEDGEMENTS

Figure 11.10 is used with the permission of the Society of Petroleum Engineers of AIME. The copyright for this figure is owned by the Society of Petroleum Engineers of AIME.

INDEX

A page number printed in italics is a reference to a glossary.

A

abandon	*103*
absorption	*123*
abstract	66
abstract company	65
acid fracture	*103*
acidize	94-95, *103*
acetic acid	95
hydrochloric acid	94-95
hydrofluoric acid	95
acquired land	64
ACT	*148*
adapter	*123*
adjustable choke	*103*
adsorption	*123*
air-actuated	*103*
air drilling	*103*
air/gas lift	*123*
Alaska pipeline	145-146
allowable	*123*
alluvial fan	*40*
alternate sources of energy	174-178
American Petroleum Institute	*103*
amorphous	*40*
angle of deflection	*103*
annual rate-of-return	158
annular blowout preventer	90, *103*
annular space	*103*
annulus	83
anomalies	55, *58*
maximum gravity	55
minimum gravity	55
anticlinal theory	46, *58*
anticline	*40*
API—see *American Petroleum Institute*	
API gravity	*139*
apparent dip	*40*
application for permit to drill, deepen, re-enter, or operate	164
application to construct/use an earthen pit	164-165
application to dispose of salt water by injection	166
apron ring	*123*
aquifer	33, *40*
argillaceous	*40*
artesian	*40*
artificial lift	12, 116-118
asphalt	28
assignee	69, 71
assignment	69, 71, 73
clauses	71
consideration	71
preparation	71
restrictions	71
assignor	69, 71
associated gas	*123*
asymmetrical fold	*40*
attic	135
automobile	4-5

B

back-in	77
back off	*103*
back pressure	*123*
back up	*103*
back-up man	*123*
baffles	*123*
bail	*103*
bailer	80, *103*
bailing line	*103*
ball	145, *148*
ball and seat	*123*
barge	*103*, 141, 142, 147
barite	82, *103*
barium sulfate	*103*
barrel	*103*, 129
barrel-mile	*148*
barrel wrench	*123*
barrier reef	*40*
base line	61
basement rock	55, *58*
basin	*40*
basket sub	*103*
batch	*123*, 145, *148*
batching	145, *148*
batch separator	145
beam	*123*
beam pumping	117
beam well	*123*
bean	*123*
bed	*40*
bedrock	*40*
bell hole	*123*
belt	*103*
bentonite	*40*
Big Inch	5, 6, 143, *148*
bird cage	*123*
bit	82, *103*
bit breaker	*103*
bit record	*103*
bitumen	27
black oil reservoir	33
blank flange	*123*
blank liner	*123*
blank-off	*123*
bleeder	*123*
bleed into	*123*
bleed off	*123*
blind ram	90, *103*
block	*103*
blooey line	*103*
blowout	89-90, *103*
blowout preventer	84, 87-88, 90, *103*
blowout prevention system	84
bob tail	*123*
boilerhouse	*123*
boll-weevil	*104*
bomb	*104*
bond	*104*
bond (land)	71
bonding requirements	164
bonnet	*123*
boomer	*123*
booster pump	145
booster station	*148*
boot	*104*

- 183 -

BOP—see *blowout preventer*
borehole..................................... *104*
bottom-hole.................................. *104*
bottom-hole choke............................ *104*
bottom-hole pressure......................... *104*
bottom-hole-pressure bomb.................... *104*
bottom water................................. *139*
bottom water drive........................... *132*
Bourdon tube................................. *104*
bowl... *123*
box.. *104*
bradenhead gas............................... *123*
brake.. *104*
break out.................................... *104*
breakout cathead............................. *104*
breakout tongs............................... *104*
bright spot technique................... 54, *58*
bring in a well.............................. *104*
BS&W........................ 12, 118, 121, *123*
bubble point............................ 131, *139*
buck up...................................... *104*
bulk plant.................................... 16
bullet perforator............................ *104*
bump down.................................... *123*
buoyancy...................................... 29
Bureau of Land Management....... 65, 71, 73, *163*
Bureau of Mines.............................. *162*
burial.. 22

C

cable.. *104*
cable tool drilling............... 11, 80-81, *104*
cable tool rig........................... 11, 80
cage... *123*
calcareous.................................... *40*
calcite...................................... *40*
capillarity................................. 29-30
caprock.................................. 30, *40*
carrier rock.................................. 29
cased.. *104*
cased hole................................... *104*
case files.................................... 65
cash flow.................................... 153
    after-tax............................ 155
    before-tax........................... 155
cash flow analysis.................... 153, 155-156
casing................................... 85, *104*
    intermediate........................ 86, *107*
    production........................... 86, 92
casing centralizer........................... *104*
casing coupling.............................. *104*
casing elevator—see *elevator*
casinghead................................... *104*
casinghead gas........................... 8, *123*
casing pressure.............................. *123*
casing shoe.................................. *104*
casing string................................ *104*
cat.. *123*
catalytic cracking............................ 14
catch samples................................ *104*
cathead...................................... *104*
catline...................................... *104*
cat walk..................................... *123*
caving....................................... *104*
cellar....................................... *104*
cementation (geology).................... 21, *40*
    secondary............................. 23
cement casing................................ *104*
cement channeling............................ *104*

cementing............................... 85-86, *104*
centrifugal pump........................ 117, *148*
centrifuge................................... *148*
chain drive.................................. *104*
chain of title................................ 66
chain tongs.................................. *104*
channel sands................................ *40*
chase threads................................ *123*
cheater...................................... *123*
check valve.................................. *104*
chemical sediments........................... *40*
chisel tongs................................. *123*
choke.. *104*
    gas.................................. *133*
choke line.............................. 90, *104*
choke manifold............................... *104*
Christmas tree.......................... *104*, 115
circulate.................................... *104*
circulation.................................. *105*
clip... *123*
closed-in.................................... *123*
close nipple................................. *123*
closure...................................... *40*
coal................................... 29, 174-175
    anthracite......................... 29, 174
    bituminous......................... 29, 174
    lignite............................ 29, 174
    mining............................ 174-175
    origin................................ 29
    subbituminous...................... 29, 174
COFCAW....................................... 136
collar....................................... *123*
combination string........................... *105*
combustion.............................. 136, *139*
    forward.............................. 136
    quenched............................. 136
    reverse.............................. 136
    wet.................................. 136
come-along................................... *123*
come out of the hole......................... *105*
commingle............................... 145, *148*
common carrier............................... *148*
communitization.......................... 71, 75
compaction............................... 21, *40*
compaction drive.............................. 34
company man—see *company representative*
company representative....................... *105*
competent.................................... *40*
complete a well.............................. *105*
completion......................... 11, 12, 92-94
    cased hole............................ 92
    open hole........................... 92-93
    sand control.......................... 93
    slotted liner......................... 93
completion/recompletion report............... 165
compound..................................... *105*
compressibility............................... 33
compressor.......................... 118, 133, *143*
computer..................................... *123*
computer control............................. *123*
computer program............................. *123*
condensate.................. 8, 132, 133, 138, *139*
conductor pipe............................... *105*
confining bed............................ 30, *40*
conformable.................................. *40*
conglomerate................................. *40*
connate water............................ *40*, 130
connection................................... *123*
conservation................................. 174
consideration.............................. 67, 69

continental shelf *40*
contour. 47, *58*
contract depth *105*
control panel *123*
core (exploration) 48, *58*, 91
core (geology) 24, *40*
core analysis. *58*
core barrel 48, *58*, 91
core bit 91
core catcher *58*
core hole. *58*
correlation *40*
correlative rights *5*
cost correlations *152*
cost estimation 151-153
    development costs. *152*
    exploration costs *152*
county records 64-65
coupling *105*
coupon *123*
crack a valve. *123*
crater. *123*
creekology 46, *58*
critical angle 54, 55, *58*
crooked hole *105*
cross section. 51, *58*
crown block. 84, *105*
crude oil 7-8, *40*
    asphalt base 8
    heavy. 8
    light. 8
    paraffin base 8
    sour. 8, *148*
    sweet. 8, *149*
crude oil production *123*
crust 24, *40*
crystallization 21
curative work. 67
cut oil *123*
cuttings 47, *105*
cycle of erosion *40*

D

damage payments 64, 67
data processing. 53-54
daylight tour *105*
deadline 84, *105*
deadline tie-down anchor. 84, *105*
dead oil reservoir. 33
dead man *123*
dead well. *124*
deadwood. *148*
debug *124*
decline curve. 154-155
deformation. *40*
degasser *105*
degradation *40*
dehydration 118
delay rental. 69
deliveryman. *148*
Department of Energy. 161-162
Department of the Interior 162-163
depletion. *124*
depletion drive. 131
deposition. 21
derrick 84, 87, *105*
derrickman. 87, *105*
desander. 83, *105*
desiccation. 21, *40*

designation of agent or operator 164
desilter 83, *105*
detrital *40*
development costs *152*
development well. *105*
deviation. *105*
deviation survey. *105*
diamond bit 82, *105*
diapir fold *40*
diapirism 23
diesel-electric power *105*
diesel engine. *105*
dip. *40*
directional drilling 91-92, *105*
discounted cash flow rate-of-return *158*
discount factor. *157*
discount rate *158*
discovery well. *105*
disintegration. *40*
dispatcher. *148*
displacement (geology) *40*
displacement fluid. *105*
disposal well. *124*
dissolved gas *124*
distillation 135, *139*
distributor 16
doghouse *105*
dog leg *124*
dolomite. 32, *40*
dome. *40*
donkey pump. *124*
doodle bug 46, *58*
door-sheet *148*
dope *124*
double. *105*
double board *105*
doughnut *124*
downcomer *124*
downhole motor 92
    positive displacement 92
    turbine drill 92
downstream functions 7, *141*
downthrow, downthrow side *40*
dozer. *124*
drag. *40*
Drake well 3, 79
drawworks. 84, *105*
dresser sleeve *124*
drifter *124*
drill bit *105*
drill collar. 82, *105*
driller 87, *105*
drilling. 11-12, 79-92
    Chinese. 1-2, 79
    costs 45, *152*
    history. 79-80
    operations. 84-87
    personnel. 86-87
drilling barge 88
drilling block *106*
drilling boom *58*
drilling contract 84-85
    daywork contract 85
    footage contract 85
    turnkey contract 85
drilling contractor. 84, *106*
drilling crew *106*
drilling engineer. 129
drilling fluid. *106*
drilling foreman. *106*

drilling line 84, *106*
drilling mud—see *mud*
drilling rate *106*
drilling report 96-102
drilling rig—see *rig*
drilling superintendent 87
drill pipe 82, *106*
drill ship 87-88, *106*
drill stem *106*
drill-stem test 49-50, *58*, 91
drill string 81-82, *106*
drip *124*
drive pipe 86
drum *106*
dry gas 8, 133-134, *139*
dry gas reservoir 33, 132-133, *139*
dry hole *124*
DST—see *drill-stem test*
dual zone completion 116
dutchman *124*
Dyna-drill *106*
dynamic positioning *106*

E

economic geology *40*
Economic Regulatory Administration 161-162
economics 151-158
edge water *139*
edge water drive 132
effective date 67, 69
electrodynamic brake *106*
elevator *106*
emulsion 119
emulsion treater 120
energy balance 129, *139*
energy consumption 169-170
    electrical generation 169-170
    industrial 169-170
    nonenergy 170
    residential-commercial 169-170
    transportation 169-170
energy demand 170
Energy Information Administration 162
energy outlook 169-178
energy source 53
energy supply 170-172
enhanced recovery 129, *139*
entrained gas *139*
Environmental Protection Agency 163
eolian 21, *40*
epoch *40*
equal opportunity 71
era 26, *40*
    Cenozoic 27
    Mesozoic 26-27
    Paleozoic 26, 27
    Precambrian 26, 27
erosion 21, *40*
erosion cycle *40*
evening tour *106*
expansion loop *124*
exploitation well *106*
exploration 9-10, 45-57
    history 45-46
exploration costs 152
exploration cycle 56-57
explorationist *58*
exploration well *58*
exposure *40*

F

facies *40*
facies changes *41*
facies map 52, *58*
    lithofacies map 52, *58*
facilities engineer 129
failure 23, *41*
farmin 75
farmout 75
fast line 84, *106*
fatigue *124*
fault 23, *41*
fault plane *41*
fault zone *41*
Federal Energy Regulatory Commission 162
federal land 64
    acquisition of rights 73
    leases 69-71
    records 65
fee land 64
fee simple estate 64
female connection *124*
fence diagram 51, *58*
field 28, *41*
field facility *124*
fill the hole *106*
filter cake *106*
fingerboard *106*
fire tube 120
fire wall *124*
fish 91, *106*
fishing tool *106*
fittings *124*
flange up *148*
flank—see *limb*
float 119, *124*
float collar *106*
floating roof *148*
float tank *148*
flood plain *41*
floorman *106*
flow a well hard *124*
flow bean *124*
flow by heads *124*
flow chart *124*
flowing well 12, 115-116, *124*
flow lines *124*, 143
flow tank *124*
flow treater *124*
flue gas *139*
fluid *41*
fluid circulating system 11, 82-84
fluid density 33
fluid expansion 33, 130
fluid injection 129, 134-138, *139*
    gas reservoirs 137-138
    oil reservoirs 134-137
fluid invasion *58*
fluid level *124*
fluid line 119
fluid movement 23-24
fluid saturation 33
flush production *124*
fluvial 21
fold 23, *41*
footage location 74
footwall *41*
force majeure 69
formation *58*

formation fracturing 95-96, *106*
formation pressure *106*
formation testing *106*
forms (state) 164-166
fossil 26, *41*
    index 26, *41*
fossil fuels 28-29
    origin 28-29
fourble *106*
fourble board *107*
frac fluid 95
fractional distillation 13
fractionating tower 13
fracture (geology) 23, *41*
fracture treating 95-96
fracturing—see *formation fracturing*
free use 69
free water 120
free-water knockout 12, 119-120
friable *41*
frost up *124*
frozen up *124*

G

gaging nipple *124*
gas cap 131
gas cap drive 131
gas cap reservoir 34, 131, *139*
gas condensate reservoir 33, 133-134, *139*
gas-cut mud *107*
gas cycling 137-138
gas exploder 53, *58*
gas injection 135, *139*
    attic gas injection 135
    dispersed gas injection 135
    gas flooding 135
gas in place 8
gas lift 118
gas lift valves 118
gas-oil ratio 131, *139*
gas-oil ratio report 166
gasoline 4-5
gasoline plant report 166
gas plant products *124*
gas processing plant *124*
gas reservoir 132-134
    dry gas 132-133
    gas condensate 133-134
    water influx 134
gas sand *41*
gas seep *58*
gas show *107*
gas well *124*
gathering lines *124*, 143
gauging *148*
genesis 20-22
geological engineer 129
geologic events 24, 25
geologic map *41*
geologic time 26, *41*
    absolute 26
    relative 26
geologist 9, *41*
geology 19-34, *41*
    early history 4
geomorphology *41*
geophone 53, *58*
geophysicist 9, 52, *58*
geophysics 52-55, *58*

geothermal energy 177-178
gilsonite 28
gin-pole truck *124*
girth *124*
go-devil 145
gone to water *124*
government 75, 161-166
    early activity 5
    federal 161-163
    legislative acts 75
    state 163-166
gpm 133
graben *41*
gradient *41*
grapple 91
grass gooser *124*
gravel packing 93-94
graveyard tour *107*
gravity drainage 34, 130
gravity meter 55, *58*
gravity methods 55, *58*
grind-out *148*
grind-out machine *148*
gross revenue 153
ground water *41*
group *41*
guide shoe 86, *107*
gun-perforate *107*
gusher *107*
guy wire *124*

H

handy *124*
hanging wall *41*
hardpan *41*
hatch *124*
hay tank *124*
heat *124*
heater-treater 12, 120, 121
high pressure zone 89-90
historical geology 24-27, *41*
historical index 65
hoist *107*
hoisting drum *107*
hoisting system 84
holiday *148*
holiday detector *148*
hook 81, *107*
hopper *107*
horst *41*
hot oil (production) *124*
hot oil (recovery) *139*
hydraulic action *41*
hydraulic fracturing *107*
hydraulic head *148*
hydraulic pumping 118
hydrocarbons 7, 27, *41*
    classification 27
hydrocracking 14
hydrodynamics 29
hydrogen sulfide 8
hydrologic cycle *41*
hydromatic brake *107*
hydrophone 53, *58*
hydro power 178

I

igneous rocks 19, *41*

extrusive	19
intrusive	19, *41*
impermeable	*41*
imports (oil)	171-172
incompetent	*41*
independents	7
Indian land	64
acquisition of rights	73
allotted	64
leases	71
records	65
tribal	64
inert gas	*139*
inhibitor	*148*
inland barge rig	*107*
inlet valve	*119*
inorganic	*41*
inorganic theories	28-29
in situ	*41*
in situ processing (oil shale)	176
in situ recovery (tar sands)	176
instrumentation	*107*
insulated flange	*124*
integrated companies	7
interest	75, 77
carried	77
net profits	75, 77
working	75
isopach map	51, *58*

J

jack board	*124*
jack-up drilling rig	6, 88, *107*
jet bit	*107*
jet gun	*107*
jet-perforate	92, *107*
joint (drilling)	82, *107*
joint (geology)	*41*
joint movement	*148*
joint tariff	*148*
joint venture	75
junk	91, *107*

K

kelly	82, *107*
kelly bushing	82, *107*
kelly hose	83
kelly spinner	*107*
kerogen	29, *41*, 175
kerosene	2, 3
KGS land	73
kick	89-90, *108*
kill a well	*124*
knockout	*124*
known geologic structure	73

L

LACT	*124*
LACT station	*148*
LACT unit	13, 121
lacustrine	21, *41*
land	10-11, 61-77
land description	61-63, 67, 69, 71
landman	10, 61
land map	64, 65
land ownership	63-67
classification	64
depth	64
divided	64
minerals	64
recording	65-66
records	64-65
surface	64
time	64
transfer	64
verification	66-67
water	64
land rights	67-73
acquisition	73
mineral	67-73
surface	67
land work	73-74
latch on	*108*
lateral changes	*41*
lay barge	144
lazy board—see *jack board*	
lead tongs	*108*
lease	10, 67-71, 73
clauses	67, 69, 71
competitive	73
early	67
federal	69-71
fee	67-69
Indian	71, 73
non-competitive	73
oil and gas	67
preparation	69, 71
state	71, 73
surface	67
lease automatic custody transfer—see *LACT*	
lease condensate	*124*
lease separator	*124*
lessee	67
lesser interest	69, 71
lessor	67
level control valve	119, 120
limb	*41*
limestone	32, *41*
limited partnership	75
liner hanger	93
lithification	19, 21, *41*
lithology	32, *41*
Little Big Inch	5, 6, 143
live oil	*124*
load binder	*124*
location	*108*
description	74
log	48-49
auxiliary	49
caliper	49
density	49
dipmeter	49
directional survey	49
dual induction focused log	48
electrical (resistivity)	48-49, *58*
gamma ray	49
induction electric	48
laterolog	48
micro-laterolog	49
microlog	49
neutron	49
radioactive	49, *58*
sonic	49, *59*
spontaneous potential	49
log book	*148*
logging	*58*
logging engineer	129

log sheet *148*
long string 92
lost circulation 89
lost circulation material 89
Lucas gusher 4

M

magma 19, *41*
magnetic brake *108*
magnetic survey............................. 55, *58*
magnetite................................... 55, *58*
magnetometer 55, *58*
main line....................................... *148*
make a connection............................... *108*
make a hand.................................... *125*
make a trip..................................... *108*
make hole................................. 81, *108*
make it up another wrinkle...................... *125*
make up *108*
make up a joint *108*
makeup cathead................................. *108*
male connection................................ *125*
manhole *125*
manifold....................................... *148*
mantle..................................... 24, *41*
marginal well *125*
marine 21, *41*
marketing..................................... 14, 16
marlstone 29, *41*, 175
marsh gas 27-28
mast 84, 87, *108*
master bushing................................. *108*
master drilling report 96
master gate.................................... *125*
master title plat 65
master valve................................... 115
material balance 129, *139*
mechanical rig *108*
member.. *41*
mercaptans....................................... 8
metamorphic rocks 20, *41*
metes and bounds............................. 62-63
methane 7
microwave..................................... *148*
migration 29-30, *41*
    buoyancy 29, 30
    capillarity 29-30
    hydrodynamics 29, 30
mill *108*
mineral 19, *41*
mineralogy *41*
mineral rights 67-73
    acquisition procedures 73
    assignment 71
    leases 67-71
    purchases 71, 73
mini-charges 53, *58*
mining (coal).............................. 174-175
    surface 174
    underground 174, 175
mining (oil shale)............................... 176
    surface 176
    underground 176
mining (tar sands) 176
miscibility................................ 136, *139*
miscible flood—see *miscible fluid displacement*
miscible fluid displacement 135, 136-137, *139*
    continuous 137
    slug 137

mix mud....................................... *108*
monkeyboard 87, *108*
monocline...................................... *41*
morning tour *108*
motorman *108*
mousehole *108*
mousehole connection *108*
moveable oil................................... *139*
mud 82-83, *108*
mud analysis *108*
mud cake *108*
mud circulation *108*
mud conditioning *108*
mud engineer......................... 84, 87, *108*
mud gun *108*
mud logging *108*
mud man—see *mud engineer*
mud pit 83, *108*
mud pump 83, *108*
mud-return line 83, *108*
mud screen *108*
multiple completion well *125*
mutliple zone completion 116

N

national security 174
natural gas 8, *42*
    dry................... 8, 133-134, *139*
    sour...................................... 8
    wet 8, 133, *139*
natural gas liquids *125*
needle valve *108*
Nelson Cost Index 152-153
net operating income 153, 155
net production 155
net revenue............................... 153, 155
nipple *108*
nipple up...................................... *108*
non-associated gas.............................. *125*
nonconformity.................................. *42*
normal circulation.............................. *108*
normal fault................................... *42*
nose... *42*
notice of intent to plug or plug and abandon 165
nuclear power............................. 176-177
    fission 176-177
    fusion 177

O

objection 66
oblique fault................................... *42*
offer to lease 73
off production *125*
offset well..................................... *125*
offshore development............................. 6
offshore drilling......................... 87-89, *109*
offshore rectangular grid system 63
oil country tubular goods *125*
oil in place 8, *139*
oil level control 119, 120
oil pool.. *42*
oil reservoirs 130-132
    combination.............................. 132
    gas cap 131
    solution gas......................... 130-131
    undersaturated............................ 130
    water influx 132

oil sand... *42*
oil shale........................... 28, 29, *42*, 175-176
    mining...................................... 176
    origin 29
    retorting 175, 176
oil shows....................................... *58*
oil smeller..................................... *58*
oil well....................................... *125*
old hand...................................... *125*
on-stream..................................... *148*
on suction.................................... *148*
on the line................................... *125*
on the pump.................................. *125*
oolitic... *42*
open.. *109*
open hole..................................... *109*
operating costs........................... 154, 155
operator............................. 75, 85, *109*
operator's monthly report 166
organic.. *42*
organic theories............................. 28-29
organization report 164
orifice.. *109*
origin of petroleum 28-29
    inorganic theories..................... 28-29
    organic theories 28-29
outcrop.. *42*
over and short station 148
overproduced................................. *125*
overshot.................................. 91, *109*
oxidation *42*
ozocerite 28

P

P&A—see *plug and abandon*
packer.............................. 49, 91, 116
    retrievable 116
paleontology *42*
panel diagram—see *fence diagram*
Pangea.................................... 25, *42*
party...................................... 67, 69
payout 156, 158
pay sand..................................... *109*
pay thickness *58*
percussion drilling............................. 80
perforate 11, 92, *109*
perforating gun *109*
period.. *42*
permeability.............................. 33, *42*
    absolute 33
    effective *40*
    relative 33
permit 67
persuader.................................... *125*
petrochemicals 14
    aliphatics 14
    aromatics 14
    inorganics................................. 14
petrography *42*
petroleum 1, 7, 27, *42*
    in ancient civilizations.................. 1-2
    in early America 2
    in medieval times......................... 2
    origin 28
    U.S. production...................... 170-171
    U.S. supply.......................... 170-171
    world production..................... 172-173
    world supply......................... 172-173
petroleum geology....................... 27-34, *42*

petroleum industry............................ 1-16
    careers..................................... 7
    early expansion and development 3-4
    early judicial and governmental activity......... 5
    historical background 2-3
    modern 7-16
    offshore development...................... 6
    people 7
    war and peace........................... 5-6
petroleum products...................... 14, 15
photogeology map 47, *58*
physical geology...................... 19-24, *42*
pig..................................... 145, *148*
pig iron..................................... *125*
pin... *109*
pipe—see *drill pipe*
pipeline 141-142, 143-146
    Alaska................................ 145-146
    construction 143-145
    gas....................................... 143
    history.............................. 141-142
    oil 143
    types 143
pipeline oil................................... *125*
pipe ram................................. 90, *109*
plate tectonics.......................... 25, *42*
platform................... 89, *109*, 121, 122
plug... 86
plug and abandon............................. *109*
plug back.................................... *125*
plugging report 165
plunger lift *125*
pneumatic air gun 53, *58*
polymer 137, *139*
polymer injection 137
pool.. 8
pooling.................................... 69, 74
    forced 74
poor boy..................................... *125*
pore space 8, *42*
porosity.................................. 32, *42*
    effective 32, *40*
    secondary 24
    total...................................... 32
porosity trap *42*
positive choke *109*
positive-displacement pump................... *148*
post-depositional phenomena................ 22-24
potential test *125*
power oil..................................... 118
power rated *125*
power system 84
power tools................................... *125*
present value............................ 157, 158
pressure...................................... *109*
pressure gauge *109*
pressure gradient *109*
pressure maintenance *139*
pressure regulator *125*
pressure-relief valve *109*
preventer—see *blowout preventer*
prices... 174
primary cementing *109*
primary recovery 129-134, *139*
    gas reservoirs 132-134
    oil reservoirs 130-132
prime mover.................................. *109*
principal meridian 61
private land 64
    acquisition of rights 73

leases	67-69
records	64
title examination	66
production	12-13, 115-122, *125*
U.S.	170-171
world	172-173, 179-180
production engineer	129
production forecast	153, 154-155
production payment	75
production testing	49-50
productive capacity	*125*
productivity test	*125*
products cycle	*148*
products line	143, *148*
profitability analysis	156-158
project	151
propellors	88
proppant—see *propping agent*	
propping agent	95, *109*
proration	*125*
proved reserves	*125*
prover	*148*
prudent operator	69
psi	*139*
public domain land	64
public land	64
title examination	66
pump	*109*, 116-118
pumping unit	12, 117
pump off	*125*
pump pressure	*109*
pump station	143, 145
purchase (land)	67, 71, 73
put a well on	*125*
put on pump	*125*

Q

quarter-quarter	62
quarter section	62
quartzite	*42*

R

rabbit	*125*
radioactive dating	26, *42*
railroad tank car	141, 142, 147
ram	*109*
ram blowout preventer	*109*
range lines	61
rathole	*109*
receiver	53, *58*
reciprocating pump	*148*
record check	66
recorder	53
record section	53-54
record title	66
recoverable oil	*139*
recovery	129-138
rectangular survey system	61-62
reeve	*109*
reeve the line	*109*
refiner-marketer	16
refiner's report	166
refining	13-14
world	179-180
reforming	135
regional dip	*42*
relative dating	26
fossil dating	26

physical feature correlation	26
relief	*42*
relief valve	*125*
remote control station	125
rental	71
report of subsurface injections	165-166
reports (state)	164-166
requirement	66
reserve forecast	153, 154
reserve pit	*109*
reserves	8-9, *139*
U.S.	171-172
world	9, 179-180
reservoir	8, 28, 29-34, *42*, 130-134
classification	33
combination	132
energy	33-34
fluids	33
gas	132-134
mechanics	33-34
oil	130-132
properties	32-33
rock types	32
reservoir engineer	129
reservoir pressure	*139*
reservoir rock	29
residual saturation	137, *139*
resistivity	48, *58*
retorting (oil shale)	175-176
retrograde condensate reservoir	133-134
retrograde condensation	133, *139*
revenue	153
revenue analysis	153
reverse circulation	*109*
rig	87-89, *109*
bottom-supported	88-89
floating	87-88
onshore	87
platform	89
rig down	*109*
right of assignment	69, 71
right-of-way	67, 143, *148*
rig up	85, *109*
riser	88, *125*
rock	19, *42*
rock a well	*125*
rock cuttings	47, *58*
rock cycle	20, *42*
rock movement	22-23
roller-cone bit	82, *109*
rotary bushing—see *master bushing*	
rotary drilling	11, 81-84, *109*
rotary helper	87, *109*
rotary hose	83, *110*
rotary rig	11, 81
rotary table	82, *110*
roughneck	87
round trip	*110*
roustabout	*110*
royalty	69, 71, 75
compensatory	74
minimum	71
overriding	71, 75
participating	75
shut-in	69
term	75
R/P ratio	45
Rule of Capture	5
run in	*110*
run ticket	121

S

saddle bearing *125*
sales 14, 16, 121-122
salt dome 23, *42*
saltwater disposal 122
sampson post 117
sand 32
sand consolidation 94
sand control 93-94
sanded up *125*
sandstone 32, *42*
scheduler *148*
scraper *125*, 145
scraper trap *148*
scratcher *110*
sea-floor spreading 25
secondary cementing *110*
secondary recovery 129, *139*
section 62
sedimentary bed *42*
sedimentary rocks 19, 20, *42*
    carbonate 22, *40*
    clastic 21, 22, *40*
sedimentation 20, *42*
seismic 52-55, *58*
    reflection 52-54, *58*
    refraction 54-55, *58*
seismograph *59*
seismology *59*
seismometer 53
semisubmersible drilling rig 6, 88, *110*
separator 119, 121
sequence *42*
serial register 65
service station 16
service well *139*
set casing *110*
settled production *125*
settling pit *110*
shake out *125*
shaker—see *shale shaker*
shaker pit *110*
shale *42*
shale shaker 83, *110*
shaped charge *110*
sharpshooter *125*
shear ram *110*
sheave *110*
Shepard's canes *148*
shot point 54, *59*
show *110*
shut down *110*
shut in *125*
shut-in bottom-hole pressure *110*
shut-in pressure *125*
sidetrack *110*
sidewall core sample 48, *59*
signature bonus 67
siliceous *42*
simultaneous filing 73
single *110*
single point mooring system 146-147
slack off *125*
sling *125*
slips *110*
slop *148*
slotted liner 93
slug 137, *139*, 145
slurry *110*

snake out *125*
snatch block *125*
soft rope *125*
solar power 177
solution gas 130
solution gas drive 131
solution gas reservoir 33, 130-131, *139*
sour 8
source rock 29
sovereign 66
spacing 74, *125*
spaghetti *125*
sparker 53, *59*
spear *110*
specific gravity *139*
sphere *148*
Spindletop 4, 80
spinning cathead *110*
spinning chain *110*
spread 143
spring pole drilling 80
spud 85, *110*
spud in *110*
squeeze cementing *110*
stab *110*
stabbing board *110*
stabilized *125*
stake a well *110*
stand *110*
standard cubic foot 129, *139*
standpipe 83, *110*
state land 64
    acquisition of rights 73
    leases 71
    records 65
state oil and gas commissions 163-164
    reports and forms 164-166
status report 66
steam injection 135-136, *139*
    drive 135
    soak 135
stimulation 12, 94-96, *111*
storage 13, 121, 122
    early history 3
    offshore 121-122
    onshore 121
storage reel 84
storer's report 166
strapping *148*
strata *42*
strat hole *59*
stratification *42*
stratigrapher 22, *42*
stratigraphic column 22, 35-39, *42*
stratigraphic trap *42*
stratigraphy 20, 22, *42*
strike *42*
string *111*
stringer bead *148*
string up *111*
strip a well *125*
stripper *125*
structural basin *42*
structural geology 20, *42*
structural high *42*
structural low *42*
structural theory 46, *59*
structure contours *42*
structure map *59*
    subsurface 50-51

surface 47
stuck pipe *111*
sub *111*
    bent 92
submersible drilling rig 88-89, *111*
submersible pumping 117-118
subsidence 22, *42*
substructure 85, *111*
subsurface contours *42*
subsurface correlation *42*
subsurface geology 47-52, *59*
sucker rods 117
suction pit *111*
sulfur 8, 14
sundry notices 164
supervisory control *148*
supply (petroleum) 174
    U.S. 170-171
    world 172-173
surface beds *43*
surface casing—see *surface pipe*
surface geology 46-47, *59*
surface pipe 85, *111*
surrender 69, 71
swab *125*
swamper *125*
sweep efficiency *139*
sweet 8
swivel 81, *111*
syncline *43*
synthetic fuels 173

T

tail out rods *125*
take a strain on *125*
take-off 66
tally *125*
tank battery *149*
tanker 141, 142, 146-147
tanker routes 146
tank farm *149*
tankship *149*
tank strapper *125*
tank truck 142, 147
tank wagon 142
tap *126*
tariff *149*
tar sands 176
    recovery 176
TD—see *total depth*
tectonic *43*
telecommunications *126*
telemetry *126*
tender *149*
term 67, 69, 71
    primary 67
terminal 16, *149*
tertiary recovery 129, *139*
test hole *59*
thermal cracking 5, 14
thief *149*
thief zone 89
thread protector *111*
thribble *111*
thribble board *111*
throw the chain *111*
thrusters 88
thrust fault *43*
thumper truck 53, *59*

tie-down *126*
tight formation *43*
tight hole *111*
tight sand *43*
time chart 27
time value of money 157
tin hat *126*
title 64
    clear 66-67
title opinion 66
tongs *111*
tool joint *111*
toolpusher 87, *111*
topography *59*
torque 82, *111*
torque converter *111*
total depth *111*
tour *111*
town lot 63
township 61-62
township lines 61
transducer *126*
transmission *111*
transportation 13, 141-147
    history 3-4, 141-142
transportation (geology) 21, *43*
transporter's report 166
trap 30-32, *43*
    structural *42*
    subsurface *42*
trapped gas saturation 137
traveling block 84, *111*
treating (oil) 12, 118-121
    chemical 121
    electricity 121
trench 144-145
trends in world production 172-173
tricone bit *111*
trip 84, 86, *111*
trunk line 143, *149*
tubing 12, 93, 94, 115
tubing job *126*
turbodrill 92, *111*

U

uintahite 28
ULCC 147
unconformity 22, *43*
unconsolidated sand 93
undersaturated reservoir 33, 130, *139*
undiscovered petroleum resources 173
uniformitarianism 20, *43*
unitization 75
unit operator *111*
universal transverse mercator system 63
uplift 22, 23, *43*
upstream functions 7, 141
uranium 177
use plat 65
U.S. gas production and supply 170-171
U.S. Geological Survey 71, 73, 162-163
U.S. oil production and supply 170-171

V

valve *111*
V-belt *111*
viscosity 33, *43*
VLCC 147

volatile oil reservoir..............................33
vug..*43*

W

waiting on cement................................*111*
walking beam..............................80, 117
wall cake—see *mud cake*
warranty..69
waste conversion.................................178
waterflooding........................134-135, *139*
water influx reservoir............34, 132, 134, *139*
    bottom......................................34, 132
    edge..34, 132
    shale..34
water injection......................134-135, 137
water well.......................................*126*
weathering...................................21, *43*
    biological.......................................21
    chemical...21
    physical...21
weevil—see *boll-weevil*
weight indicator.................................*111*
weighting material...............................*112*
welder...*149*
wellbore...*112*
well completion..................................*112*

wellhead............................*112*, 115-116
well samples..................................47, *59*
well stimulation.................................*112*
wet gas...............................8, 133, *139*
wet gas reservoir................................133
whipstock....................................92, *112*
widow maker.....................................*126*
wiggle stick......................................46
wildcat well......................................46
winch...*126*
windfall profits tax..............................155
wind ring..*149*
wing valve.......................................116
wireline...*112*
wireline log......................................*59*
    see *log*
wire rope..*112*
WOC—see *waiting on cement*
words of grant................................67, 69
working pressure................................*126*
work over..*126*
workover rig......................................86
world production and supply of petroleum.....172-173
worm...*112*

Z

zone..*59*